IEE TELECOMMUNICATIONS SERIES 6

SERIES EDITORS: PROFESSOR J. E. FLOOD AND C. J. HUGHES

Signalling in Telecommunications Networks

Previous volumes in this series

Volume 1   Telecommunications networks
J.E. Flood (Editor)

Volume 2   Principles of telecommunication-traffic engineering
D. Bear

Volume 3   Programming electronic switching systems
M.T. Hills and S. Kano

Volume 4   Digital transmission systems
P. Bylanski and D.G.W. Ingram

Volume 5   Angle modulation: the theory of
system assessment
J.H. Roberts

Volume 6   Signalling in telecommunications networks
S.Welch

Volume 7   Elements of telecommunications economics
S.C. Littlechild

Volume 8   Software design for electronic switching systems
S. Takamura, H. Kawashima and H. Nakajima

Signalling in Telecommunications Networks

SAMUEL WELCH, O.B.E., M.Sc(Eng.), C.Eng., F.I.E.E.
Telecommunications Consultant *formerly*
Head, Signalling,
  UK Post Office
Technical Director,
  AEI Telecommunications Ltd.

PETER PEREGRINUS LTD.
on behalf of the Institution of Electrical Engineers

Published by The Institution of Electrical Engineers, London, and New York
Peter Peregrinus Ltd., Stevenage, UK, and New York

ISBN: 0 906048 04 4 (Casebound)
ISBN: 0 906048 46 X (Paperback)

Typeset by The Alden Press, Oxford, London and Northampton
Printed in England by A. Wheaton & Co., Ltd., Exeter

Contents

		Page
	Preface	xi
1	**General**	**1**
	1.1 Introduction	1
	1.2 Signalling functions	3
	1.2.1 Supervisory functions	3
	1.2.2 Selection functions	3
	1.2.3 Operational functions	4
	1.3 Factors influencing the telephone signalling technique	5
	1.3.1 General	5
	1.3.2 Influence of type of switching system	5
	1.3.3 Influence of type of speech transmission system	7
	1.3.4 Influence of facilities to be given	10
	1.3.5 Part of network treated differently from signalling aspect	10
	1.4 Signalling for pulse code modulation transmission systems	12
	1.5 Common channel signalling	13
	1.6 International signalling	14
	References	14
2	**Subscriber line signalling**	**16**
	2.1 General	16
	2.2 Signalling from dial telephone sets	17
	2.3 Pushbutton telephone sets	20
	References	23
3	**Junction network signalling**	**24**
	3.1 General	24
	3.2 Typical forms of junction network d.c. signalling	27
	3.2.1 Loop d.c. signalling	28
	3.2.2 Other forms of junction network d.c. signalling	33

		3.2.3	Low-frequency a.c. signalling	37
		3.2.4	Proceed-to-send	40
		3.2.5	Analysis of continuous and pulse signalling	43
		3.2.6	Junction network signal repertoire	45
	3.3	Signalling for direct control and common control switching in junction networks		47
		References		49
4	**Pulse repetition and pulse distortion**			**50**
	4.1	General		50
	4.2	Pulse repetition		53
		4.2.1	Line resistance	54
		4.2.2	Line leakance	55
		4.2.3	Line capacitance	55
		4.2.4	Relay performance variation	55
		4.2.5	Voltage variation	56
	4.3	Pulse correction		56
	4.4	Pulse regeneration		59
		References		59
5	**Long-distance d.c. signalling**			**60**
	5.1	General		60
	5.2	Typical l.d.d.c. systems		61
		5.2.1	UK DC2 system	61
		5.2.2	UK DC3 system	67
		5.2.3	Release guard and blocking signals	71
		5.2.4	Bell system CX and DX systems	73
		5.2.5	Bell system Simplex (SX) system	80
	5.3	Duplex signalling		81
		References		82
6	**P.C.M. signalling**			**83**
	6.1	General		83
	6.2	P.C.M. signalling 24-channel p.c.m. systems		86
		6.2.1	Bell system D1 24-channel system	86
		6.2.2	UK Post Office first-generation 24-channel system (short-haul)	86
		6.2.3	Bell system D2 24-channel system	88
	6.3	P.C.M. signalling 30-channel p.c.m. system		91
	6.4	Comments on built-in p.c.m. signalling		92
		References		93
7	**Influence of transmission on signalling**			**94**
	7.1	General		94
	7.2	4-wire circuits		94
	7.3	Echo suppressors		95
	7.4	Long propagation time		97
	7.5	Satellite circuits		98
	7.6	Time assignment speech interpolation equipment		99
	7.7	Compandors		102

7.8 Signal power 102
7.9 Bothway operation of circuits 103
  7.9.1 General 103
  7.9.2 Double seizure 105
7.10 E- and M-lead control of signalling 107
References 115

8 Voice frequency signalling **117**
8.1 General 117
8.2 Speech immunity 118
8.3 Signal information transfer on multilink connections 124
  8.3.1 General 124
  8.3.2 End-to-end signalling 124
  8.3.3 Link-by-link signalling 124
  8.3.4 V.F. signalling line splits 125
8.4 V.F. receiver signal-guard 126
8.5 General arrangement v.f. line signalling 132
8.6 Type of v.f. signal 134
  8.6.1 Continuous v.f. signalling 135
  8.6.2 Pulse v.f. signalling 137
8.7 Bell SF 1 v.f. line signalling system 137
  8.7.1 Analogue version line signalling 138
  8.7.2 Comments on Bell SF system 154
  8.7.3 Digital version SF line signalling 157
8.8 Pulse v.f. line signalling 157
  8.8.1 General 157
  8.8.2 UK/AC9 and AC11 1 v.f. pulse line signalling systems 158
  8.8.3 Other national network v.f. line signalling systems 167
  8.8.4 Comments on pulse v.f. line signalling 169
References 170

9 Outband signalling **172**
9.1 General 172
9.2 Consideration of the signalling mode 175
9.3 Application constraints of outband signalling 177
9.4 Typical outband line signalling systems 178
9.5 CCITT (CEPT) R2 system outband line signalling 190
  9.5.1 General 190
  9.5.2 R2 analogue outband line signalling 191
  9.5.3 R2 line signalling digital version 196
References 199

10 Interregister multifrequency signalling **201**
10.1 General 201
10.2 Interregister signalling modes 203
10.3 Link-by-link and end-to-end interregister signalling 212
10.4 Control of address information transfer 216

10.5	General arrangements interregister m.f. signalling system	219
10.6	Bell system interregister signalling (R1)	220
10.7	CEPT interregister signalling (R2)	223
	10.7.1 General	223
	10.7.2 R2 interregister signal code	224
	10.7.3 Basic philosophy of the R2 interregister signal code	232
	10.7.4 Operational features	235
	10.7.5 R2 interregister signalling on satellite circuits	238
	10.7.6 General comment on R2 interregister signalling	239
10.8	UK MF2 interregister signalling system	240
	10.8.1 General	240
	10.8.2 Guarded pulse m.f. signalling	241
	10.8.3 MF2 interregister signal code	245
	10.8.4 Outline of MF2 operation	248
10.9	French Socotel interregister m.f. signalling system	253
	References	257
11	**Common channel signalling**	**259**
11.1	Introduction	259
11.2	Basic common channel signalling	260
11.3	Association between c.c.s. and speech (or equivalent) networks	264
11.4	Network centralised service signalling	267
11.5	CCITT No. 6 signalling system	269
	11.5.1 Basic concepts	269
	11.5.2 Signal codings	270
	11.5.3 Error control	282
	11.5.4 Analysis of the system 6 error control method	289
	11.5.5 Synchronisation	290
	11.5.6 System 6 analogue and digital versions	294
11.6	Common channel signalling loading	296
11.7	Signalling link security and load sharing	299
	11.7.1 Security	299
	11.7.2 Load sharing	304
11.8	Changeover, retrieval and changeback	306
11.9	Continuity check of the speech path	309
11.10	Signal priority	309
11.11	CCITT No. 7 optimised digital common channel signalling system	311
	11.11.1 Basic concepts	311
	11.11.2 Signalling bit rate	314
	11.11.3 Error control	314
	11.11.4 Message structure	320
	References	321

12   **CCITT international signalling systems**   **322**
     12.1   General   322
     12.2   CCITT signalling system 3   323
     12.3   CCITT signalling system 4   324
     12.4   CCITT signalling system 5   335
     12.5   CCITT signalling system 5 bis   348
     12.6   International signalling: sending sequence of
            numerical information   361
            References   364

     Glossary of terms   **365**

     Appendix to Chapter 11   **377**

     Index   **381**

Preface

Ease of communication is a vital necessity both nationally and internationally, and perhaps the most important contribution to this is made by the various telecommunication services, particularly the telephone service. Need generates the means, and this has resulted in the now commonplace direct dialled telephone service in national networks, and as these networks are interconnected by an international network, in the present rapidly expanding international world-wide direct dialled telephone service. This book deals with the telephone service in the main since this is the most important aspect of telecommunications, and the dominant service.

Signalling systems link the various switching machines in a network to enable the network to function as a whole, and must be compatible with the different types of switching and transmission systems equipped. They also link the various national networks to the international. A good understanding of the signalling art is thus essential in network design. The approach of the book has been to deal with the principles of the various signalling techniques as understanding of the principles involved is particularly important in the evolving signalling art in view of the impact of new techniques of the other disciplines such as stored program control, integrated digital switching and transmission, etc., may have on networks. The detailed design of signalling systems is not covered, as, for a given technique, such detail varies greatly depending on different designers' approaches and administration preferences, but, where necessary, simplified design information is included to enable the principles employed to be discussed.

As most administrations and operating organisations adopt their own designs of signalling systems, and thus there are many different system designs, selection was necessary in view of the limitation of text, but adequate reference is made to the literature for further reading. The selection adopted is from the UK and Bell system (USA) sources in the main, but as the book is concerned with principles, it is my view, supported by an experience in signalling of some 40 years, that such a selection is adequate for the purpose of assessing different approaches. The book analyses the selected techniques and preferences are proposed, which preferences also reflect the conclusions reached during my signalling career.

While the book deals with signalling on the public switched telephone network in the main, the principles discussed are equally applicable to other services such as private networks, data networks, etc.

On-speech path signalling is the main technique in switched telephone networks at present, and will continue to be used for many years to come, but the technique is relatively expensive in total network signalling cost owing to the lack of time-sharing. For economic reasons, highly desirable signalling rationalisation has not been achieved with the on-speech path signalling technique, which accounts for the variety of different signalling systems in any given network, with the consequential administrative and operational problems. With the advent in networks of stored program control of switching, both analogue and digital, and thus, logically, for the adoption of time-shared common channel signalling, the opportunity exists for a rationalised signalling system which would be economic in terms of total network signalling cost. It is hoped, and there is every possibility, that this will be achieved. Further, with common channel signalling, the previous distinction between national and international network signalling would not be necessary, and the rationalisation of signalling method so extended to embrace both national and international. It is thought that the common channel signalling technique is perhaps the most dramatic advance in signalling system evolution and that there is little doubt that it will be the main signalling method of the future.

There is an extensive literature on the description of various signalling systems, but as far as is known, no publication exists which presents and analyses in one place, the various signalling techniques and approaches which may be applied in networks. This book is intended to fill that gap, and to provide telecommunication network designers, telecommunication engineers, and university undergraduates and postgraduates reading telecommunication technology, with a knowledge of signalling principles in the concept of a complete network. It is hoped that the book will also be of value as a source of reference information on the particular signalling systems treated in some depth.

Chapter 1 is devoted to a general survey of the signalling covered in the book and is intended as a general introduction easy-reading aid to the more detailed treatment in the subsequent chapters. Chapter 2 gives an account of the signalling on subscriber lines and, from this, develops the basic signalling requirement for the switched network. Chapters 3 and 4 deal with the signalling in local exchange area junction networks, including treatment of the various problems arising from address pulse distortion. This leads to Chapter 5 dealing with long distance d.c. and low-frequency a.c. signalling. Chapter 6 discusses built-in signalling on p.c.m. transmission systems applied point-to-point, the CCITT specified 24- and 30-channel systems being given the main treatment. The importance of the influence of various transmission features on signalling is treated in some detail in Chapter 7 as a lead in to Chapters 8–10 dealing with the trunk (toll) network signalling, d.c., voice frequency, outband, and interregister multifrequency, the influence of the switching on signalling being discussed where relevant. Chapter 11 deals with the important common channel signalling technique in both the analogue and digital environments, covering the requirements and preferred approaches in various problem areas. The treatment is in some depth as little, if any, published information of analytical nature is available. The final Chapter 12 deals with the CCITT specified international signalling systems, covering their evolution, field of application, and the reasons why the respective systems were produced.

I am sincerely grateful to Professor J. E. Flood of the University of Aston in Birmingham and the series editor of the IEE Telecommunications Series, for his helpful criticism of the work, and for his invaluable suggestions for the preparation of the text. His encouragement eased the long drawn-out effort involved in the writing of the book.

S. Welch

Chapter 1

General

1.1 Introduction

Signalling in a telecommunications network is the interchange of information between different functional parts of the telecommunications system. The present book concerns signalling in telephone networks in the main, and, in view of the many signalling systems involved, it is perhaps useful to present a broad survey, in general terms, in this chapter as an introduction and aid to general understanding of telephone network signalling before the later detailed treatment.

The total signalling requirement in a telephone network may be broken down into three well defined areas:

(*a*) the transfer of information from subscribers, calling and called, to the switching machines
(*b*) signalling within an exchange
(*c*) the transfer of information over the switched network.

Signalling (*a*) is performed over the subscriber lines and reflects the on-hook and off-hook conditions of the subscriber's telephone set. The on-hook condition is said to apply when the telephone set is in the idle condition (the handset resting on the cradle to open the subscriber's loop, or an equivalent condition), and the off-hook when the telephone set is in the active condition (the handset removed from the cradle to close the subscriber's loop, or an equivalent condition). The exchange equipment detects for the absence or presence of current on the subscriber's line resulting from the on-hook and off-hook conditions, respectively, and it is clear that all telephone call signalling emanates from these conditions on the subscriber lines.

Subscriber line signalling may be regarded as being basic function signalling, signalling which is independent of the type of switching system and type of switched network, the basic requirements being:

Calling subscriber's line: seizure (off-hook) to indicate to the exchange that a call is initiated

dialled address information (sequential on-hook/off-hook conditions for dial telephone sets)

clear forward (on-hook) to indicate to the exchange that the call is terminated by the caller

Called subscriber's line: answer (off-hook)

clear back (on-hook) to indicate to the exchange that the call is terminated by the called party

Signalling (*b*) is internal with the switching equipment and is of a type depending on the particular switching machine arrangements. This signalling does not involve signalling systems of the type required on the switched network and will not be considered further.

Signalling (*c*) is performed between exchanges, information being transferred forward and backward over the switched network to control call connection and clear down. It is this signalling which involves the use of signalling systems of the type to be discussed in this book. Signalling over the switched network is usually more complex than its subscriber line signalling counterpart, owing to the concept of machines 'speaking' to each other, the inclusion of signals to ensure efficient operation of the network and the need to take account of additional signalling requirements, as discussed later.

The various signalling system techniques have evolved over the years in step with advances in the telephony art, embracing both switching and transmission, starting from the simple manual service and evolving to the present highly sophisticated automatic telephony networks incorporating national and international subscriber dialled services. Thus many early type signalling systems are of historical interest only and will not be considered; only signalling techniques for automatic telephone switched networks will be discussed.

1.2 Signalling functions

The basic signalling on the subscriber lines may be considered as comprising two distinct signalling functions: supervisory and selection. The signalling on the switched network must also comprise these two functions as a minimum requirement, but it is preferable and more usual for switched network signalling to incorporate a further signalling function, which may be termed 'operational'. Thus any telephone switched network signalling facility should perform supervisory, selection and operational functions.

1.2.1 Supervisory functions
The supervisory conditions serve to detect or change the state or condition of some element (in general, subscriber and network lines) of the network and reflect the subscribers' on-hook/off-hook conditions. They involve

> off-hook seizure
> off-hook answer
> on-hook clear back
> on-hook clear forward

detection and any consequential changes of state of lines from the idle to the busy condition, and vice versa.

1.2.2 Selection functions
These are concerned with the call connection set-up process and are initiated by the caller sending the called party's address information, this information (or part of it) being transferred between exchanges. The selection information so transferred may also include signals additional to the dialled address information to enable the switching process to be carried out satisfactorily, e.g. proceed to send, number received, request for the sending of digit(s) etc. There may also be a requirement for other signals which serve to acknowledge receipt of signal(s), according to the specific system.

Selection function signalling, being concerned with call connection set-up, has possible influence on the magnitude of the postdialling delay (the interval between the completion of dialling by the caller and the receipt of ring tone). The postdialling delay is an obvious criterion on which subscribers tend to assess the efficiency of the telephone system, as a long delay would give rise to subscriber anxiety as to possible system failure, which could result in an abandoned call and a repeated attempt. Thus, in addition to the requirement that the selection function signalling between exchanges be performed efficiently and reliably to ensure that the switching correctly functions, the form and speed of the signalling should minimise the postdialling delay.

1.2.3 Operational functions

While call processing could be performed by the supervisory and selection functions only, further functions, which may be termed 'operational' are necessary to enable the best use to be made of the network and to cater for system and administration facilities. The operational requirements are variable, depending on the signalling system capability and administration policies, typical features being:

(*a*) to detect, and transfer information regarding, congestion in the network; the information may be used, typically to prompt a repeat attempt, initiate clear down, to inform the caller of the congestion condition etc.
(*b*) to inform of the nonavailability of equipment, or of circuits, owing to breakdown or outages for maintenance
(*c*) to provide call charging information, e.g. metering over junctions
(*d*) to provide a means of extending fault-alarm information from unattended exchanges etc.

The above discussion concerns 'functions' rather than 'signals', and is thus in terms of the philosophical concept of signalling systems. A function could be represented by one or more signals in a signalling system; also, a specific signal could well serve to carry out one or more separate functions.

It is usually convenient to classify signals as 'line (supervisory) signals' and 'selection (address) signals'.[1] This classification has much in common with the function classifications 'supervisory' and 'selection', respectively, but this is not always completely so. The classification 'line' and 'selection' signals tends to differentiate signals between supervisory units (auto-to-auto relay sets etc.) equipped on the speech path at exchanges, and those between control units (registers), while the functional classification emphasises the different missions the signalling system carries out. For example, the proceed-to-send signal is a selection function, and, while transferred by the selection (address) signalling in some networks, is transferred as a line signal in others. Most operational functions tend to be incorporated in the selection signalling facility, but, here again, there is no firm rule; some operational functions are carried out by selection signals, e.g. congestion, and others by line signals, e.g. metering over junctions.

1.3 Factors influencing the telephone signalling technique

1.3.1 General

The telephone network signalling technique divides into two broad areas: (i) on-speech path signalling, which may be referred to as being 'conventional' and (ii) common channel signalling, the signalling not being performed over the speech path. All present-day telephone networks are equipped with on-speech path signalling systems, but, with the advent of computerlike techniques for the control of switching (stored program control), however, many administrations are at present planning the application of common channel, instead of on-speech path, signalling systems. Networks have a considerable inertia to change, however, and, because of this, on-speech path signalling will continue to apply for many years to come, but there is little doubt that many networks will gradually evolve to common channel signalling.

Unlike common channel signalling, on-speech path signalling systems do not employ time-shared techniques, and are thus relatively expensive. The consequent anxiety to minimise total signalling cost results in every effort being made by administrations to take advantage of any particular network conditions to reduce the cost. As a consequence, conventional on-speech path signalling tends to be designed for particular network application conditions, which accounts for the variety of different signalling systems in any given network. Adoption of this philosophy results in various factors influencing the on-speech path signalling technique, the main being:

> the type of switching system
> the type of transmission system
> the facilities to be given
> part(s) of a network being treated differently from other part(s) from the signalling aspect

In general terms, the influence of these various factors on the design of on-speech path signalling systems is as follows:

1.3.2 Influence of type of switching system

There are two main types of telephone switching system: direct control (noncommon control) and indirect control (common control).[2] Common control systems may be wired logic control, i.e. nonstored program control, or stored program control, and the switching analogue (space division) or digital. The signalling implications of digital switching are considered in Section 1.5.

Direct control switching systems, typically step-by-step, do not allow for separation of the supervisory and selection signalling functions. The dialled address positions the switches directly, the numbering scheme controls the routing of the call connection and the speed and break/make pulse ratio of the decadic address signals are conditioned to the speed and characteristics of the switching machines. The supervisory and selection signalling functions are combined in the line signalling system, and, as the address selection signals are decadic, the selection process is slow, but, owing to the direct control, the post-dialling delay is negligible.

In the absence of registers, the facilities, and thus the signal repertoire, permitted by direct control switching systems are modest. While this does not meet the requirements of modern networks, it does, however, permit the signalling systems to be relatively simple, which has economic merit in view of the per-speech circuit signalling system provision.

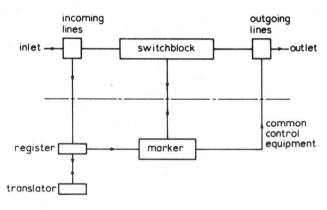

Fig. 1.1   *Generalised principle of common control switching.*

In common (indirect) control switching systems, the common control equipment deals with the call connection set-up, being released from a call on connection establishment to deal with other call connections. The common control equipment receives the dialled address information, processes it and then controls the switching process. In principle, the calling line gives an incoming mark at the switchblock and the information (the translation) resulting from the processing of the address information, an outgoing mark. Typically, the marker controls the switching process to connect the two marks together through the switchblock. Fig. 1.1 shows common control switching in generalised form. The translation facility permits alternative routing, and the routing of the connection is thus divorced from the numbering scheme. The processing of the address information delays the start of switching, which increases the postdialling delay.

Early forms of common control switching continued to rely on decadic address signals, these being transferred between the register functions of the common control equipments of the various switching machines involved in the connection, the relevant line signalling systems carrying the decadic address signals. This concept permits little, if any, facility exploitation on a network basis.

With the appearance of the more advanced forms of common control switching, e.g. crossbar, attention was given to the separation of the supervisory and selection signalling functions and the address information transferred between registers in a form, and at a speed, independent of the form and speed of the common control switching machines, the transmitting register functioning as a coder and the receiving register as a decoder. The preferred form of interregister coded (nondecadic) signalling is multifrequency (2/6 m.f.), which, in addition to giving coding possibility for a wide signal repertoire, contributes to the speeding up of the connection set-up process, particularly on multilink connections. This minimises any increase in the postdialling delay due to processing of the address information. Enhanced facilities in networks requires registers, or equivalent functions, and thus interregister m.f. signalling gives a convenient arrangement to enable the network to be exploited for facilities.

The above usually applies for wired-logic common control analogue switching. Should the common control be stored program type, there is considerable merit in applying common channel signalling between the common control equipments, the common channel carrying all the signalling requirements for a large number of speech circuits. Common channel signalling, which has potential for a considerable signal repertoire, could be applied to wired logic common control, but is usually uneconomic in this situation.

1.3.3 Influence of type of speech transmission system
Transmission plant may vary in type: audio, frequency division multiplex (f.d.m.), digital.[3-5] As conventional signalling is conveyed on the speech path used for the call, it is clear that the type of transmission plant can have a profound influence on the signalling, since particular types of transmission media may permit, or preclude, the application of particular signalling methods. Signalling using a signal frequency (or frequencies) within the voice band (300–3400 Hz) could, of course, be applied to any type of transmission media on the logic that such

signals may be transferred in any situation affording adquate speech transmission. This is called voice frequency (v.f.), and, being inband, i.e. within the voice band, is completely flexible in application, admitting the possibility of the very desirable concept of a rationalised signalling method in networks. However, this particular rationalisation of signalling methods is uneconomic. To economise in per-speech

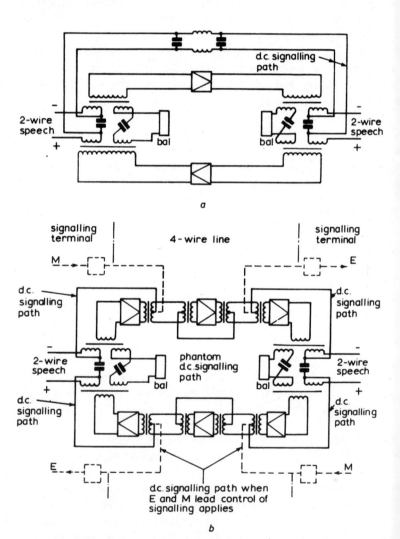

Fig. 1.2   *Amplified audio lines d.c. signalling paths*
   (*a*) 2-wire line
   (*b*) 4-wire line

circuit equipped signalling systems, since the cost of the various signalling methods varies greatly, conventional signalling in analogue
networks tends to be designed for different types of transmission media
to reduce total signalling cost.

Audio line plant, unamplified (2-wire) and amplified (2-wire or
4-wire), permits a metallic signalling path per speech circuit and d.c.
signalling on the speech path is usually applied. This is particularly
attractive to administrations as d.c. signalling is relatively simple,
reliable and cheap. Special arrangements are required to obtain the
d.c. signalling path on amplified audio circuits (Fig. 1.2). The d.c.
signalling may be of simple 'local' form, or long-distance d.c. when
the local signalling limit is exceeded. Low-frequency a.c. signalling,
which also usually requires a metallic signalling path, is sometimes
applied instead of long-distance d.c.

A d.c. signalling path is not available when circuits are provided by
means of f.d.m. transmission systems, d.c. signalling is not possible
and a.c. line signalling – inband v.f. or outband – is usually applied
in analogue networks. Inband v.f. signalling must be protected against
false operation by speech or other interference, at some cost and
complexity. Signalling during the speech period is precluded. Despite
these disadvantages however, v.f. signalling is the most widely applied
system in the long-distance network of any telephone network, owing
to its flexibility of application.

Outband signalling uses a signalling frequency above the voice
band (above 3400 Hz), but below the upper limit of 4000 Hz of the
nominal voice channel spacing. The system can only be used on f.d.m.
transmission circuits, is integrated with the transmission equipment
and is controlled by d.c. conditions (E and M lead control) to and
from the relevant switching equipment. It is not complicated by measures to protect against false operation by speech and thus has potential for simpler signalling arrangements compared with v.f. signalling.
Signalling during the speech period is possible. Unfortunately, however,
various constraints preclude large-scale application of outband signalling in existing analogue networks.

Interregister m.f. signalling is inband and, when applied is applied
regardless of the type of transmission media. Thus the variation in
signalling method as a consequence of the type of transmission system
in analogue networks concerns line signalling systems only, d.c., v.f.,
outband etc. systems being applied as appropriate when interregister
m.f. signalling is applied (Fig. 1.3). Should the line signalling incorporate decadic address signalling (interregister m.f. signalling not
applying), as in direct control switching networks, and in common

control switching networks should decadic address signals be transferred between registers, again, the type of line signalling system may vary according to the type of transmission system.

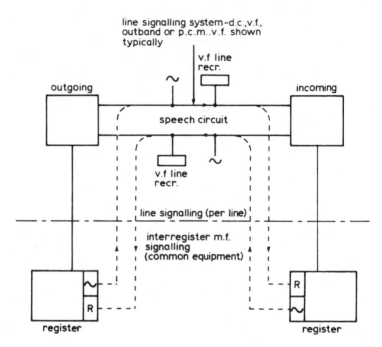

Fig. 1.3   *Line and interregister signalling*

1.3.4 Influence of facilities to be given

Simple signalling systems, such as d.c., have a limited signal repertoire. The trend in modern networks is for enhanced facilities (subscriber, system and administration), and thus increased signals. In the main, enhanced facilities are concerned with the connection set-up process and require registers or the equivalent (and thus common control switching), the signalling sophistication being given by an interregister m.f. signalling system. This approach permits all per-speech circuit line signalling to be simple, dealing with the relatively few supervisory signals in the main.

1.3.5 Part of network treated differently from signalling aspect

National telephone networks consist of many multiexchange local area (junction) networks and a long-distance (trunk, toll) network (Fig. 1.4).[6] The numerous local area networks represent a considerable cost investment and, as the traffic is not so concentrated as in the long-distance network, measures are adopted to reduce local network cost. To this end, the line plant in analogue networks is usually 2-wire unamplified audio and the line signalling local d.c., loop signalling being preferred. Local d.c. loop signalling has a limited range and some local d.c. signalling methods use single wire earth return signalling to increase the signalling limit.

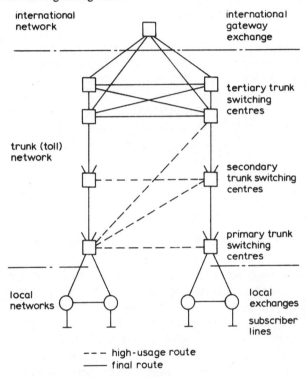

Fig. 1.4 *Typical national telephone network*
3-level trunk hierarchy

Local d.c. loop signalling is relatively cheap, requiring one signalling unit only per circuit; it is the simplest signalling method and no transmission equipment is involved. It can be used for supervisory signalling or for supervisory plus decadic address signalling, in local areas, and, even when the interregister signalling is m.f., local d.c. line signalling is preferred when the line plant and signalling limit permit. Low-frequency a.c., or long-distance d.c., signalling may be adopted when the local d.c. signalling limit is exceeded, both these methods requiring outgoing and incoming signalling terminals, which increase the cost relative to local d.c. signalling.

To contribute to simple and cheap local area signalling, the practice is often adopted to avoid incorporating certain 'system' signals which would normally be included in the more sophisticated line signalling systems in the long-distance network. Analogue long-distance network circuits are invariably 4-wire, mostly high frequency, the line signalling systems thus requiring both outgoing and incoming signalling terminals. Such signalling is more costly and, as stated, usually more sophisticated than local d.c. signalling.

All telephone signalling is required to be duplex in the sense that both forward and backward signalling applies without mutual interaction. The duplex requirement may be achieved by (*a*) simultaneous forward and backward signalling on the one signalling path, or (*b*) the use of two (forward and backward) simplex signalling paths. (*a*) is adopted for such techniques as local d.c. loop signalling on 2-wire circuits. (*b*) is adopted when separate forward and backward speech transmission paths, and thus signalling paths, are available, as on 4-wire circuits (and also on 2-wire circuits when, typically, band splitting is applied, e.g. interregister m.f. signalling with 2-wire switching or 2-wire circuits).

1.4 Signalling for pulse code modulation transmission systems

Pulse code modulation (p.c.m.) transmission is already extensively used in analogue networks for junction circuits, short-haul trunk routes and is beginning to be applied on long-haul trunk routes, the application being point-to-point in the analogue environment.[7] The inherent nature of p.c.m. allows a convenient way of transferrring signalling information as a built-in facility in individual p.c.m. systems without requiring the speech circuits to be equipped with analogue signalling systems.

With built-in p.c.m. signalling, d.c. signals from the analogue switching machines are converted into bit coded form at the transmit end of a p.c.m. channel and reconverted back to their original d.c. form at the receive end. The signalling is implicitly 4-wire. The signalling bit(s) may be inslot or outslot, the slot being the time slot identifying a specific p.c.m. speech channel, depending on the particular type of p.c.m. system, but, in either case, the signalling is time assigned to the individual speech channels and circuit labelling of signals is not required. Each speech channel has exclusive use of its own signalling facility, the total p.c.m. system signalling facility being time shared between the speech channels in the system.

All built-in p.c.m. signalling is continuous, the signal meaning being given by the bit-coded pattern of the signalling bit(s). When the code pattern identification starts at any point in the continuous signal code pattern sequence, such codes are called 'comma free'.

Built-in p.c.m. signalling replaces all conventional analogue line signalling systems in the relevant application, but not interregister m.f. signalling, for which analogue signalling systems continue to be applied with point-to-point p.c.m. transmission, the analogue m.f. signals being bit coded as for speech.

1.5 Common channel signalling

Stored program (processor) control (s.p.c.) of switching has prompted a reappraisal of the network signalling technique.[8-10] With s.p.c., it is inefficient for the processor, which works in the digital mode, to deal with analogue signals on the speech path. A much more efficient way of transferring information between s.p.c. exchanges is to provide a bi-directional high-speed data link between the two processors over which they transfer signals in digital form by means of coded bit fields. A group of speech circuits (many hundreds) can thus share a common channel signalling (c.c.s.) link in the time-shared mode, which has significant potential for signalling economy.[11] In preferred c.c.s., all the signals for the speech circuits served are passed over a signalling link which is separate from speech, c.c.s. thus replacing all conventional on-speech path signalling systems, with considerable potential for signalling rationalisation in networks.[12] While c.c.s. could be applied to non-s.p.c. exchanges, it is uneconomic in this situation.

Unlike built-in p.c.m. signalling, individual speech channels do not have exclusive use of a signalling facility with c.c.s., the signalling is not time assigned and each signal transferred must carry a bit coded circuit label which identifies the particular speech channel the signal is for.

C.C.S. may be applied in analogue or digital networks, the c.c.s. bit stream being obtained from modems in the analogue environment, and from the p.c.m. multiplex in the digital case.

Many administrations are presently planning for integrated digital networks, with integrated digital switching and transmission, which is of particular interest in regard to c.c.s.[13,14] Digital switching has potential for significant switching economy relative to space division switching.[15] Digital switching does not necessitate s.p.c., although this is preferred; wired logic control could apply. Digital switching implies high-speed digital transmission and thus the convenient availability of a high-speed bearer for digital c.c.s. Logically, time shared digital switching should be combined with time shared switching control, and thus s.p.c., and, equally logically, with the availability of s.p.c., c.c.s. should apply. Thus, in digital networks, all functions, switching, control, transmission and signalling can be time shared, which has potential for significant reduction in total network cost relative to that of an analogue network. Current planning for digital networks has adopted this approach.

1.6 International signalling

International gateway exchanges interface the international network and the various national networks, allowing the respective networks to evolve without mutual interaction. The international network has some similarity to national long-distance networks, and, in basic concept, the signalling problems are much the same. The international network, however, requires features not required in national networks, e.g. special operators, language-speaking operators, and the network includes features not normally included in many national networks, e.g. time assignment speech interpolation, satellites, echo suppressors etc., but, of course, which may be used in some. As a consequence, international signalling systems tend to be somewhat more complex than the national, to cater for such features, but the basic signalling concepts are much the same.

As with national systems, the international signalling systems have evolved in step with the evolving international network and a number of different designs of analogue signalling systems exist, v.f. line signalling being the main technique, with interregister m.f. signalling in some systems. Also, as with the national systems, there is little doubt that c.c.s. will be applied as the next step in the evolution of international signalling, which will offer the possibility of signalling rationalisation on the international network. Further, c.c.s. also offers the possibility of a rationalised signalling method for both the international and the national networks.[12]

References

1 FLOOD, J. E. (Ed): 'Telecommunication networks' (Peter Peregrinus, London, 1975), chap. 5
2 op. cit., chap. 4

3 op. cit., chap. 2

4 op. cit., chap. 3

5 op. cit., chap. 7

6 op. cit., chap. 1

7 CATTERMOLE, K. W.: 'Principles of pulse code modulation' (Iliffe, 1969)

8 *Bell Syst. Tech. J.* 1964, **43**, (5); 1969, **48**, (8); 1970, **49**, (10)

9 HILLS, M. T., and KANO, S.: 'Programming electronic switching systems' (Peter Peregrinus, London, 1975)

10 ADELAAR, H. H.: 'The 10c system – a stored program controlled reed switching system', *IEEE Trans.*, 1969, **COM-17**, pp. 333–339

11 'Specification of signalling system no. 6. CCITT: Green Book 6, Pt. 3, Recommendations Q251–Q295 ITU, Geneva, 1973

12 WELCH, S.: 'Common channel signalling – a flexible approach'. International Switching Symposium Record, Kyoto, Japan, 1976

13 MORNET, P., CHATELON, A., and LE CORRE, J.: 'Application of PCM to an integrated telephone network', *Electr. Commun.*, 1963, **38**, p. 23

14 DUERDOTH, W. T.: 'Possibility of an integrated switching and transmission network'. Colloque Internationale de Communication Électronique (Éditions Chiron, Paris, 1966), p. 64

15 ALEXANDER, S. P.: 'A multipurpose digital switching subsystem'. International Switching Symposium Record, Kyoto, Japan, 1976

Chapter 2

Subscriber line signalling

2.1 General

Subscriber equipment (telephone set, subscriber line and exchange termination) is usually considerably underutilised, and, as it represents a significant capital investment on a network basis, the subscriber's line is usually 2-wire audio for economy. A subscriber's line

(a) transmits speech in both directions
(b) is normally required to carry power from the exchange to the telephone set
(c) carries the appropriate audible tones to the caller (dial tone, busy tone, ring tone etc.)
(d) carries ring current to the called party
(e) transfers the necessary control signals from the subscribers (calling and called) to the exchange.

All these conditions are relatively independent and our present interest is in the control signalling requirement, the basic signalling from the telephone set being d.c. for simplicity and economy.

As discussed in Section 1.1, the basic signals are derived from the on-hook/off-hook conditions of the telephone set, or a sequential combination of these two conditions. When on-hook, the line loop between the telephone set and the exchange is open and no current flows. When off-hook, the line loop is closed and d.c. flows, also powering the telephone set. Thus the on-hook/off-hook conditions on the subscriber lines determine the required minimum conditions from telephone sets to the switched network (Fig. 2.1).

2.2 Signalling from dial telephone sets

The basic signals on the subscriber lines are:

forward (from calling subscriber):　seizure (off-hook), dialled address (sequential off-hook/on-hook)

clear forward (on-hook)

backward (from called subscriber):　answer (off-hook)

clear back (on-hook)

No current flows on the subscriber lines in the idle (on-hook) condition of the telephone sets.

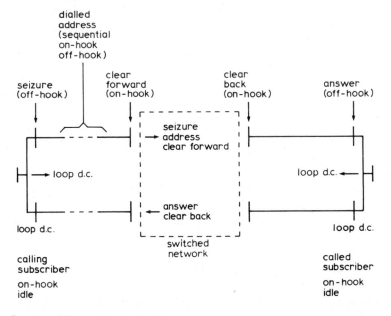

Fig. 2.1　*Minimum basic signalling*

The subscriber local exchange detects the caller's loop d.c. flow on seizure and connects appropriate equipment capable of receiving address information to the subscriber's line. As such equipment (a switch in direct control switching systems, and a register in indirect common control) is traffic provided, a slight delay occurs before it is connected, which necessitates a return of the dial tone when the equipment is connected, to inform the caller that dialling may commence.

The dialling interrupts the caller's loop d.c. in decadic form, the interruption occurring on the dial break (on-hook) condition, a decimal series of interruptions signalling the dialled digit value to the exchange, and similarly for other dialled digits. Thus a dialled digit consists of a sequence of breaks (on-hook) and makes (off-hook), the number of breaks indicating the digit value. An off-hook interdigital pause generated by the dial, and of duration much longer than the longest off-hook dialled make, serves to indicate the end of each dialled digit train to the exchange receiving equipment. This dialled address information, the called subscriber's number, achieves connection set-up.

The dial speed and break/make pulse ratio characteristics reflect the characteristics of the dialled address receiving equipment at the exchange, and, historically, by the characteristics of direct control switches. A subscriber dial speed of nominal 10 pulses per second (p.p.s.) is usual, but the nominal break/make ratio varies, reflecting the type of dialled address pulse receiving equipment used in particular networks, typically UK 66/33, the Bell system 61/39 and Germany 60/40. Tolerances vary, typically UK 9–11 p.p.s. 63–72% break and the Bell system 8–11 p.p.s. 58–64% break. The dial break/make ratio is deliberately not made 50/50 to compensate for the characteristics of relays, switches etc., and the signal transmission systems, which differ substantially, and to make the most advantageous use of circuit conditions occurring during the dial pulse break and make time intervals.

The on-hook clear forward is signalled to the calling subscriber's local exchange by cessation of the loop d.c. on the calling subscriber's line. This condition normally releases the connection and achieves release at any time in the call sequence. Timing discriminates between the on-hook dialled address break pulses and the on-hook caller's clear, the signal recognition duration of the latter being much longer than that of the longest dialled break pulse.

The off-hook answer condition results in loop d.c. flow on the called subscriber's line, and, when detected at the called party's local exchange, ceases the ring current (and the backward ring tone). The answer condition initiates call charge.

The on-hook clear back signal to the called subscriber's local exchange is given by cessation of the loop d.c. on the called subscriber's line. In the more usual calling party release networks, the clear back signal has no direct connection release function in the automatic service, but, in certain circumstances, may be used to cease the call charge and sometimes to initiate connection release, should the caller's clear forward signal be unduly delayed. In first party release networks, the clear back signal releases the connection should it occur before the clear forward signal.

All these conditions occur on the subscriber lines. The signalling is continuous', the signal meaning being given by the presence or absence of the loop d.c., and thus 2-state signalling applies in the forward direction (from the calling subscriber) and in the backward (from the called subscriber). A signal condition persists during on-hook or off-hook, being changed by change from on-hook to off-hook, and vice versa. The direct relationship between the signal current conditions and the on-hook/off-hook states results in simple signalling arrangements, but the signal repertoire is clearly limited. It is convenient to refer to this arrangement as the 'd.c. signalling condition' and as being a reference to assess signalling simplicity of other signalling arrangements.

Signalling on the subscriber lines incorporates the supervisory (seizure, answer, clear back, clear forward) and the selection (address information) signalling functions, both these being conveyed by the one signalling arrangement. It will be noted that the impedance at the signal sending end of the line is relatively low for the off-hook condition and substantially infinite for the on-hook, these conditions also applying for the sequential off-hook/on-hook dialling. As the line is alternately looped and disconnected to perform the signalling, this arrangement is sometimes referred to as being the 'loop-disconnect' technique. The varying send-end impedance is of no significance when supervisory signalling, but has significant pulse-distortion reaction when detecting d.c. address pulse signals, as discussed in Section 4.1.

The above minimum signals are also required to be transferred over the switched network on relevant calls (Fig. 2.1).[1] In some networks, the switched network conventional line signalling systems are based on merely reproducing the d.c. signalling condition (or its equivalent) and thus giving the minimum signals only. Other networks require more than the basic minimum signals on the switched network, typically requiring additional signals of the 'system' type, and often do not reproduce the equivalent of the d.c. signalling condition on line signalling.

With 4-wire circuits, having independent forward and backward signalling paths, it is a simple matter to obtain the 2-state signalling condition in each direction when signalling on the switched network. Complications arise when independent forward and backward signalling paths are not available on the switched network (typically when loop signalling on 2-wire circuits). Here, typically, the equivalent of the d.c. signalling condition may be given by loop d.c. on/off in the forward direction and the 2-state in the backward direction by reversals of the loop d.c. flow.

2.3 Pushbutton telephone sets

With pushbutton telephone sets ('touch tone',[2] 'keyphone'), the seizure, answer, clear back and clear foward d.c. signals are given in the same way as by dial sets. The selection function address signals, however, are transmitted on the caller's line in a coded form, the digit value being denoted by a single electrical signal transmitted to the exchange when a button is depressed. Compared with decadic dial pulsing, the method reduces fatigue (and errors) when a large number of address digits are to be sent [subscriber trunk dialling (s.t.d.), direct distance dialling (d.d.d.), international subscriber dialling (i.s.d.)], speeds up the address information transfer from the telephone set to the exchange and is highly compatible with common control nondecadic switching systems.

Although pushbutton telephone sets transmitting d.c. signals are known, and, in certain circumstances, may be economic, it is desirable that, to achieve an internationally standardised pushbutton signalling arrangement, the signalling should not be conditioned by transmission considerations of the subscriber lines. For this reason, it is preferred that the address signalling be a.c., the frequencies being within the voice range (inband), and thus capable of being conveyed by any subscriber line giving adequate speech transmission.[3] Assuming this approach, single signal address digit signalling implies a multifrequency arrangement, signals being frequency coded to indicate the address digit value.

The simplest multifrequency concept is a coding of two frequencies out of five (2/5 m.f.) to give the ten possible values of a digit, a signal thus consisting of two frequencies compounded, appropriate push-button m.f. receivers being equipped at the exchange. Early experimental work on subscriber pushbutton signalling adopted five frequencies of 700, 900, 1100, 1300, 1500 Hz, the signalling being 2/5 m.f., but this approach was abandoned, owing to the signal imitation problems which arose. When pushbutton signalling, depression of a button disconnects the telephone set microphone from circuit during the sending of an m.f. signal, but, with some arrangements, the microphone is restored to circuit between successive button depressions, and, of course, when address signalling is complete. As the signalling is inband, signal imitation and signal interference problems (discussed later in regard to inband voice frequency signalling systems) between button depressions can occur, the telephone-set microphone picking up background speech, music, noise etc., during the pushbutton address signalling process, with the possibility of false signalling.

As discussed later in connection with voice frequency signalling systems, a guard circuit in the receiver is a powerful feature which may be exploited to minimise signal imitation when signalling inband, the guard relying on frequencies outside the signalling bandwidth. The pushbutton signalling frequencies, however, being relatively numerous, embrace that part of the frequency spectrum which normally makes effective guard, and, as preferred frequencies are not available for guarding, it is difficult to guard efficiently by using guard frequencies when pushbutton signalling. With the nonavailability of frequency band guarding, it was established that 2/5 m.f. (and 2/6 m.f.) signalling was not satisfactory from the signal imitation aspect. For this reason, the CCITT* internationally recommended pushbutton signalling is of 2(1-out-of-4) [2(1/4) m.f.] form, a signal consisting of two frequencies compounded, one frequency from each of two separated blocks of four frequencies each, the frequency spacing within a block being narrower than the 200 Hz spacing of the early 2/5 m.f. (and 2/6 m.f.) experiments.[4] The 2(1/4) type of m.f. signal is less prone to imitation by voice etc., compared with 2/5 (or 2/6) m.f. and the signal imitation performance is satisfactory, despite the signalling frequencies spanning the voice band, no guard frequency bandwidth being available. Further protective safeguards, such as limiting and persistence checks are generally incorporated.[5] A 2-and-2-only frequency persistence check, typically for 70 ms, applied at the receiver checks for valid signal, the busy tone being returned if this condition is not met. The protection against signal imitation is not so powerful as in the inband v.f. signalling systems, but this is acceptable with pushbutton telephones, as, while speech etc. may be picked up by the microphone, unlike v.f. signalling, it is not normally present.

The CCITT recommended pushbutton frequencies:

> lower block: 697, 770, 852, 941 Hz
> upper block: 1209, 1336, 1477, 1633 Hz

are allocated (two frequencies per signal pulse, one from each block) to digits as in the matrix of Table 2.1. The physical disposition of the buttons is as in the Table.

This matrix gives 16 possibilities, ten being used for the digit value. Combinations 11 and 12 are often used for additional facilities transmitted from the telephone set, and combinations 13–16 are spare, with potential for future use for any desired purpose. The general use of pushbutton sets for purposes other than telephone address

---

* The International Telegraph and Telephone Consulative Committee.

Table 2.1

Frequency Hz	1209	1336	1477	1633
697	digit 1	2	3	(13) spare
770	4	5	6	(14) spare
852	7	8	9	(15) spare
941	(11) spare	0	(12) spare	(16) spare

signalling is a clear future possibility, e.g. the sending of slow-speed data from subscriber's telephones, and the potential for 16 signals has merit in this respect.

The CCITT recommendation includes the following technical detail:

(*a*) each transmitted frequency to be within ± 1·8% of the nominal frequency

(*b*) the total distortion products (resulting from harmonics or inter-modulation) must be at least 20 dB below the fundamental frequency.

It is not practicable to standardise the sending signal level, as these level conditions would depend on national network transmission plans, which are not the same in all countries. The CCITT recommends, however, that, for crosstalk reasons, the absolute power level of each component part of a short-duration signal in signalling systems should not exceed the values stated in Table 2.2, and that the sending level of the m.f. signals from the CCITT recommended pushbutton telephone set should not exceed the values stated.[6]

Table 2.2 *Maximum permissible value of power at a zero relative level point for pulse a.c. signals*

Signal frequency, Hz	Maximum permissible power for a signal at a zero relative level point, μW	Corresponding absolute power level, dB referred to 1 mW (dBmO)
800	750	− 1
1200	500	− 3
1600	400	− 4
2000	300	− 5
2400	250	− 6
2800	150	− 8
3200	150	− 8

When signals are made up of two different frequency components compounded, the maximum permissible values for the absolute power levels of each frequency are 3 dB below the values given in Table 2.2.

Pushbutton signalling may be applied for direct control switching systems without the provision of m.f. receivers by arranging for the pushbutton set to transmit decadic d.c. pulses at the desired speed when a button is depressed. This method is not internationally standardised, but is sometimes used as expediency in certain direct control switching situations when economic provision of pushbutton telephone sets is desired.

References

1   WELCH, S.: 'The fundamentals of direct current pulsing in multiexchange areas'. Institution of Post Office Electrical Engineers, Printed Paper No. 184, 1944
2   HAM, J. H. and WEST, F.: 'A touch-tone caller for station sets', *IEEE Trans.*, 1963, CE-9, p. 17
3   SCHENKER, L.: 'Pushbutton calling with a two-group voice frequency code', *Bell Syst. Tech. J.*, 1960, 38, p. 254
4   CCITT: Green Book, 6, Pt. 1, Recommendations Q11 and Q23, ITU, Geneva, 1973
5   WELCH, S.: 'The influence of signal imitation on the design of voice frequency signalling systems', Institution of Post Office Electrical Engineers, Printed Paper No. 206, 1953
6   CCITT: Green Book, 6, Pt. 1, Recommendation Q16, ITU, Geneva, 1973

Chapter 3

Junction network signalling

3.1 General

A telephone exchange connects two of the exchange terminations for a call, a termination being a subscriber's line, an external line connecting with another exchange or an access to a service (operator, directory enquiry etc.). Means must be provided to indicate a demand for a call, to receive the instructions for a connection and to supervise the connection so that it can be cleared down when the call is terminated. Provisions for calling a wanted subscriber (ringing) and for indicating the chargeable duration of a call are also frequently associated with the supervisory arrangements.

These basic functions must be carried out by any telephone exchange switching system. The means for indicating that a subscriber's line is calling must be individual to that line, and is called the subscriber's line circuit. This line circuit must register the following conditions to indicate them to the switching equipment:

> line idle
> line calling (but call not yet attended to)
> line engaged on a connection.

In basic concept, these conditions are registered on two relays (or the electronic equivalent), known as the line relay (L) and the cut-off relay (K). The L relay operates to the subscriber's loop when the subscriber goes off-hook to initiate a call, and this signals to the exchange that a connection is required. The K relay operates when the line is connected to the switching equipment and this cuts off the calling signal produced by the L relay. The K relay, but not the L relay, operates on an incoming call to the subscriber to prevent originating call conditions being set up when the subscriber goes off-hook to answer. In systems that use meters (message registers) for recording the charge of calls, these are also associated with the subscriber's line circuit.[1]

Equipment for supervising calls is only needed for call connections, which are much fewer than the total number of subscriber line terminations on an exchange. Consequently, the supervisory functions are not performed by the subscriber line circuits, but by equipment accessed by the switching stage(s) to concentrate calls from a large number of subscribers onto a smaller amount of supervisory equipment, thus minimising the amount of supervisory equipment required (Fig. 3.1).[2]

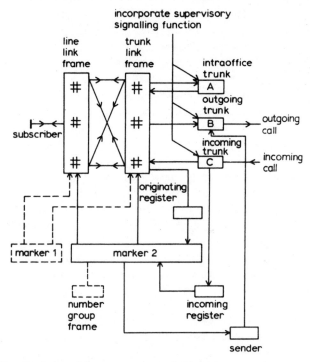

Fig. 3.1   *Typical trunking common control switching*

The supervisory equipment (generic term 'supervisory unit' and variously called supervisory link, junctor circuit, trunk circuit, auto-to-auto relay set, outgoing dial pulse repeater etc.) is thus needed at each exchange:

      to send and receive supervisory signals
      when applicable, to repeat decadic address d.c. pulse signals
      to incorporate a transmission bridge which supplies microphone feed current to power the telephone sets.

On an own exchange call, the supervisory unit supplies d.c., from the respective sides of the transmission bridge, to power the calling and called telephone sets, respectively. On calls extended over external line(s), the supervisory unit at the originating exchange powers the calling subscriber's telephone, that at the terminating exchange powers the called subscriber's telephone and the supervisory unit at tandem exchanges receives and retransmits the supervisory signals transferred in the respective directions between the originating and terminating exchanges. Thus the form of the supervisory unit varies and depends on its location, function and call connections (Figs. 3.2–4).

In direct control switching systems, the last switch in the connection, for example, the final selector in the UK direct control step-by-step system, the connector in the Bell step-by-step system, usually incorporates the supervisory equipment (Fig. 3.4). In this event, the supervisory equipment for an own exchange call is the same type as that in the terminating exchange on a call connected over the external line(s), and the supervisory equipment at the originating exchange is the same type as that at the tandem exchange(s) (Fig. 3.4).[3,4]

In common control switching systems, the supervisory equipment is not built-in with switches and separate units of appropriate type, typically equipment A, B and C of Fig. 3.1, are used, depending on the type of call: own exchange, outgoing or incoming.[3,4]

Line units associated with external lines enable electrically independent signal conditions to be used in the exchanges and on the external lines, and the supervisory signalling function is often combined with such line units (typically units B and C Fig. 3.1). This is of particular significance with an incoming line unit, as external lines in local areas are much more heavily loaded than subscriber lines, and it is therefore unnecessary to provide a switching stage to connect incoming calls from external lines onto fewer supervisory units.

It is convenient to refer to the external line speech circuit groups in multiexchange local telephone areas as junction circuits (trunks). The cost of local area equipment (switching, junctions and subscriber lines) is a significant part of the total network cost, owing to the relatively large quantities involved. For economy, the junction circuits are usually 2-wire unamplified audio, although, exceptionally, 2- or 4-wire amplified circuits may be used, and the line signalling is d.c. on the speech path; d.c. signalling is preferred in local junction networks because it is simple and relatively cheap. Point-to-point p.c.m. transmission with analogue switching is being increasingly applied in junction and short-haul networks, the signalling being the built-in p.c.m. types as discussed in Sections 6.2 and 6.3.

D.C. signals may be obtained in many different ways: opening and closing of a loop or a wire, high/low-resistance loop, application or removal of potentials, potential reversals, continuous or pulse signals etc., and thus there are many different forms of junction network d.c. signalling systems; the choice depends on the conditions to be met, particularly on the number of signals to be given and the required signalling line limit.[5-7] Loop d.c. signalling, which aims primarily at reproducing the basic d.c. signalling condition, is simple, reliable and is preferred when the signals to be given are relatively few, the continuous-type signalling permitting simple arrangements.

Both the supervisory and selection (address) signalling functions (Section 1.2) must be transferred to other exchanges on junction calls, and the arrangements for the signal transfer differ, depending on the type of switching system: common control or direct control.[3,4,6] In common control switching, the address information is stored in a register and all (or part) is transferred between exchanges as inter-register signals; this signalling may be by decadic d.c. pulses or an interregister m.f. signalling system. The line signalling system transfers the supervisory signals (Figs. 3.2–3). Should the interregister signalling be decadic d.c., arrangements are often adopted to avoid pulse repetition of the d.c. address pulses by the relevant supervisory units. In direct control switching systems, step-by-step, for example, while differentiation between the supervisory and selection signalling functions exists, it is not practicable to separate the two types of signal and in this event the supervisory unit repeats the decadic address d.c. pulse signals on junction calls, and the line signalling system transfers both the supervisory and address signalling, as, for example, in the auto-to-auto relay set in the UK step-by-step system (Fig. 3.4).

3.2 Typical forms of junction network d.c. signalling

There are many different forms of junction d.c. signalling, of which loop d.c. signalling is the simplest, and certainly the most widely applied. Loop d.c. signalling is based on the lines being successively looped and disconnected; there are many different designs, varying in detail, but the common principle may be illustrated by the typical examples of the Bell system 'loop reverse battery' and the UK 'loop-disconnect' methods.

3.2.1 Loop d.c. signalling

Bell system loop reverse battery d.c. signalling:[8,9] Fig. 3.2 shows the basic principle in simplified form, as applied to an interexchange line between, typically, common control crossbar exchanges. Direct control step-by-step switching exchanges employ similar circuits, except that relay A repeats the decadic address d.c. signals. The originating exchange relay A responds to the caller's off-hook seizure condition and A1 and A2 repeat this condition to the distant exchange. The polarised relay CS is polarised so that it does not operate to this condition. The terminating relay A operates to the incoming seizure loop current, A1 and A2 extend the connection to the called party and the ring current is applied. The operation of the respective A relays holds the connection.

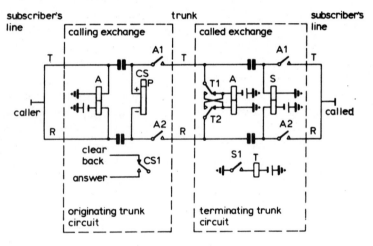

Fig. 3.2   *Loop reverse battery d.c. signalling*

On answer, the called party's off-hook condition operates S, which trips the ring, operates T, and T1 and T2 reverse the junction line loop current to operate CS, CS1 setting up the answer condition at the originating exchange. The called party's on-hook clear releases S, and the consequent release of T restores the polarity of the junction loop current to the incoming seizure direction (assuming that the clear back is prior to the clear forward), to release CS, which sets up the clear back condition at the originating exchange. The caller's on-hook clear forward, in releasing relay A, repeats the clear forward to the distant exchange by A1 and A2 disconnecting the line loop to release A at the

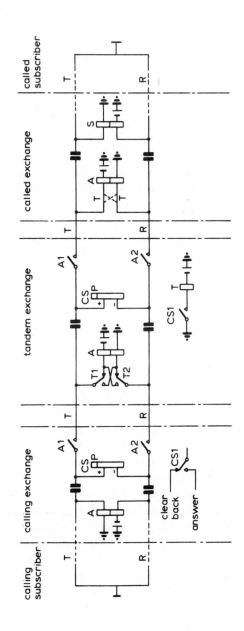

Fig. 3.3 *Repeated reverse battery signalling*

terminating exchange, the connection being released in sequence, starting at the calling exchange. The clear forward signal achieves connection release in any condition of the connection, and is thus overriding.

Fig. 3.3 illustrates the principle of repeated reverse battery signalling at a tandem exchange. Tandem relay CS operates to the reversed line loop current from the called exchange on called party answer. T, operated by CS1, reverses the incoming line loop current by changeover of T1 and T2 to repeat the answer current reversal to the calling exchange. Tandem relay CS releases to the restoration of the line loop current polarity from the called exchange on the called party's on-hook clear, and the release of T repeats the clear back from the tandem to the calling exchange by the restoration of the incoming line loop current to the seizure (pre-answer) polarity.

UK loop-disconnect d.c. signalling:[6,10] Fig. 3.4 shows a typical direct control switching single junction link condition, the receive circuit being part of the final selector switch in the UK step-by-step direct control system. The supervisory auto-to-auto relay set interface, outgoing on the junction, converts the, typically, 3-wire $(-, +, p)$ exchange condition to the 2-wire $(-, +)$ junction condition. Some administrations designate the wires as *a, b, c* or T, R, S, respectively. No incoming signalling interface equipment is provided on the junction, which contributes to the signalling economy.

Assume that the number 34567 is dialled; the typical connection is shown in Fig. 3.4. The first digit 3 steps the switch directly in exchange A over the calling subscriber's line, and the switch seizes an outgoing auto-to-auto relay set on the junction to exchange B. The auto-to-auto relay set is now held to the off-hook caller's telephone and supplies d.c. to the caller via relay A which operates. A1 operated at exchange A extends the loop seizure over the junction to seize the incoming switch terminating the junction at B. Relay D at A, polarised by the rectifiers, does not operate to the seizure junction loop d.c. polarity. The further dialled digits 4567 step the relevant switches at B, the address signals being repeated over the junction in the loop-disconnect mode by A1 at A. Arrangements are included to short circuit the transmission bridge during this address signalling repetition, the short circuit being removed during each interdigital pause and at the end of address signalling.

The step-by-step final selector switch at B incorporates the transmission bridge to supply d.c. to the called party, and ring current is applied should the called line be free, busy tone being returned if it is engaged. The operation of the A relays at A and B holds the connection.

Relay D at B operates to the subscriber's loop current when the called party goes off-hook to answer. D1 and D2 reverse the polarity of the loop d.c. over the junction to operate D at A, this reversal thus transmits the answer signal on the network. Relays A and D at A and B are in the operated condition during the speech period. Should the call be to a noncharged nonelectrical answer signal condition service, the D relay(s) may not be operated to the off-hook answer, but this does not negate speech transmission.

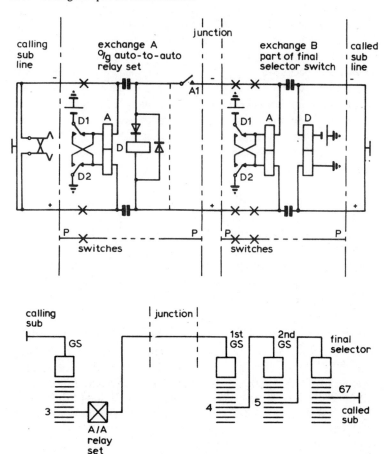

Fig. 3.4   *UK loop d.c. signalling principle*

Relay D at B releases to the on-hook clear from the called party, and, assuming the called party's clear occurs before the caller's, this restores the loop d.c. on the junction to the pre-answer polarity to release D at A. This restoration of the junction loop d.c. polarity to the pre-answer condition thus transfers the clear back on the network.

Relay A at A releases to the caller's on-hook clear, the connection release condition. This, at A1, disconnects the junction loop to signal the clear forward over the junction to B, to release relay A. The release of the A relays at A and B release the equipment at A and B, respectively. The clear forward signal achieves connection release in any condition of the connection and is thus overriding. Timing, with a generous margin, discriminates between the caller's on-hook dialled address pulse break and the on-hook clear forward.

Similar arrangements apply on multilink connections, an auto-to-auto relay set being equipped outgoing on each junction, the signals, forward and backward, being repeated at the tandem switching points.

Common control switching systems employ much the same basic arrangements for loop-disconnect d.c. signalling (electromechanical or electronic) as described for the direct control switching for the supervisory signalling, whose signalling reception is by a supervisory unit at the terminating exchange and not by a switch; the address signalling is interregister.

Comments on loop d.c. junction signalling:
(*a*) It will be noted that the basic 2-state signalling condition, on-hook/off-hook, is reproduced forward and backward on the network. All telephone call processing signalling is required to be duplex, and, when signalling on 2-wire circuits, this requires loop current forward for the forward signals and loop current backward for the backward signals, should loop signalling be the requirement. Both the forward and backward signals are required to operate independently, and both loop current conditions to apply simultaneously, to meet the duplex requirement. In loop d.c. signalling, this is achieved by the loop current operated by disconnect and loop in the forward direction to give the forward signalling, and by change of loop current polarity in the backward direction, to give the backward signalling. This accounts for the reverse battery signalling in both the typical Bell and UK loop d.c. signalling methods, which, while differing in design detail, are the same in principle. The 'reverse battery' in the Bell loop d.c. signalling simply denotes the called party's off-hook/on-hook reversals of the loop d.c.
(*b*) A transmission bridge separates an input from an output from the d.c. aspect and may be transformer or capacitor type (Figs. 3.2–4 show the capacitor type). On junction calls, the bridge at the originating exchange supplies d.c. to the caller's telephone (relay A, Figs. 3.2–4)

and that at the terminating exchange supplies d.c. to the called telephone (relay S, Figs. 3.2–3 and relay D, Fig. 3.4). On an own exchange call, of course, the transmission bridge powers both the calling and called telephones, d.c. being supplied from each side of the bridge.
(*c*) All supervisory units incorporate a transmission bridge and all d.c. signals are repeated across the bridge. All local d.c. signalling, and, indeed, all d.c. signalling, is thus implicitly link-by-link.
(*d*) All the signals associated with loop d.c. signalling are balanced and do not give rise to interference, as induced noise voltages affect the two wires symmetrically and cancel. Both the battery and earth potentials are applied at the same exchange, giving immunity from earth potential differences.
(*e*) On release, outgoing equipment must be maintained in a busy state to prevent follow-on calls while equipment at the distant end of the junction is releasing. In junction network signalling, this is usually achieved by a locally applied busying feature at the outgoing end, the time being preset to cover the normal release time of the incoming equipment.
(*f*) The basic signal repertoire given by 2-state local loop d.c. signalling is not always adequate for junction network signalling, and further signal(s) may be obtained within the basic 2-state concept. Typically, when, in the UK network, meter pulse signalling during the speech period is applied on junction circuits for bulk billing (nonitemised) call charging, the meter pulses are given by short loop d.c. reversals in the backward direction, the pulse shapes being gradual (softened) to ensure that the pulses are inaudible. This signalling conforms to the basic 2-state condition.

3.2.2 Other forms of junction network d.c. signalling
Loop d.c. signalling, reproducing the basic d.c. signalling condition, has limitations in signal repertoire and in signalling ohmic limit, and some administrations adopt different d.c. junction signalling methods to overcome these limitations as far as their networks are concerned. The following gives brief general information of some typical alternative arrangements to illustrate the basic differences:

Single wire d.c. signalling:[5,6] Earth current signals are transmitted on the individual wires of the junction. A battery (negative potential) or an earth can be applied to any of the wires for signalling, earth or battery, respectively, being present at the distant end. The signalling is not balanced, and the immunity from the effects of inductive interference (noise etc.) and earth potential differences is less than that for loop signalling. Table 3.1 shows a typical arrangement adopted, the three wires *a*, *b* and *c* being run on the junction circuit between exchanges A and B. Other arrangements are possible.

Table 3.1   *Typical earth-return signal code*

| Signal | Form of signal | | Signalling direction, → forward ← backward |
	At outgoing exchange A	At incoming exchange B	
Seizure	Earth on *c*-wire		→
Address information	Earth pulses on *a*-wire during dial breaks		→
Clear forward	Disconnection of earth from *c*-wire		→
Number received		150 ms earth pulse on *b*-wire	←
Answer		Battery on *a*- and *b*-wires	←
Meter pulses		Pulses of battery (augmented current on *b*-wire)	←
Clear back		Disconnection of battery from *a*-wire and change from battery to earth on *b*-wire	←

Combination of single-wire and pulse signalling:[5-7] This is sometimes adopted to extend the signal repertoire still further by a combination of single wire and pulse signalling. It employs pulse signalling to a greater extent than the arrangement of Table 3.1. Table 3.2 shows a typical case, the three wires *a*, *b* and *c* being run on the junction circuit between exchanges A and B. Other arrangements are possible.

High/low-resistance d.c. signalling: To extend the signal repertoire above that given by the loop d.c. signalling basic d.c. signalling condition, high/low-resistance signalling may be employed as high/low-resistance loops, high/low-resistance wire(s) or a combination of both. A typical high/low-resistance loop discrimination is high-resistance 22 000 Ω loop, low-resistance 440 Ω loop.[8] It is clear that such a technique complicates the junction network signalling and departs

Table 3.2   *Typical earth-return plus pulse d.c. signal code*

| Signal | Form of signal | | Signalling direction, → forward ← backward |
	At outgoing exchange A	At incoming exchange B	
Seizure	Earth on c-wire		→
Address information	Earth pulses on a-wire during dial breaks		→
Trunk-call indicator	70 ms battery pulse on b-wire		→
Trunk offering	60 ms earth pulse on a-wire		→
Coin-box line	55 ms earth pulse on a-wire		→
On-hook signalling (for malicious-call tracing)	200 ms earth pulse on a-wire		→
Clear forward	Disconnection of earth from c-wire		→
Number received		Earth on b-wire for time of last digit plus 130 ms	←
Answer		130 ms earth pulse on a-wire, battery on b-wire	←
Subscriber busy and congestion		Battery on a-wire, earth on b-wire	←
Clear back		Continuous earth pulsing on a-wire (155 ms make, 475 ms break), battery on b-wire	←
Blocking		Disconnection of c-wire	←

from the basic simplicity of loop d.c. signalling by the opening and closing of the loop or by current reversals.

Battery and earth pulsing: This is adopted by the Bell system to increase the signalling limit, particularly the decadic address pulsing limit, above that given by the loop reverse battery d.c. signalling method of Figs. 3.2–3, but without the use of an incoming signalling interface.[8,9] The T and R speech conductors only are run on the junctions. In one arrangement, termed 'battery and earth pulsing, loop supervision', the extended limit is achieved by applying battery and earth at both ends of the line T and R conductors, but with opposite polarities at each end (Fig. 3.5) for the decadic address signalling. This nearly doubles the current flow over the line loop, which improves the decadic address pulsing limit and the pulse distortion. Contacts A1 and A2 open and close both line conductors to furnish forward on-hook/off-hook signals.

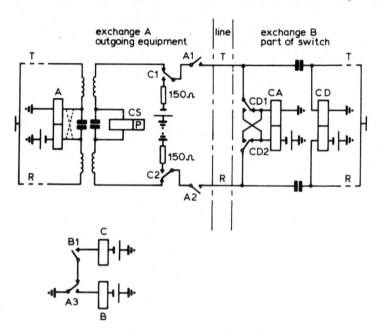

Fig. 3.5   *Principle of Bell system battery earth pulsing with loop supervision*

The response of CA at the incoming end serves to control the incoming equipment corresponding to the on-hook/off-hook conditions from the outgoing end. Relays B and C are slow release, B being maintained operated for the call duration and during all decadic address signalling, and C being maintained operated during each digit, but releasing during the interdigital pauses and on completion of dialling. When released, C1 and C2 substitute the transmission bridge supervisory relay CS for the pulsing battery and earth to detect the backward loop current signals and thus the on-hook/off-hook conditions of the called party. A1 and A2 operated to the caller's off-hook extend seizure via relay

CS to the distant equipment. The polarised relay CS does not operate to this loop current direction. C1 and C2 operate on the first break pulse of a digit to give the two single conductor conditions and A1 and A2 repeat the address pulses forward as battery and earth pulses on the respective conductors. The pulsing is effectively loop-disconnect d.c. in the sense that the sending-end impedances are not equal during the address pulse make and break periods. CD operates to the called party's off-hook answer to reverse the line loop current to operate CS. Subsequently CS releases to the restoration of the loop current direction on the called party's on-hook clear. A releases on the caller's on-hook clear to initiate release of the connection, A1 and A2 releasing repeating the clear forward to the distant equipment. The clear forward is over-riding to release the connection.

When a further extended limit is required, 'battery and earth pulsing, battery and earth supervision' may be employed, the supervisory signalling being earth return instead of loop.

Pulsing noise interference can arise in some circumstances, owing to the two single conductor pulsing condition.

3.2.3 Low-frequency a.c. signalling
This technique is defined as that used by systems using a signalling frequency below 300 Hz and thus below the speech band, typical frequencies used being 25, 50, 80, 135, 150 or 200 Hz.[5] The technique (Fig. 3.6) is usually referred to as simply 'a.c. signalling' to imply a distinction between this and voice frequency and outband signalling, which are also a.c. D.C. signals are converted to low-frequency a.c. signals for transmission over the line, the a.c. signals being converted to d.c. at the distant end. Both outgoing and incoming signalling interface terminals are required and the signalling is link-by-link on multilink connections.

The system may be applied to metallic circuits above the range of junction network d.c. signalling and to circuits isolated for d.c., where d.c. signalling is not possible. The technique is not applied to f.d.m. carrier circuits because of the very high attenuation at the low signalling frequency. When adopted, low-frequency a.c. signalling fills the gap between local d.c. signalling and carrier circuits with v.f. or outband signalling. Some administrations adopt long-distance d.c. signalling (l.d.d.c.) to fill this gap, either by preference for the simplicity and reliability of l.d.d.c. or because of inability of equipment in the network to pass the low a.c. frequency.

Since the more usual application of low-frequency a.c. signalling is on 2-wire circuits, and thus only one transmission path being available for the two signalling directions, the signals are generally pulsed. The phase position at the start of a pulse may significantly reduce the transmitted a.c. pulse length. To overcome this distortion, particularly when decadic address pulses are transferred, some administrations incorporate pulse correction, which complicates the system; others arrange that the signalling terminals themselves generate the necessary a.c. to ensure a correctly timed start of signalling.

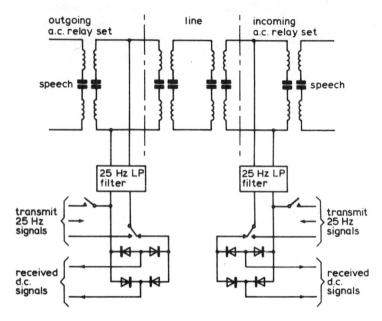

Fig. 3.6   *Simplified low-frequency a.c. signalling*

Assuming that some precaution against excessive pulse distortion is adopted, then, within reason, the line limit is not so constrained by this factor. As the attenuation to the low-frequency low-level signal is relatively high, which would normally call for a high-level transmitted signal, perhaps the main factor in determining the line limit is the permissible difference between the transmit and receive signal levels. The transmit level is limited by crosstalk and the receive level by the sensitivity of components. In practice, the permissible line length is limited, being of the order 80–100 km maximum.

The technique permits signalling during the speech period, e.g. meter pulses over junctions; this signalling should be inaudible to subscribers. The lower the frequency, the more certain it is that the lower-order harmonics, with a still relatively high energy content, do not extend to the speech band. For this reason, a low-frequency (say 25 Hz) is preferred when meter pulse transmission during speech is required. Signal frequencies of up to 200 Hz can be used in applications not requiring meter pulses.

In the pulse signal codes usually adopted, it is important that important signals are significantly longer than any less important signal(s) in the opposite direction to ensure that the more important signal(s) gets through to be effective. Also, relative to more precise signalling methods, two signals transmitted consecutively in the same direction must be well spaced to ensure clear recognition as distinctive signals. For example, in the typical signal code of Table 3.3, if the backward number received signal and the answer signal merged into each other for any reason (such as adverse timing tolerances) the condition could be falsely recognised as busy.

Table 3.3 *Typical pulse signal code, low-frequency a.c. signalling*

Signal	Duration, ms
Forward	
Seizure	40
Decadic address pulsing	40
Clear forward	1600
Backward	
Number received	150
Answer	150
Busy	750
Meter pulses	140
Clear back	Pulse 150, gap 475 repeated
Release guard	750
Blocking	Continuous

Although a number of national networks, particularly in Europe, have significant provision for low-frequency a.c. signalling, the technique is declining in importance:

(*a*) it is not suitable for the long-distance network.
(*b*) the increasing application of p.c.m. transmission systems in junction and short-haul networks reduces low-frequency a.c. signalling provision, even in the shorter line ranges.

(*c*) in some situations, the technique competes with l.d.d.c., which is often preferred, as, apart from reliability and signalling precision, l.d.d.c. has application in both the local junction and the trunk long-distance networks.

3.2.4 Proceed-to-send

In common control switching, time is required to associate a traffic provided register (sender), usually accessed by switching stage(s), with an incoming line.[11] Similarly, in direct control switching, time is required to seize a traffic provided incoming switch over a line. Thus the sending of address signals over lines to such equipment must be delayed to accommodate for this time, which gives rise to a proceed-to-send philosophy.

Dial tone from a register, or a switch, is a form of proceed-to-send indication, in that the start of dialling, or keying, is required to be delayed until the register, or switch, is associated with the calling line at the calling subscriber's exchange and is prepared to accept address signals. On calls extended over external lines, other means must be adopted for proceed-to-send indication on the switched network.

As no storage exists in direct control switching systems, there is no facility to automatically delay sending address signals over the network, and proceed-to-send indications are not given. The line and the distant switch terminating the line are seized, and the switch conditioned to accept address signals, during the interdigital pause of the dial, which is normally adequate. Thus the problem of delaying the sending of address signals over the switched network concerns common control switching.

Address information storage exists in common control switching systems, which permits delayed outpulsing, and a number of possibilities exist to control the outpulsing from an outgoing register (or sender):

(*a*) After outgoing seizure of the line, a fixed time delay is used before outpulsing, to cover the normal time required for incoming register association. A backward proceed-to-send indication on the network is not given.

(*b*) Liberal register provision to minimise the above outpulsing fixed time delay.

(*c*) The outpulsing is delayed until a backward proceed-to-send signal is received, which is returned when an incoming register is line associated and prepared to accept address signals. This is a positive indication and is preferred.

Network proceed-to-send signals are transmitted in different ways depending on the circumstances and arrangements of the common control switching system and are typically:

(*a*) Modern analogue networks transfer the address (and other relevant) signals by interregister m.f. signalling systems in both the local junction and the long-distance networks, the line signalling systems transferring the supervisory signals. When the interregister m.f. signalling system incorporates both forward and backward signalling, which is preferred, the proceed-to-send indications are given by backward interregister m.f. signals, and not by signals in the line signalling systems. This approach has the merit that the m.f. coding of the proceed-to-send signal can give additional information of the request nature, e.g. requesting the sending of a particular address digit, or group of digits, which is of significance when the interregister m.f. signalling is operated in the end-to-end signalling mode. This approach is adopted by the CEPT* CCITT R2 and the UK MF2 interregister m.f. signalling systems (Sections 10.7 and 10.8).

(*b*) When interregister m.f. signalling with forward signalling only is used (as in the Bell system R1), the lack of backward m.f. signalling precludes a backward proceed-to-send signal in the interregister signalling system, and the signal must be included in the relevant line signalling systems. It should be noted that the Bell interregister m.f. signalling system is operated in the link-by-link signalling mode, and not end-to-end, and it is not required to utilise coding of proceed-to-send signals for additional information of the request nature. The requirement is for a signal with a proceed-to-send meaning only.

(*c*) When the interregister address information transfer is by decadic d.c. pulses (and not coded m.f.), no backward interregister signalling exists, and the proceed-to-send signal must be included in the line signalling systems.

(*d*) Decadic address d.c. outpulsing from common control exchanges to direct control exchanges (an interworking condition) does not usually necessitate proceed-to-send indication, it being sufficient for the register (or sender) to delay sending the first pulse for a short time after outgoing circuit seizure to allow time for the distant switch to be seized and prepared to accept address signals. Special circumstances however, may require the outpulsing to be controlled by a backward line signal from the direct control exchange (as instanced by the 'stop/go' method of operation described in Section 5 of Reference 8).

---

* European Conference of Posts and Telecommunications Administrations.

When required to be conveyed by the line signalling systems in junction networks, the backward proceed-to-send indication can be given in various ways; for example:

(*a*) A pulse proceed-to-send signal in pulse-type d.c. (and other pulse-type) line signalling systems.

(*b*) With loop d.c. signalling (and other Bell signalling systems), the Bell system 'delay dialling' method of controlling the outpulsing. On seizure detection, the incoming exchange immediately returns an off-hook delay dialling signal for at least 140 ms until a register is associated and ready to accept address signals and, when ready, the incoming exchange returns an on-hook 'start dialling' (proceed-to-send) signal.[8]

(*c*) With loop d.c. signalling (and other Bell signalling systems), the later Bell system 'wink' method of controlling outpulsing. On seizure detection, the incoming exchange does not immediately send an off-hook delay dialling signal; the idle backward on-hook condition from the incoming exchange is maintained until a register is associated and ready to accept address signals. When ready, the backward on-hook condition is changed to off-hook for a period of some 140–290 ms, the on-hook condition being then restored. This off-hook interruption (wink) of the backward idle on-hook condition is the proceed-to-send, and outpulsing commences on termination of the wink.[8]

Note: In the Bell delay dialling and wink, the backward signals cannot be defined in absolute terms of polarity because of the many variations in practice; it is always true that on-hook is the polarity existing when awaiting the called party answer, while off-hook is the polarity when the called party has answered. In local d.c. signalling, the calling end receives signals transmitted by the called end by operation of the polarised relay CS (Figs. 3.2–3). Trunks not requiring the delay dialling signal are in the on-hook condition when idle, whereas trunks requiring the delay dialling signal may be in the off-hook condition when idle. Trunks arranged for the wink proceed-to-send signal are in the on-hook condition when idle.[8]

Both the 'delay dialling' and 'wink' line signalling methods of controlling register outpulsing, while involving timing, are compatible with the 2-state on/off line signalling philosophy. On the other hand, however, any additional signal(s) by timing begin to erode the basic simplicity of 2-state signalling and gives rise to other problems; for example, both the delay dialling and wink involve a short off-hook period, the double seizure (glare) on bothway working of circuits involves detection by simultaneous recognition of both outgoing and incoming off-hook seizures at the same end, and, to avoid confusion, the recognition time of the off-hook glare must be longer than the off-hook delay dialling or wink. This is not a significant problem, but it serves to illustrate the type of situation which can arise when the basic 2-state signalling concept is departed from. Bothway working is usually limited to long-distance network circuits, and is not applied in local junction networks.

The various proceed-to-send conditions have been discussed in relation to local junction d.c. signalling, but the same considerations hold for the long-distance (trunk, toll) signalling systems. While the main function of the proceed-to-send signal is to control outpulsing, the signal is also an acknowledgment of a previous forward signal, such as seizure. Thus the combination of seizure and proceed-to-send signals performs a per-call continuity and signal integrity check of the line. The arrangement gives assurance that the line is capable to conveying address information satisfactorily. Without the proceed-to-send signal, the possibility could arise that an outgoing register could outpulse to an unsatisfactory line and the caller be left high and dry.

3.2.5 Analysis of continuous and pulse signalling
Consideration of choice of continuous or pulse signalling applies to both d.c. and a.c. signalling and to both local junction and long-distance network signalling systems. In modern common control switching analogue networks, the signalling in both the local junction and long-distance networks has much in common, the main variant being in the type of line (supervisory) signalling system. Local d.c. line signalling is the main, but not exclusive, application in local junction networks, and it is appropriate to examine the continuous as opposed to pulse signalling problem in relation to local d.c. signalling at this stage. The problem for the v.f., outband, and interregister m.f. signalling is considered later in relation to these systems.

Continuous d.c. signalling: Here, in the theoretical concept, the significant merit is in the simple recognition of the 2-state on-hook/off-hook condition in each direction, no timing being required to complicate the system. In reality, however, a measure of timing is necessary, as appropriate signal recognition delays are necessary to safeguard against false signalling conditions, due, for example, to line interruptions when the signal meaning is given by d.c. cessation. Further, appropriate delay in signal recognition is a normal and necessary feature in some cases, for example, d.c. cessation on the on-hook clear forward is relatively long and longer than the on-hook dial break pulse d.c. cessation. It should be noted that, should the decadic address pulsing be incorporated in the line signalling system, the break/make pulse times are generated outside the signalling system. Thus, in the consideration so far, the necessary timings are in no sense transmitted pulse timings and the continuous signalling gives highly desirable signalling simplicity with good operating margins, combined with reliability in that transient interruptions do not normally cause loss of signal, as may arise with pulse signalling.

Signals additional to the basic on-hook/off-hook conditions are often required, e.g. proceed-to-send, meter pulses over junctions, etc. This additional signal requirement can be conveyed in the continuous 2-state philosophy as distinctive signals by time coding the on/off intervals, but this could imply departure from the theoretical merit of 2-state signalling in that pulse timing the on/off conditions could apply for certain of the additional signals. Thus, depending on the number of additional signals to be given, the boundary between the continuous and pulse signalling techniques could become blurred.

Modern signalling systems are related to common control switching systems, which allows the line signalling system to convey relatively few signals, which is preferred. In this situation, there is considerable merit in the adoption of continuous 2-state d.c. signalling for junction networks, even if this includes a modest pulse timing requirement for any additional signal(s). The same conclusion could hold for the d.c. signalling for direct control switching systems where limited facilities apply.

Pulse d.c. signalling: Here, the signal meaning is given by timed pulses. A total signalling requirement may be given by a d.c. signalling system combining both continuous and pulse signals (Tables 3.1–2). Pulse signalling allows an extensive signal repertoire, more than that permitted by continuous signalling, as more information than the on-hook/off-hook conditions can be conveyed. The signalling terminals are more complex, owing to the requirement to generate and receive timed pulses, and the reliability is less than that of continuous signalling, owing to the possibility of a signal being lost or clipped when coinciding with a line interruption. With common control switching, the modern requirement, it is considered that continuous 2-state d.c. signalling gives an adequate repertoire and is thus preferred to pulse d.c. signalling.

Should low-frequency a.c. signalling be applied in the 2-wire junction networks, pulse signalling is the usual arrangement adopted, owing to the nonavailability of separate forward and backward a.c. signalling paths.

3.2.6 *Junction network signal repertoire*

The minimum signalling requirements are seizure, address, answer, clear back and clear forward. When interregister m.f. signalling is used, which is preferred, the address (and other relevant signals such as class of service etc.) is conveyed by this signalling system, whose coding possibilities permit an extensive signal repertoire, and the immediate interest is the line signalling system repertoire. Mention has been made of the requirement to include proceed-to-send line signals in some circumstances. Should l.d.d.c., v.f. or outband line signalling systems be equipped in local junction networks, the requirement for further signals such as release guard and blocking may well arise. Thus the line signalling repertoire in junction networks varies, being determined by the type of signalling system, the type of switching system and the administration's policies.

There is a traditional distinction between local junction network and trunk (toll) network signalling, which arises for a variety of reasons, probably because of the type of transmission plant and the preference for simple local d.c. line signalling on the 2-wire circuits in junction networks in the main. Certain system signals (typically, release guard and blocking, as mentioned), highly desirable in trunk network signalling, when both outgoing and incoming signalling terminals are equipped on the 4-wire circuits, are not always given in local junction network signalling. On the other hand, junction networks may require signals, typically meter pulses over junctions, not required in the trunk network. This distinction may well largely disappear, if not completely, in future evolution of networks, but, owing to the inertia of networks to change, the various differences will apply for a long time.

As national networks vary in their signalling requirements, and, to simply illustrate a particular network example, it is convenient to examine the local junction network signalling requirement of the present-day UK analogue network, the switching in the local areas being a mixture of direct and common control. The present signal repertoire requirement is seizure, address, answer, clear back, clear forward, metering over junctions, trunk offering, manual hold and coin and fee checking. This does not imply that any particular signalling system in the UK junction network must always carry the complete repertoire, as signalling terminals are often designed for particular application, e.g. metering over junctions, trunk offering, manual hold and coin and fee checking would not be required on a subscriber to subscriber dialled local call. It does imply however, that a particular signalling method, e.g. loop-disconnect d.c., should have the potential within its basic concept to incorporate the complete repertoire to enable relevant facilities (and thus signals) to be used when appropriate in particular applications.

The seizure, decadic address, answer, clear back and clear forward signals have been discussed. For the others:

(*a*) Metering over junctions arises on trunk traffic with the nonitemised charging (bulk billing) technique adopted, the charge pulse rate being determined at the lowest level trunk exchange and meter pulses passed back on the junction to operate the meter at the subscriber local exchange; achieved by reversals of loop d.c. backward over the speech leads in loop-disconnect d.c. signalling.

(*b*) The trunk offering facility enables an operator to offer a long-distance incoming call to a busy subscriber; achieved by earth connected to both speech leads forward in loop-disconnect d.c.

(*c*) The manual hold enables an operator to have control of the clear down of a call; achieved by reversal of loop d.c. backward on the speech leads in loop-disconnect d.c. signalling.

(*d*) Calls originated from a pay-on-answer coinbox are routed via a coin and fee checking (c.f.c.) equipment at the local exchange. Where the operator has control of call charging, a c.f.c. signal is used to control the operation of the c.f.c. equipment; achieved by a pulse of $+ 50\,\text{V}$ d.c. on the positive speech lead, together with a disconnection on the negative speech lead, in loop-disconnect d.c. signalling.

The signalling techniques in the UK junction networks are loop-disconnect d.c. and p.c.m. signalling in the main, but long-distance d.c. and outband signalling may be applied when relevant transmission systems are equipped in the junction networks. Thus all the junction network signalling systems must have the potential within their basic concepts to give additional signal(s) to the basic seizure, decadic address, answer, clear back and clear forward signals, depending on the application condition. Inband v.f. signalling is not at present applied in the UK junction network, owing to its inability to signal during speech and thus its inability to convey meter pulses.

Signal discrimination: The signal repertoire and the signalling mode (continuous, pulse etc.) of a signalling system are collectively referred to as the 'signal code'. A number of factors help in signal discrimination in the signal code, and, typically, in junction network signalling:

(*a*) The transmission direction of a signal, e.g. seizure — forward, answer — backward.

(*b*) Duration. The on-hook dialled break pulse is relatively short and is distinguishable from an on-hook clear forward signal, which is transmitted in the same direction, but for a longer period. Typically, the on-hook clear forward should be at least 400 ms to cause clear down, but the clear down should not occur on, say, 140 ms, to cover the extreme case of a dialled break pulse.

(*c*) Relative time of occurrence. In continuous signalling, and when line signalling conveys the backward proceed-to-send signal, a proceed-to-send off-hook occurs before any address information has been sent on that link, while the off-hook answer signal occurs after the sending of address information. Although both signals are backward off-hook condition, they are distinguished by their relative time of occurrence. In pulse signalling, the answer (off-hook) and clear back (on-hook) signals may be of the same duration, on the logic that an answer signal always precedes a clear back.

There are other possibilities, and such ways of signal discrimination aid the logic design of the signalling terminals.

3.3 Signalling for direct control and common control switching in junction networks

The junction network line signalling with both direct and common control switching is of a type depending on the type of transmission plant. Local d.c. signalling is the preferred, and most widely applied, type, but p.c.m., l.d.d.c., outband and sometimes v.f., but noting the limitation of v.f. in its inability to signal during speech, may be used, depending on the type of transmission plant when this plant does not permit local d.c. signalling.

Direct control switching: Signals are repeated stage by stage over the connection and the decadic address pulse signalling is included in the line signalling system. Proceed-to-send signals are not normally given. The address pulsing performance must conform to the general pulsing plan. As in any transit switched connection, the object is to operate a switch in the terminating exchange, either directly over an incoming junction or via an incoming signalling interface on the junction; the multilink connection is an onerous condition, owing to the multiplicity of address pulse repetitions, and the pulsing requirements of the terminating exchange switches determine the pulse distortion margins.[10] This has some constraint on the maximum length of individual junctions, the limit being influenced by the type of switch and pulsing circuit. Typical limits for a 3-link connection with local d.c.

signalling and step-by-step 2-motion direct control switches are first and second junctions of 800 Ω maximum each and the terminal junction of 1500 Ω maximum, but the typical is stressed, as these limits vary a great deal. Pulse correctors and pulse regenerators may allow an increase in the limits. The pulse distortion may also limit the maximum number of links in a multilink connection, three links being a typical maximum, but, here again, the typical is stressed. For economic reasons, per-speech circuit line signalling systems should be as simple as possible, which, combined with the lack of registers, results in limited facilities in the signalling systems in direct control switching networks, which normally do not meet modern requirements.

Common control switching: This may be:

(*a*) Common (indirect) control, e.g. crossbar. Here the common control equipment at each exchange translates to switch the connection, the original address information (or part) being transferred between exchanges on multilink connections. Alternative routing is possible.
(*b*) Common (direct) control, e.g. UK director system. Here the common equipment (director) at the originating exchange translates for the complete routing of the call, including the transit switching, i.e. originating register control of routing.[1] In the UK director system, the switches are direct control, and relevant translation and address information is transmitted from the director in the decadic d.c. pulse form with controlled interdigital pauses. Alternative routing is difficult and not adopted. Signalling for common (direct) control is usually the same as that for direct control.

In common (indirect) control — or more simply common control — the interregister address signalling may be decadic or m.f. In either case, the line signalling is of type depending on the type of transmission media (Fig. 1.3). Proceed-to-send signals are normally required, which may be line signals or interregister m.f. signals, depending on the circumstances.

Decadic address interregister signalling is slow and not error detect-able. The increased postdialling delay on multilink connections could be tolerated if few links apply in the worst connection, and overlap input and output of the address signalling at the registers (senders) helps in this regard.

Owing to storage, each register functions as an address pulse regenerator and the d.c. junction limits on multilink connections are not so constrained by pulse distortion as they are in direct control switching systems. Typically, each junction in a multilink connection may be some 2000 Ω. Should the line signalling system by of a type requiring both outgoing and incoming signalling terminals per-speech circuit (l.d.d.c., v.f., outband), the decadic address interregister signals (loop-disconnect d.c. at the registers) are conveyed by such line signalling systems with appropriate signal conversions. In the interests of simple line signalling, the facility potential normally permitted by registers (or the equivalent) is not fully exploited.

In common control switching, it is not desirable to combine all the interexchange signalling in one signalling system. Separation of supervision and selection signalling often permits economies, promotes flexible application and simplification of line signalling and permits adaption to different signalling languages for the selection information, this being possible because of the coding and decoding possibilities in the registers. Thus an interregister m.f. signalling system, separate from the line signalling system, is preferred to deal with the transfer of address, and other relevant, information in fast coded (nondecadic) form. The line signalling system is correspondingly simplified. The interregister m.f. signalling system also gives potential for enhanced facility exploitation on a network basis, and the fast signalling reduces the postdialling delay relative to decadic signalling. Economically, the interregister signalling system, common with the registers, may be more sophisticated than per-speech circuit provided signalling systems.

References

1 FLOOD, J. E. (Ed.): 'Switching systems' *in* 'Telecommunication networks' (Peter Peregrinus, London, 1975) Chap. 4
2 FLOOD, J. E. (Ed.): 'Signalling and call establishment', op. cit., Chap. 5
3 JOLLEY, E. H.: 'Introduction to telephony and telegraphy', (Pitman, London, 1968)
4 SMITH, S. F.: 'Telephony and telegraphy' (Oxford University Press, 1969)
5 CCITT: 'National automatic networks for the automatic service' chap. VI, ITU, Geneva, 1964
6 CCITT: 'Local telephone networks', ITU, Geneva, 1968
7 FUHRER, R.: 'Landesfernwahl – Bands 1 und 2', (R. Oldenbourg Verlag, Munich, 1962)
8 Blue Book 'Notes on distance dialling'. (AT&T Co., 1975)
9 BREEN, C., and DAHLBOM, C. A.: 'Signalling systems for control of telephone switching', *Bell Syst. Tech. J.,* **39**, pp. 1381–1444; and Bell System Monograph 3736, 1960
10 WELCH, S.: 'The fundamentals of direct current pulsing in multiexchange areas'. Institution of Post Office Electrical Engineers, Printed Paper No. 184, 1944
11 BEAR, D.: 'Telecommunication traffic engineering' (Peter Peregrinus, London, 1975)

Chapter 4

Pulse repetition and pulse distortion

4.1 General

In direct control systems, each dialled address pulse operates a switch, typically a 2-motion step-by-step switch making one step per address pulse in the process of selection, each dial break causing switch magnet energisation. The subscriber's decadic dial break pulse is an interruption of the subscriber's line current; the corresponding pulse to the switch magnet being current applied during the break. The magnet current is relatively heavy and cannot be line current, so a relay, or equivalent device, is inserted into the line to convert subscriber line pulses to magnet pulses, the relay repeating the pulses. The same technique is used when pulsing over a junction when the distant switch is directly operated, the pulsing contact being at the outgoing interface on the junction (Fig. 3.4), and, on multilink connections, pulses are repeated by the various outgoing interfaces. Should both outgoing and incoming interface signalling terminals be equipped on the junction, a pulse receiving relay in the incoming interface repeats the pulses to the switch pulse receiving relay.

The same basic considerations obtain for decadic d.c. pulsing in common control switching, but here register storage of the address information is involved, as distinct from switch operation.

The dialled d.c. pulsing over the subscriber line, and over a junction terminating on a switch, is in the so-called loop-disconnect mode (Fig. 4.1a). Here, the send-end impedance is substantially zero on pulse make and is infinite on pulse break, the significant point being that the impedance is not equal during the make and break periods, and, because of this, the pulsing technique is far from precise, as the pulsing waveforms are not symmetrical. This technique is always used for dialling over subscriber lines and is the more usual for d.c. signalling

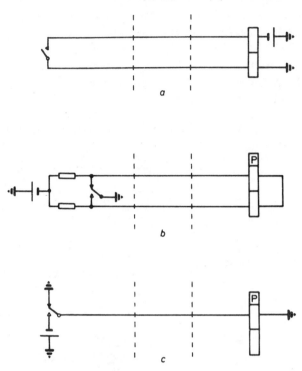

Fig. 4.1   *Decadic pulsing modes*
     a Loop-disconnect
     b Loop-battery
     c Earth-battery

over junctions. It is not used for l.d.d.c. over junctions, as this requires a more precise pulsing arrangment.

     There are many different designs of direct control switches and pulsing circuits. Fig. 4.2 shows a generalised 2-motion step-by-step switch pulsing circuit, this being adopted for the present description for clarity of exposition of the principles. The pulse receiving relay A is bridged across the speech wires, and operates to the off-hook closure of the loop and releases to pulse the switch magnet M to the on-hook disconnect of the loop. The slow release relay B operates on seizure, holds in operation during pulsing, and does not release until clear down. The slow release relay C is required to hold in operation during each pulse digit train and release at the end of each train. Depending on the particular switch arrangement, C may be preoperated prior to a digit train (Fig. 4.2), or not preoperated, in which case C is operated on the first received pulse break of the digit train.

On seizure, A1 operates B and B2 operates C on winding 2. During pulsing, B is energised during the pulse make period of A1. C and the switch magnet M are energised in series during the pulse break periods of A1. On the first step of the switch to the first break pulse of the digit, the preoperate circuit of C is disconnected at the switch off-normal contact N1, but C maintains operation during the digit train by pulses of current in winding 1 on pulse breaks of A1. C releases at the end of the digit train to disconnect the magnet circuit at C1, in addition to its other switching functions.

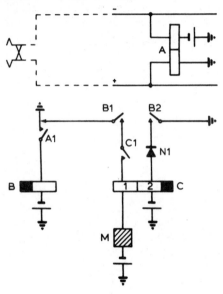

Fig. 4.2   *Typical loop-disconnect pulsing circuit*

Pulse distortion is liable to occur on the A relay pulse repetition. Distortion resulting in a reduced make period for A1 tends to cause failure because

(*a*) B is not receiving sufficient energisation to enable it to hold in operation during pulsing
(*b*) there is insufficient time for the magnet to release during pulsing.

Distortion resulting in a reduced break period for A1 tends to cause failure because

(*a*) C is not receiving sufficient energisation to enable it to hold in operation during the digit train
(*b*) there is insufficient time to operate the magnet.

In direct control switching systems, this pulse distortion on repetition constrains the maximum permissible limit of junction lengths and the number of links in multilink connections, as discussed in Section 3.3. The pulse distortion can arise from a number of causes; typical are the electrical properties of the line (resistance, leakance and capacitance), battery voltage variations within tolerances, and mechanical adjustment differences ('light' or 'heavy' adjustment within the tolerances) of the pulse repetition A relay, which functions in the single current manner.[1,2]

The dial characteristics, the speed and the break/make ratio, reflect the type of switch and pulsing circuits used in particular networks. A subscriber dial speed of a nominal 10 pulses per second (p.p.s.) is usual, but the nominal break/make ratio varies; typically, UK 66/33, Bell system 61/39, Germany 60/40. Tolerance vary; typically, UK 9–11 p.p.s. 63–72% break, Bell system 8–11 p.p.s. 58–64% break. The dial break/make ratio is deliberately not made 50/50 to compensate for the characteristics of relays, switches and signal transmission systems, which differ substantially, and to make the most advantageous use of circuit conditions occurring during the break and make time intervals. Signal circuits can be designed to shift the break ratio of received pulses if necessary to a value better suited to the requirements of the circuit to which they deliver the pulses, or, alternatively, the make ratio, i.e. pulse correction.

It should be noted that the term break/make ratio (or break ratio, make ratio, break period, make period) always implies the presence of a dial contact, a switch contact, or a relay contact, at the point where the break ratio is specified. The term has no meaning apart from such a contact.

4.2 Pulse repetition

Decadic address pulses are repeated by each pulse receiving relay. In direct control switching systems without pulse regeneration, the pulses are immediately retransmitted, as no storage exists. In common control switching systems, register storage exists, direct operation of switches not being involved. An efficient pulse repetition is one which will repeat, with negligible distortion, pulses received in the presence of various factors which give rise to distortion.[1] Fig. 4.3 shows the generalised waveforms resulting from the loop-disconnect d.c. pulsing mode. Practical single current pulse repetition relays cannot exactly translate changes of current in their windings into changes of state of their contacts. The important factor is the state of the contact, as it is on this that the subsequent activity of the circuit depends. This is why break/make ratio, or more simply the break ratio, refers to the state of the pulsing contacts and not to the current flow or any other feature of the circuit. Pulse distortion exists when the duration of the output pulse break or make from the receive device is different from that at the input (Fig. 4.3). It will be understood that the pulse speed is not distorted.

Various factors influence the pulse distortion:[1, 2]

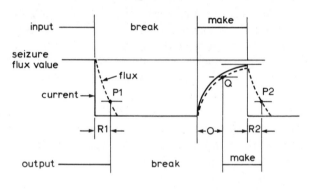

P1 and P2 = release value
Q = operate value
R1 and R2 = release time
O = operate time

Fig. 4.3   *Typical current/flux waveforms for loop-disconnect d.c. pulsing*

4.2.1 Line resistance

This reduces the current and flux in the pulse receiving relay and slows the waveforms. This increases the relay operation time to the make pulse, and hence the lower is the point on the arrival wavefront at which the relay operates; that is, operating on a small fraction of the available current, the less the effect. The effect of eddy currents in the relay magnetic circuit increases rapidly with resistance, to increase further the operation time. As the maximum current and flux decrease with increasing line resistance, the release flux is reached more quickly and the relay release time to the break pulse decreases. With the usual type of repetition relay, and at the higher values of line resistance, the operation time tends to increase more rapidly than the release time decreases. The net result is that the output break ratio measured at the relay contacts increases, which tends to failure of the B relay to hold during pulsing in the typical pulsing circuit Fig. 4.2.

4.2.2 Line leakance

Leakance provides a current path to preflux the relay, and the operation time to make is decreased. On release, three factors tend to maintain the flux and thus to increase the relay release time to break:

(*a*) eddy currents in the relay magnetic circuit
(*b*) the pulsing battery circulating a current through the leakance path
(*c*) the leakance functioning as a noninductive shunt, and the inductive discharge of the relay on the input break circulating by this path.

The relay release time increases at a far more rapid rate than the operation time decreases at the higher values of leakance. The net result is that the output break ratio measured at the relay contacts decreases. This tends to failure of the magnet to step and the C relay to hold in operation during the pulse train in the typical pulsing circuit (Fig. 4.2).

Line resistance offsets the effect of line leakance on pulse distortion, but this is significant only when the resistance is relatively high. A fast dialling speed is an onerous condition on short leaky lines, particularly when a number are in tandem on a multilink connection.

4.2.3 Line capacitance

The line *CR* time constant determines the steepness of the pulsing waveforms. Increased time constant degrades the arrival wavefront, thus tending to increase the output break. On sending the break pulse, the send end is on open circuit and the line capacitance charges to the battery behind the receive relay to delay the relay release to break. On long highly capacitive lines, the decay wavefront may be extremely gradual, the relay release time increases rapidly to result in a considerably reduced output break and, at high capacitances, this reduction is far greater than the increased break on relay operation. The net result is that the output break ratio measured at the relay contacts decreases, which tends to cause failure of the switch magnet to step and the C relay to hold in operation during the pulse train. Fig. 4.4 shows a general characteristic, and the decrease in the break ratio on repetition will be noted. The pulse distortion due to high line capacitance, e.g. that on 4-wire audio circuits and the phantom used for signalling, is a particular constraint on local loop-disconnect d.c. signalling and is one of the reasons, operating sensitivity being the other, for the requirement for l.d.d.c. signalling.

4.2.4 Relay performance variation

Relays can reproduce distortionless pulse signals on repetition only if the operation and release times are equal. Varying pulse waveshapes due to the various line characteristics produce distortion, as discussed above. Additionally, variation in relay performance affects the operation and release times. The operation currents of practical single current relays are greater than those for release, which can result in distortion. Variation in the operation and release currents due to varying mechanical adjustment of the relay within its tolerances may vary the distortion. Relative to the nominal, heavy relay adjustment causes the relay to operate later and release earlier, increasing the output break ratio. Light adjustment has the opposite effect. It is obviously good practice to ensure a rapid rate of change of current about the operation and release points to minimise the effect of relay variations.

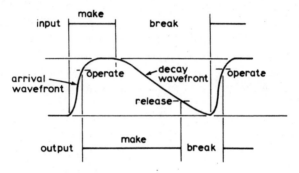

Fig. 4.4   *Typical waveforms for loop-disconnect d.c. pulsing over highly capacitive lines*

4.2.5 Voltage variation

The battery voltage may vary within the accepted tolerances. A high voltage reduces the operation time and increases the release time, thus decreasing the output break ratio. A low voltage has the opposite effect. The combination of a high voltage and a light adjustment presents difficulties for leaky and highly capacitive lines, while the converse condition presents difficulties for high-resistance lines. In practice, the variations within tolerance occur in a random manner; the probability of several adverse conditions being present simultaneously is remote.

4.3 Pulse correction

Pulse correction and pulse regeneration are methods for converting distorted pulses into pulses of sufficient quality to actuate switches.

Pulse correction, as distinct from pulse regeneration, may be defined as a process in which the components concerned with pulse repetition are influenced directly by the received pulses, but also have a subsidiary control which tends to correct the ratio.[1,3] No pulse storage is involved. In general, any method other than complete digit train regeneration can only correct for ratio, and cannot alter the pulse speed. As the pulse speed cannot change in transmission and repetition, it may be argued that there is no need for it to be corrected. The pulse target diagram (Fig. 4.5) shows the basics of pulse correction. Pulse correction in which a certain fixed speed is assumed and a new break (or make) period is generated for each pulse has the obvious objection that different pulse speeds will produce output pulses of widely varying ratios.

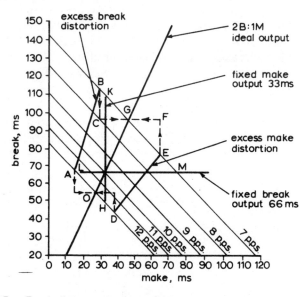

Fig. 4.5   *Target diagram pulse correction*

We assume a dial speed of 10 p.p.s, a 2B/1M ratio and a speed variation 7–12 p.p.s. and that the ideal repetition ratio output over the speed range is 2B/1M (line OG). If the input at all speeds is at a 2B/1M ratio and the repetition is such that excess break distortion results, this can be represented by the line AB, or, if excess make results, by the line DE. If the output break is of fixed duration (say, 66 ms), the output ratio over the speed range is represented by the line LM, and the divergence between LM and the ideal ratio output OG is clearly seen. Similarly, a fixed make output (say, 33 ms) would give an output represented by the line HK, which is again widely different from OG. Thus widely varying output ratios would occur over the speed range.

Various methods involving measurement of both the received break and make have been proposed to overcome the pulse correction problem over a speed range, but none is really satisfactory. A typical proposal is to measure one component, say, the break, and then determine the correction to be applied to give a 2B/1M ratio output at all speeds. Each component break could be corrected in two stages, first by an adjustment depending on the preceding make condition, and secondly on an adjustment dependent on itself. With this method, the break at point B at 7 p.p.s. (Fig. 4.5) would be decreased by BC, and the make thus increased by CG (CG = BC), thus moving B to G to give a 2B/1M ratio. Similarly, the break at E would be increased by EF, and the make decreased by FG (FG = EF) to move E to G. The effect is similar at other speeds. With the first break of a train, there is no preceding make pulse component from which to make the first stage of correction and assumptions must be adopted (say, 33 ms) and first pulse distortion would result at all speeds other than 10 p.p.s. Correction devices aiming to give a 2B/1M output ratio over the speed range are complex and are not satisfactory. Those aiming to give a fixed break or make output by a fixed time base are relatively simple and are adopted when pulse correction is required, but reasonable performance can only be obtained with closely controlled pulse speeds.

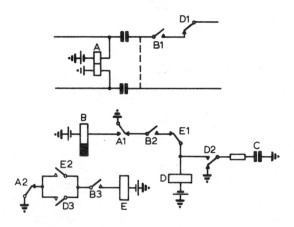

Fig. 4.6   *Fixed break output pulse correction*

Fig. 4.6 shows a typical fixed break pulse corrector. A1 on break operates D, and D1 starts the output break. D3 operates E, and E1 disconnects D from the A1 circuit, but D2 in operation allows the capacitor C to charge through D in the D hold in operation direction, and the termination of the output break is delayed by the charging of the capacitor. When D releases, C discharges to D2. Thus D and C form a fixed timebase. The weakness of the method is that the capacitor timebase circuit is ineffective when the received break exceeds the charge time constant of the capacitor circuit.

4.4 Pulse regeneration

Pulse regeneration is implicit in the registers in common control switching. Various types of pulse regenerators, electromechanical and electronic, are available for speed and ratio correction in direct control systems.[1,4] As the regenerator output is completely independent of the input, storage is required, as, clearly, output pulses cannot be transmitted more slowly than they are received, and it is usually arranged for a digit train to be completely stored before the transmission of that digit train commences.

The main objection to pulse regenerators lies in the delays associated with the storage, at least one digit delay at each regenerator, these delays increasing progressively if the number of tandem regenerators increases on multilink connections. Against this, however, the advantages of storage must be compared, these being:

(*a*) good output speed and ratio
(*b*) the output interdigital pause can be controlled to any desired value. This has the merit in direct control switching systems of allowing more time than is given by normal dial interdigital pauses for various functions to be performed during the interdigital pause.

References

1 WELCH, S.: 'The fundamentals of direct current pulsing in multiexchange areas'. Institution of Post Office Electrical Engineers, Printed Paper No. 184, 1944
2 'Notes on distances dialling'. Blue Book, Section 5, 'Signalling' (AT & T Co. 1975)
3 FUHRER, R.: 'Landesfernwahl – Bands 1 und 2' (R. Oldenbourg Verlag, Munich, 1962)
4 LOCK, P. J., and SCOTT, W. L.: 'The regenerator 5A – a microelectronic project for Strowger exchanges, Pt. 1 – development of the mark 1', *Post Off. Electr. Eng. J.*, 1978,71, Pt. 2, pp. 110–116

Chapter 5

Long-distance d.c. signalling

5.1 General

Long-distance d.c. signalling (l.d.d.c.) is employed on lines beyond the limits of local d.c. signalling when a through metallic circuit is available; it is adopted by many operating administrations (as in the UK and the Bell systems) either through preference for the simplicity and reliability of d.c. signalling or because of an inability to use low-frequency a.c. signalling. Compared with local d.c. signalling, l.d.d.c. extends the signalling limit by the use of sensitive receive devices, such as polarised relays, and extends the decadic address pulsing limit by a more efficient pulsing technique than the loop-disconnect system. With loop-disconnect pulsing (Fig. 4.1a), the send-end impedance varies during the pulsing process, being substantially zero on pulse make and infinite on break, and this variation contributes to the asymmetric pulsing waveforms and thus to pulse distortion on repetition. Further, the single current receive device, with fixed operation and release currents, gives varying pulse distortion with varying signal level when operating on asymmetric waveforms. Loop disconnect is thus unsatisfactory on long audio circuits, particularly when the decay wavefront is gradual because of high line capacitance. For this reason, l.d.d.c. pulsing is usually based on the symmetrical-waveform principle (Fig. 5.1), the shapes depending on the microfarad–ohm figure for the line. Here, the input pulse time T is reproduced between A and B on the zero datum. With a receive device having operation currents [points C, D and E, F in each direction (Fig. 5.1)] set equally about the zero datum, T is also reproduced between these points. This applies at all signal levels and the fixed operation currents do not result in varying pulse distortion with varying signal level.

Symmetrical pulsing waveforms are achieved by double current working (or the equivalent). This requires:

(*a*) the send-end impedance to be substantially equal during the pulse make and break periods

(*b*) the pulsing battery to be at the send-end (it is at the receive end with loop-disconnect pulsing)

(*c*) the receive device to be bidirectional in operation (polarised relay or the equivalent), operating in one direction for the make and in the other direction for the break pulse.

These requirements require the provision of both outgoing and incoming signalling terminal interfaces, and l.d.d.c. systems are thus more costly than local d.c. signalling systems when the latter require only one interface per circuit.

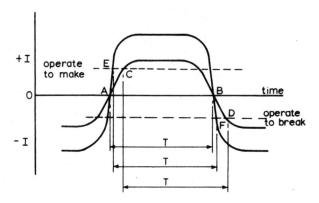

Fig. 5.1   *Symmetrical current pulse waveforms*

There are different types of l.d.d.c. system, but the use of polarised relays for signalling sensitivity and for bidirectional operation to minimise decadic pulsing distortion is a reasonably common principle.

5.2 Typical l.d.d.c. systems

5.2.1 UK DC2 system

This system is used in the UK local junction and trunk networks when decadic address signalling is required, and may be used on 2-wire and 4-wire audio circuits, the phantom being used for signalling in the latter case (Fig. 1.2*b*).[1] A single send-end battery is commutated by a single changeover contact, the polarised relay responding to the symmetrical loop current reversals to repeat the pulse makes and breaks (Fig. 4.1*b*). The send-end impedance is substantially equal during the pulse make

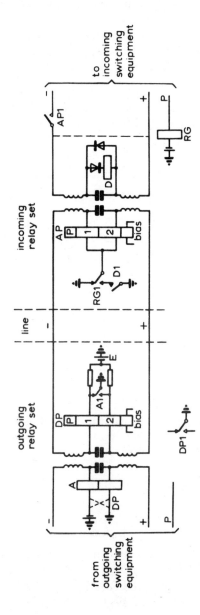

Fig. 5.2   *Principle of UK DC2 signalling system*

and break periods. This technique is conveniently called loop-battery as distinct from loop-disconnect.

Fig. 5.2 shows the principle. The polarised relay DP, with windings 1 and 2 in series opposition, responds to earth, but not to loop, currents. The polarised relay AP, with windings 1 and 2 in series aiding, is responsive to loop currents and, when operated to a forward loop current in either direction, is unresponsive to backward earth currents in either direction. Duplex signalling is used; in general, loop current signals are in the forward direction (A1 commutating battery E) and earth current signals are in the backward direction.

Table 5.1 shows the signal code for the trunk network. The local junction network application includes the additional signal potential:

Table 5.1   *UK DC2 system signal code for trunk network*

Signal	Condition at outgoing equipment	Condition at incoming equipment
Idle	Negative battery positive wire Earth negative wire	Earth centre point AP (incoming equipment free condition)
Forward		
Seizure		
(a) unidirectional	Negative battery negative wire Earth positive wire	On seizure recognition, earth centre point AP disconnected and loop applied
(b) bothway	Negative battery negative wire Positive battery positive wire	
Decadic address	Loop current reversals in step with dial pulses	
Clear forward	Negative battery positive wire Earth negative wire	
Backward		
Answer		Earth centre point AP
Clear back		Loop at AP
Release guard		Loop at AP (earth centre point AP – the incoming equipment free condition – terminates the release guard signal)
Blocking		Disconnection of the earth from centre point AP

forward: trunk offering (positive battery positive wire, disconnect, negative wire)

backward: manual hold (earth negative and positive wires)

coin and fee check (battery negative and positive wires)

metering over junction (loop reversals)

Corresponding relay conditions

Idle (A1 released): The earth current positive wire holds AP 'released' on winding 2 aiding the bias. DP is 'operated' on winding 2 opposing the bias. The earth centre point AP at the incoming end is the incoming equipment free condition and DP1 operated indicates the free condition at the outgoing end.

Seizure (A1 operated): The earth current negative wire operates AP on winding 1 opposing the bias. DP maintains operation on winding 1. When seizure is effective at the incoming end (some 20 ms) and the switching equipment is seized, an earth is returned on the P-wire to operate RG. RG1 removes the earth centre point AP and applies loop. AP maintains operation to the loop current; DP releases.

Address signals (A1 pulsing): AP responds to the loop current reversals; DP does not. AP1 repeats the pulses forward in the loop-disconnect mode.

Answer (D operated): D1 operated (recognition time 30 ms) applies earth centre point AP. The earth current negative wire maintains AP in operation on winding 1 opposing the bias. DP operates on winding 1 opposing the bias. DP operated repeats the answer condition.

Clear back (D released): D1 releasing applies loop at AP. AP maintains operation. DP releases to the loop current to repeat the clear back.

Clear forward (A1 released): The clear forward signal achieves release of the connection in any condition:

(a) Clear forward after clear back: Loop conditions AP already exist and DP is already released, owing to the clear back. A1 released on the caller's clear reverses the loop current to release AP. AP1 extends the clear forward signal forward. DP released constitutes the release guard signal condition, as this indicates that the equipment free condition does not exist at the incoming end. RG1 releases when the

incoming equipment release is complete and applies earth centre point AP. The earth current positive wire maintains AP released on winding 2 and operates DP on winding 2. DP1 operated terminates the release guard condition at, and unbusies, the outgoing end. Clear forward signal recognition is 300 ms.

(*b*) *Clear forward before clear back – release from the answered condition:* AP and DP are operated (as for answer) prior to the clear forward. A1 released to the caller's clear reverses the loop current to release AP on winding 2 aiding the bias. DP maintains operation. AP released with D still operated applies the clear back loop condition at AP (circuitry not shown). AP maintains released and DP releases to the loop current. The subsequent release guard sequence is the same as for (*a*). Thus, when releasing after the clear back, the normal clear back loop condition is extended back to give a release guard indication. When releasing prior to clear back, a clear back loop condition is manufactured by the system to give the release guard indication.

Blocking: Cessation of the idle condition earth current positive wire, by removal of the earth centre point AP by maintenance or by line disconnection due to fault, releases DP to give the backward busying. The outgoing end returns to the unblocked free condition when the earth current positive wire reappears.

If desired, the signal repertoire could be extended by the addition of a battery centre point AP condition at the incoming end and the inclusion of another series opposition connected polarised relay in series with DP at the outgoing end.

Performance

> signalling limit: some 8000 Ω loop
> pulse distortion: some 2–3 ms on maximum circuit
> some 76 000 μFΩ

Extraneous earth currents
Unless compensated for, unwanted earth currents, which may arise from differences in earth potential (e.p.d.) and by induced longitudinal currents by extraneous sources such as alternating e.m.f. from 50 Hz power lines, are constraints on earth current signalling systems. System DC2, which does not incorporate compensation, is tolerant to extraneous potentials not exceeding some $E/2$, where E is the send battery potential, which is judged to be acceptable in the network concerned.

Bothway working

When idle, earth current flows on the positive wires of the equipment owing to the earth on the positive wire at the incoming equipment and the negative battery on the positive wire at the outgoing equipment. This is the normal condition of undirectional working, and extends the equipment free indication to the outgoing equipment. In the idle condition on bothway working, both the negative and positive line wires must be used to convey these currents to the respective ends. To achieve this, the positive wires of the equipment are connected to both line wires in the idle condition, as shown in Fig. 5.3, the negative wires of the equipment not being connected to line.

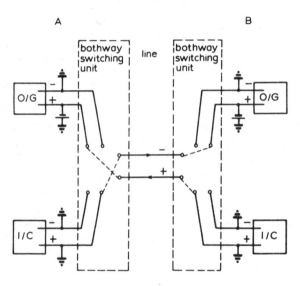

Fig. 5.3   *UK DC2 signalling system bothway working idle condition*

The undirectional seizure is the earth current negative wire of the equipment, but, as the negative wires are not line connected in bothway working idle, the unidirectional seizure cannot obtain for bothway working. Seizure must be performed by earth currents on the positive wires of the equipment – these wires carry the incoming equipment free earth currents – and AP is operated over these wires. For this reason, the seizure signal in bothway working is a positive battery earth current pulse of 50 ms. This reverses the direction of earth current on the equipment positive wires to operate AP (on winding 2 instead of winding 1 as in the unidirectional mode) in the incoming equipment.

Assuming seizure from A, when the positive battery seizure pulse is transmitted (on the positive line wire in the condition shown in Fig. 5.3), the negative line wire is switched from the incoming equipment to the associated outgoing equipment at A. AP operating on receipt of the seizure at B causes the line wire connected to the associated outgoing equipment to be switched to the incoming equipment at B, both line wires now being connected to the outgoing equipment at A and to the incoming equipment at B. The remaining signals operate in the same way as for unidirectional working. Similar arrangements occur when seizing from B.

Double seizure is detected at an end when both the outgoing and incoming equipment at that end are in the seized condition. The outgoing call is not completed, the seizure signal is disconnected and the incoming equipment is connected to both line wires. An equipment engaged tone is returned to the outgoing caller, and the release of the outgoing equipment is by the subscriber clear. The incoming equipment at that end is now available for incoming calls, and, should the double seizure not have been detected at the other end, the setting up of the incoming call proceeds.

Should the double seizure be detected at both ends, both line wires will terminate on incoming equipment at both ends. Both calls are lost and the equipment engaged tone is returned to both callers. When one of the callers clears, the outgoing equipment at that end is freed and a follow-on call on this equipment is completed via the distant incoming equipment already connected to both line wires. The outgoing equipment at this distant end becomes free and available for further calls when its caller clears and the associated incoming equipment is free. The general bothway working philosophy is discussed in Section 7.9.

5.2.2 UK DC3 system

This system is used when the l.d.d.c. line signalling is not required to carry decadic address signalling, and thus when common control switching and interregister m.f. signalling are used.[2] The only requirement is for increasing the signalling limit (5000 Ω loop) for the supervisory signals, pulsing limit requirements not being involved. This allows the DC3 system to be simplified in comparison with the DC2 system, and, for this reason, DC3 operates on a forward and backward loop signalling basis; earth current signalling, and the constraints of this, being eliminated. Table 5.2 shows the signal code.

Fig. 5.4 shows the simplified elements. Seizure, release guard and blocking are signalled by the connection of earth and battery potentials. The other conditions are signalled by an increase or decrease in the loop current by changing the resistance at the incoming end. Line relays DP (answer detector), AP (seizure detector) and BGP (release guard detector) are polarised for sensitivity and are held over by bias in the no-signal condition. The signalling path resistance at incoming equipment under idle, seized (but unanswered) and clear back conditions is some 19·5 kΩ and some 1·7 kΩ in the answered condition.

Table 5.2   *UK DC3 system signal code for trunk network*

Signal	Loop current in signalling path
Idle	No current
Forward	
Seizure	Low current
Clear forward	No current
Retest	Low current
Backward	
Answer	High current
Clear back	Low current
Release guard	High current (reverse direction)
Blocking	
(a) unidirectional	High current
(b) bothway	Low current

Seizure: The outgoing relay A operated to the caller's off-hook operates B, which causes loop current to flow via DP, but is insufficient to operate it. The incoming relay AP operates to this loop current (recognition 20 ms) to seize the incoming equipment.

Answer: D operates to the called-party's off-hook answer (100 ms recognition) and operates DD, which reduces the incoming-end signalling loop resistance. The increased loop current operates the outgoing relay DP to repeat the answer condition. The incoming relay AP remains in operation in the seized condition.

Clear back: D releases to the called-party's on-hook clear and

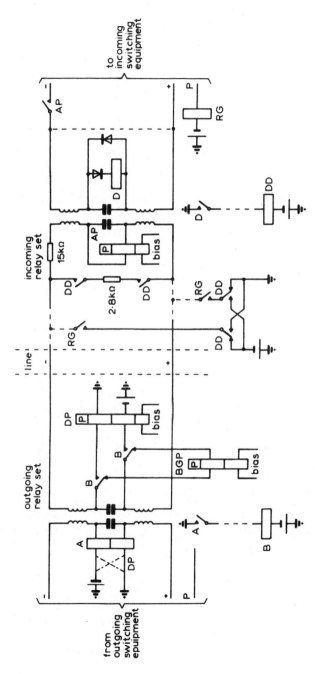

Fig. 5.4 *Principle of UK DC3 signalling system*

increases the incoming end signalling loop resistance, which releases the outgoing DP to repeat the clear back. The incoming relay AP remains in operation in the seized condition.

Clear forward: A and B release to the caller's on-hook clear. B disconnects the seizure line current to release AP in the incoming equipment (clear forward recognition cessation of loop current for 100 ms or more). B also connects BGP across the line. As RG remains in operation during the release of the incoming equipment, the release guard is signalled (maintained for at least 400 ms) in the form of a reversed battery and earth connection via RG and DD contacts to the line in the incoming equipment to operate BGP at the outgoing end. BGP releases to terminate the release guard and free the circuit when the incoming equipment is completely released; RG releases and the battery and earth potentials via DD are disconnected from the line. The direction of the release guard current, connected via DD contacts, is opposite to that of a seizure signal, which ensures that a spurious release guard signal occurring on a bothway circuit does not cause seizure of the incoming equipment and so cause a circuit interaction.

This clear forward release guard sequence is the same regardless of whether the clear forward occurs before or after the on-hook clear back. Nonreceipt of the release guard signal within 250 ms, owing to a malfunction, brings the retest feature into operation. Cyclic retransmission of the seizure and clear forward signals, once every 50 s occurs to invite repeatedly the return of a release guard signal, the condition being alarmed after 6–12 min should a release guard signal not be received within this time. The reception of a retest seizure by the incoming equipment does not cause a register to be associated or switching equipments. It should be noted that this retest feature is not an essential item for the DC2 system; it is not provided, as it is not necessary to cover line malfunction. In the DC2 system, the termination of the release guard is the idle condition on the earth centre point of AP, and thus earth current on the positive wire. This condition will automatically exist, and the circuit be freed, when the line malfunction which prevented its transmission eventually ceases, the condition being alarmed after 6–12 min.

Blocking: A unidirectional line, in the idle state, is terminated at the outgoing end by the release guard detector BGP and at the incoming end by the seizure detector AP. Blocking is conveyed by disconnecting AP by operating RG, which transmits loop current backward to operate BGP, which, in these circumstances, is used to detect blocking signals.

Bothway working

Current feed to line arrangements are adopted on both 2-wire and 4-wire (phantom signalling) bothway circuits so that, on double seizure, the seizure signal battery and earth connections of both the outgoing relay sets connected to the line result in earth currents in each wire of the signalling path. The resultant currents flowing operate the DP relays at each end, simulating a false answer. False answer signals evoked by double seizure are ineffective, as the circuit is associated with an outgoing register at this stage of the call; an incoming register is not associated. An equipment congestion tone is returned to both callers, both calls are lost and the double seizure conditions persist until either a calling party clears, or an outgoing register sets up a forced release on time-out.

As the bothway line is terminated at both ends by the seizure detector AP of the incoming equipment, blocking is conveyed in either direction as required by disconnecting the relevant AP relay and substituting a battery and earth connection by RG operating. The resulting loop current operates the distant AP relay which functions under this condition to busy the outgoing path at that end.

5.2.3 Release guard and blocking signals

Mention has been made in Sections 5.2.1 and 5.2.2 of release guard and blocking signals, which are of a 'system' nature, and it is appropriate to discuss these signals at this stage. When included in the signal repertoire, these signals are associated with signalling systems having both outgoing and incoming signalling terminals per circuit, e.g. l.d.d.c., v.f., outband etc. The signals are used in international and in most national long-distance network signalling systems, and are not usually incorporated in the simple types of signalling system such as loop d.c. in local junction areas.

Release guard: Electromechanical switching equipment is relatively slow to release, and, on connection release by the clear forward signal, the outgoing equipment on a circuit must be guarded against reseizure by a follow-on call while the incoming equipment terminating the circuit is releasing. A number of techniques are used:

(*a*) The outgoing equipment is held busy for a guard time, typically 1000 ms, after sending the clear forward signal. This is the usual arrangement in local junction networks, whether direct or common control switching.

(*b*) With common control switching, the outgoing equipment may be reseized by a follow-on call, but the sending of the seizure signal to the line is delayed for a time (typically 700 ms) or until a test of the line indicates that release has taken effect. This method makes efficient use of lines, as a follow-on call, which may otherwise be refused, is accepted. The method cannot be used for bothway working, as here the seizure signal must be sent immediately, nor can it be used with direct control switching, as the interdigital pause may not always be adequate.

(*c*) A 'release guard' line signal is returned when the incoming equipment has released, the outgoing equipment being maintained in a busy condition until the release guard is received. The release guard signal technique has the significant merit in that it may be regarded as being an acknowledgement of the important clear forward signal. Nonreceipt of the release guard signal within a certain time would indicate that the clear forward signal had not been effective in releasing the incoming equipment, and gives opportunity for appropriate action to be taken; typically, the outgoing end is kept busy and the circuit passed to maintenance attention, or, with certain signalling systems, automatic repeat attempts are made to clear down, as instanced by the retest feature of the UK DC3 signalling system (Section 5.2.2).

Electronic switching equipment is fast release, and here it could be reasoned that there is no necessity to guard the outgoing equipment on clear down, as the probability of a follow-on call arriving during the short release time is remote. Despite this, however, it is good practice to adopt a short time guard when the release guard signal technique is not used, but the usual preference is for the release guard signal because of its clear forward signal acknowledgement merit, despite the fast release of electronic equipment. The release guard signal technique finds its main application in trunk network signalling systems in many national, and in the international, telephony networks. Although not usual, it could be applied in local junction networks in appropriate equipment conditions.

Blocking: This signal busies an outgoing signalling terminal when an incoming signalling terminal on that circuit is placed out of action, or under maintenance attention. In continuous signalling, this usually involves transmitting a continuous off-hook condition signal in the backward direction, in, for example, the UK DC2 and DC3 systems

(Sections 5.2.1 and 5.2.2). This is also the preferred approach in pulse signalling systems, but, alternatively, discrete blocking-on and blocking-off signals could be used, both being pulse. Blocking signals find their main application in trunk network signalling systems and in international signalling systems. They could be used in local junction networks in appropriate equipment conditions.

5.2.4 Bell system CX and DX systems

The acceptance of 2-state line signalling in each direction has had an important bearing on the evolution of the Bell system signalling, and all Bell system conventional line signalling systems are based on 2-state continuous signalling, all having the same signalling repertoire. So-called derived signalling links are used for the longer circuits (toll and longer junctions), the signalling connections between the trunk relay circuits and the derived signalling links being obtained via a uniform system of E and M leads (Section 7.10). The Bell Composite (CX) and Duplex (DX) l.d.d.c. systems are applied on 2-wire and 4-wire audio circuits when d.c. signalling is required and the decadic address pulsing and/or the signalling limit is beyond the ranges of Bell loop d.c. signalling. Both systems conform to the 2-state continuous and the E- and M-lead control philosophies.

The principles of the CX and DX systems are much the same. The extended signalling limit is obtained by the use of sensitive polarised relays and the extended pulsing limit by earth-battery pulsing (as distinct from loop-disconnect) with polarised relay reception on single conductor working (Fig. 4.1c). Earth-battery pulsing, earth for pulse break and battery for pulse make on the M-lead, ensures equal send-end impedance during the pulse break and make periods to give symmetrical pulse waveforms. Balanced polar relay sets in a symmetrical arrangement at each end permit the duplex operation, and the symmetry eases the equipment problems of bothway working. Balancing networks must be adjusted for each signalling circuit according to the impedance of the line signalling conductor. Both the CX and DX systems employ a single line conductor with earth return for signalling for each speech circuit. Both incorporate arrangements to compensate for a.c. induction, earth potential differences and battery voltage variations. Since the line signalling conductor and the compensation line conductor both carry the unwanted earth currents, the compensation conductor cancels (in a selfcompensating manner) the effect of the unwanted earth currents on the signalling conductor.

Composite (CX) signalling:[3,4] This system employs a single conductor with earth return for each signalling channel (Fig. 5.5).

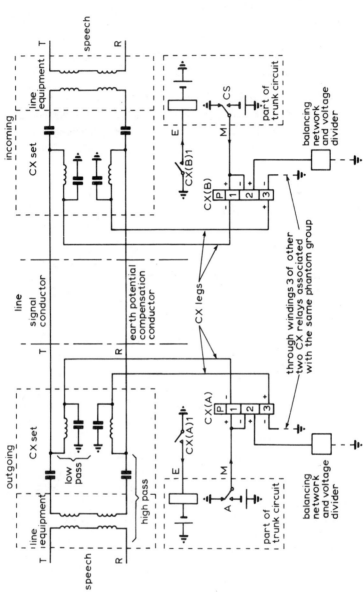

Fig. 5.5   *Bell system Composite (CX) d.c. signalling system*

The higher-frequency speech currents are separated from the low-frequency currents arising from the d.c. signalling by a highpass–lowpass filter arrangement called a CX or 'composite' set, the crossover frequency being about 100 Hz. Two CX signalling legs can be derived from a pair of wires and four from a phantom group. All these legs can be used to signal independently with earth return, but, in most cases, one leg is used for an a.c. or d.c. earth potential difference compensation path on a common basis for three signalling conductors. The signalling channels can be assigned independently of the speech channels with which they are physically associated because of the isolation provided by the CX sets. The composite path can be used only for d.c. flow.

The circuit operation is much the same, and the signalling code is the same as for the DX system described below.

Signalling limits: short-haul, 4800 Ω loop
                             long-haul, 12 000 Ω loop
Decadic pulse distortion: some ± 4% on maximum circuit

Duplex (DX) signalling:[4–6] This later, and preferred, system is also based on a symmetrical and balanced arrangement that is identical at each end, with earth-battery pulsing. It is patterned after the CX signalling system, but does not require a composite set to separate the low frequencies arising from the d.c. signalling from the speech transmission. The system uses the same conductors as the speech circuit for signalling. One conductor of the pair is used for signalling and the other for unwanted earth current compensation (Fig. 5.6). Table 5.3 shows the signal code.

When idle, the E-leads at both ends are open, the M-leads are earthed and DX(A) and DX(B), appropriately biased, are in the released position. Relay A operates to the incoming off-hook seizure and the A1 battery to the M-lead and the T signal conductor operates DX(B) on winding 1 to the incoming M-lead earth. DX(A) stays released, the balance, balancing the impedance of the signal conductor circuit, ensuring that the net energisation in DX(A) due to the M-lead battery is virtually zero, winding-1 energisation opposing winding-2 and winding-3 energisation. DX(B)1 earths the incoming E-lead to repeat the seizure condition at C1.

The delay dialling backward off-hook condition, battery on the incoming M-lead, results in battery on both the M-leads at the outgoing and incoming ends, outgoing and incoming referring to the traffic direction of the call, and the net current flow on the signal line

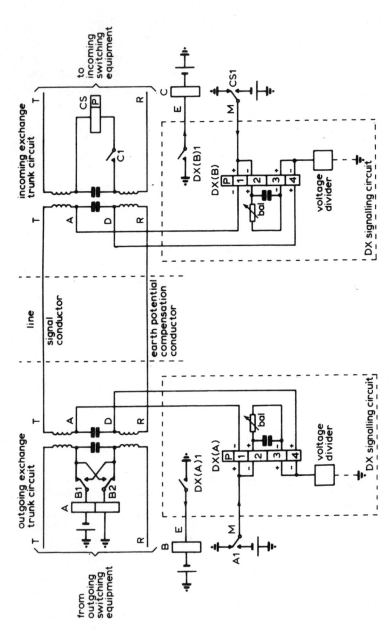

Fig. 5.6   Bell system Duplex (DX) d.c. signalling system

Table 5.3   *Bell DX system signal code*

Signal condition	D.C. on signal conductor $\longrightarrow$	$\longleftarrow$	State DX(A)	DX(B)
Idle	Ceased	Ceased	Released	Released
Forward				
Seizure (off-hook)	Continuous	Ceased	Released	Operated
Decadic address pulsing (sequential on-hook, off-hook) when applicable	Continuous seizure d.c. interrupted during break pulses	Ceased	Released	Responds to pulsing
Clear forward (on-hook)	Ceased	Ceased	Released	Released
Forward transfer (on-hook)	Pulse 65–135 ms cessation of d.c.	Ceased	Released	Released to the pulse
Backward				
Delay dialling (off-hook) when applicable[4]	Continuous[1]	Continuous[1,3]	Operated[2]	Operated[2]
Start dialling proceed-to-send (off-hook) when applicable[4]	Continuous	Ceased[3]	Released	Operated
Answer (off-hook)	Continuous[1]	Continuous[1]	Operated[2]	Operated[2]
Clear back (on-hook)	Continuous	Ceased	Released	Operated
Blocking (off-hook)	Ceased	Continuous	Operated	Released

Notes:
(1) As continuous d.c. obtains in each direction on the signalling conductor, the resultant d.c. flow on the conductor is zero.
(2) The DX relays at each end hold in operation to local off-hook arrangements when condition (1) above holds.
(3) The durations of these signals are variable, depending on when the succeeding signal occurs, and, to ensure proper registration, the transmitted signal durations should not be less than 140 ms.
(4) The delay dialling and proceed-to-send signals are not applied in direct control switching system applications.

conductor is zero. DX(B) remains in operation on windings 2 and 3 to the incoming-end M-lead battery. DX(A) operates on windings 2 and 3 to the outgoing-end M-lead battery to repeat the delay dialling condition. The backward on-hook proceed-to-send, by earth on the incoming-end M-lead, restores the line signal conductor current to release DX(A) to give the proceed-to-send indication at the outgoing end. The proceed-to-send and clear back signals, both backward on-hook, are recognised as such by their sequence.

When decadic address pulsing transfer is used, A1, responding to the incoming pulses, applies earth (on-hook break pulse) and battery (off-hook make pulse) pulsing. DX(B) and C via the incoming E-lead respond, and C1 repeats the pulsing forward in the loop-disconnect mode. CS1 operating to the called-party's off-hook answer substitutes battery for earth on the incoming M-lead. In this, the speech condition, both ends are off-hook, both M-leads are battery, and the condition is the same as for the delay dialling condition, both DX(A) and DX(B) being in the operated position. DX(A)1 earths the outgoing-end E-lead to operate B, which reverses the incoming loop current to repeat the off-hook answer. The delay dialling and answer signals, both backward off-hook, are recognised as such by their sequence. During speech, both E-leads are earthed as both DX(A) and DX(B) are operated, and both M-leads are battery. The on-hook clear back earths the incoming-end M-lead at CS1 released to release DX(A) to open the outgoing-end E-lead, which releases B to repeat the clear back. DX(B) remains in operation to the signal current on the T-line conductor.

On clear forward prior to the called-party clearing, the caller's on-hook releases A, which earths the outgoing-end M-lead. This releases DX(B) to open the incoming E-lead. C and CS release, and the incoming M-lead will be earthed, effectively short circuiting winding 1 of DX(A), which releases. C1 released repeats the clear forward signal forward. The outgoing E-lead will be open on DX(A)1 released, and, with both M-leads earthed and both E-leads open, an idle trunk condition exists. On clear forward after the clear back, the outgoing-end M-lead earth stops the T-line conductor current, and so release DX(B), which releases C on the incoming-end E-lead, C1 repeating the clear forward signal. The clear forward signal achieves clear down of the connection from any condition of the connection.

An e.p.d. current flows over the R-line conductor between windings 4 of DX(A) and DX(B) to compensate for the e.p.d. current flow on the T-line conductor between windings 1 of DX(A) and DX(B).

In semiautomatic service, the forward transfer signal (not required in automatic service), is sent by an originating operator to recall an incoming operator after a connection has been established. Both ends are off-hook when the signal is to be sent and thus the forward transfer can only be given by an on-hook condition. This should not be a continuous on-hook signal, as this would conflict with the continuous on-hook clear forward. For this reason, a third state, that of time, is used in the basic 2-state signalling, and the forward transfer signal is a timed on-hook pulse. The clear forward signal recognition time (300 ms or longer) is longer than the forward transfer pulse (65–135 ms) to achieve the necessary discrimination to avoid conflict between these two on-hook condition signals. Although this case is not particularly significant in itself, it serves to illustrate the problem of continuous 2-state signalling when more than the basic d.c. signals are required, the basic simplicity being eroded.

Bell system signalling systems do not provide the release guard signal. The outgoing end equipment is maintained guarded by a locally applied busying time feature of a period (typically 1000 ms) assessed to cover the normal release time of the incoming equipment, and the release is not sequenced. Blocking is given by applying the off-hook condition at the incoming end, the resulting backward d.c. signal on the signal conductor operating the DX relay at the on-hook outgoing end, to busy that end for as long as the incoming end is off-hook.

Signalling limit: 5000 Ω loop
Decadic pulse distortion: $\pm 4\%$ on the maximum circuit.

On bothway working, double seizure results in each end of the circuit receiving an immediate and sustained off-hook signal ('glare'), each end sending and receiving off-hook seizure, which enables the double seizure to be detected. When delay dialling signals are used, the incoming seizure signal is recognised at each end as a delay dialling signal. If a proceed-to-send on-hook signal is not received within a certain time-out period (typically 5 s) double seizure is assumed. When delay dialling signals are not used, double seizure detection is immediate. The incoming off-hook seizure does not start the charging process as, owing to the signal sequence, there is no conflict between this signal and an incoming off-hook answer signal. In the event of double seizure, depending on the arrangements in the particular equipment:

(*a*) an automatic repeat attempt is made to set up the call

or

(*b*) no automatic repeat attempt is made, reorder conditions are applied and the engaged tone sent to the callers.

With either method, the double seized circuit is released.

The DX system is a particularly simple and attractive arrangement, but the simplicity reflects the limited signal repertoire given. The per-line balance adjustment is, perhaps, a slight penalty, but this is not particularly significant. It will be noted that, when both ends are off-hook, e.g. during speech, and thus no line signal current flows, a coincident persistent malfunction line break in the signal conductor, a remote possibility, would not cause a change in signalling condition, both ends remain off-hook to the subscribers and the signalling system would not monitor the speech circuit. The signalling system would not clear down until the on-hook clear by the subscribers on an abandoned call. Again, this is not a particularly significant penalty.

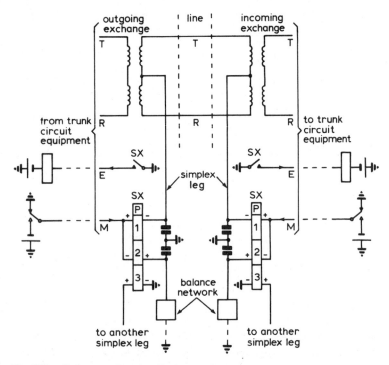

Fig. 5.7   *Bell system Simplex (SX) d.c. signalling*

5.2.5 Bell system Simplex (SX) system

Simplex (SX) signalling[3] feeds signalling currents through the centre points of the line transformers to the balanced paths furnished by the line conductors (Fig. 5.7). With SX, the line resistance is halved by paralleling the two conductors, thus extending the range relative to loop d.c. signalling. Earth potential compensation requires the use of one conductor of an additional pair for each five signalling channels. Thus only five SX signalling circuits can be derived from six physical pairs. The signalling currents on the line side induce no voltage in the equipment, since they flow in opposite directions in the two halves of the line transformer winding, and, conversely, speech currents in the equipment cause no current flow in the simplex leg. As in the CX and DX systems, earth-battery decadic address pulsing is used, giving equal send-end impedance during the pulse makes and breaks. Simplex signalling has been largely superseded for new work in the Bell system by the DX signalling system.

5.3 Duplex signalling

All supervisory signalling associated with telephony call processing involves duplex operation, in the sense that both forward and backward signalling conditions must exist simultaneously. Typically, in the speech condition, both the forward and backward off-hook conditions must occur simultaneously; when applicable, the backward off-hook delay dialling condition in the proceed-to-send sequence must exist at the same time as the forward off-hook seizure, and so forth. This duplex requirement may be achieved in different ways, d.c. signalling typically by:

(*a*) Standing forward and backward signal currents on the line conductor(s), one signalling path only being available for the two signalling directions; e.g. the forward loop-disconnect and the backward loop reversals existing at the same time in loop d.c. signalling, the forward loop current and the backward earth current signals at the same time in the UK DC2 system and the loop currents in the UK DC3 system. (*b*) No line signal current, one signalling path only being available for the two signalling directions, but the off-hook signal condition is maintained at each end of the signalling system by local arrangements, depending on the absence of line signal current, whose absence, in turn, depends on valid off-hook conditions being signalled from each end; e.g. the Bell system CX and DX systems.

When signalling on 4-wire circuits with systems having independent forward and backward signalling paths, e.g. v.f., outband and p.c.m. signalling, the duplex condition is greatly eased, being given by the two simplex signalling paths.

Pulse signalling requires the complications of memory logic to achieve the duplex condition. Compared with pulse signalling continuous 2-state signalling in each direction greatly simplifies the duplex signalling operation; this is true with one signalling path and with two independent paths. Further, it facilitates signalling systems which are symmetrical, balanced and identical at each end, as instanced by the Bell CX and DX signalling systems. In Bell system practice, the additional signalling and trunk-switching complexity for bothway working is absorbed in the trunk equipment at each end, and the line signalling system arrangements are simple, owing to the symmetrical operation at each terminal. With symmetrical operation, each of the input/output functions is applied, conveyed and delivered in an identical manner for each direction of operation. It will be appreciated, however, that this symmetry is possible only when the supervisory signalling functions are reduced to continuous 2-state conditions in either direction, which is often acceptable.

References

1   WELCH, S., and HORSFIELD, B. R.: 'The single commutation direct current signalling and pulsing system', *Post Off. Electr. Eng. J.*, 1951 **44**, Pt. 1
2   MILLER, C. B.,   and   MURRAY, W. J.:   'Transit-trunk-network   signalling systems. Line signalling systems', *ibid.*, 1970, **63**, Pt. 3, pp. 159–163
3   BREEN, C., and DAHLBOM, C. A.: 'Signalling systems for control of telephone switching', *Bell Syst. Tech. J.*, 1960, **39**, pp. 1381–1444; and Bell System Monograph 3736, 1960
4   'Notes on distance dialling'. Blue Book Section 5, 'Signalling' (AT & T Co., 1975)
5   NEWELL, N. A.: 'DX signalling – a modern aid to telephone switching', *Bell Labs. Rec.*, 1959, 37
6   NEWELL, N. A.: 'DX signalling', *ibid.*, 1960, **38**, p. 216

Chapter 6

P.C.M. signalling

6.1 General

In pulse code modulation (p.c.m.) transmission systems, speech or other information is transmitted in digital coded form, typically 8-bit encoding, the p.c.m. system being synchronised. The speech input is sampled at twice the top frequency to be transmitted, i.e. sampling at 8 kHz, and the original speech wave is reproduced at the receive end from this coded information. By this technique, a number of circuits, typically 24 or 32, can be derived from, typically, two 2-wire circuits by using one pair for the go and the other for the return. The 8 bits (8-bit octet) for each p.c.m. channel are associated into a time slot, and the number of channels in the particular p.c.m. system make up a frame, i.e. a sampling period of 125 μs, each channel time slot being sampled every 125 μs, which corresponds to the sampling frequency of 8 kHz (Figs. 6.1 and 6.2). As the sampling frequency is 8 kHz, and 8-bit encoding is used, each p.c.m. channel has a 64 kbit/s capability. The system line bit rate is therefore $64N$ kbit/s, where N is the number of time slots. N may be the same as the number of speech channels in the p.c.m. system, but, in some systems, one or more time slots are used for control purposes and not for speech.[1,2]

In itself, some signalling over p.c.m. systems could be achieved by sampling and coding a.c. analogue signals in the same way as for speech and thus a built-in p.c.m. signalling arrangement is not required. At present, this is so for some signalling systems; for example, there are no plans to specify modified versions of the CCITT International analogue Signalling Systems 4, 5 and 5bis for application to p.c.m. transmission systems. The inherent nature of p.c.m., however, allows a convenient way of transmitting signalling information as a built-in function in individual p.c.m. transmission systems without requiring the speech circuits to be equipped with conventional analogue signalling systems. Further, switching signals, owing to their relatively low requirements for signal distortion and speed, do not have to be sampled

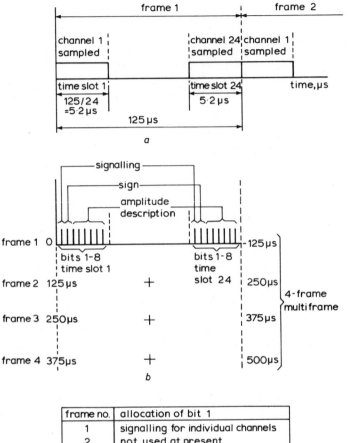

frame no.	allocation of bit 1
1	signalling for individual channels
2	not used at present
3	signalling for individual channels
4	frame and multiframe alignment

bits 3-8 amplitude description of speech sample

Fig. 6.1 *UK 24-channel p.c.m. system*
(*a*) Before coding
(*b*) After coding; each time slot consists of 8 bits

as frequently as speech signals, and advantage can be taken of this, typically in a method where the signals of a number of speech channels are combined.

Any t.d.m. (time division multiplex) transmission system must include some redundant bits, that is, bits not used for p.c.m. speech information, to define the frame of the system and enable constituent channels to be identified. Further, some bit capacity must be provided for signalling. With built-in p.c.m. signalling, analogue d.c. signals are converted into coded bit form at the transmit end of a channel and reconverted back to their original form at the receive end. With this built-in p.c.m. signalling, the bit signalling is time assigned, meaning that the identity of the speech channel for which the signal applies is given on a time basis. The bit signals may be in-slot or out-slot, depending on the particular p.c.m. system, but, in either case, the time assignment implies that the bit signals are 'associated' with the speech channel.

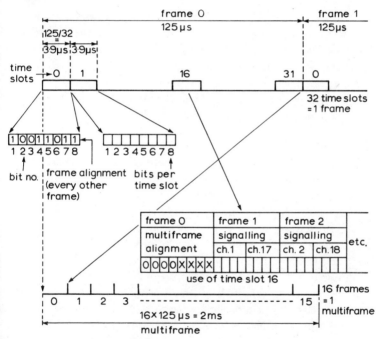

Fig. 6.2   *30-channel p.c.m. system*

Two p.c.m. transmission systems, the Bell D2 24-channel and the CEPT 30-channel, are at present specified by the CCITT. The 24-channel system uses all 24 channel time slots for speech. The 30-channel system has 32 time slot channels, 30 being used for speech, one (time slot 0) for frame alignment and one (time slot 16) for signalling. Built-in p.c.m. signalling is used when p.c.m. transmission is applied in the analogue environment, the switching being analogue. The p.c.m. transmission system is then said to be applied point-to-point.

6.2 P.C.M. signalling 24-channel p.c.m. systems

6.2.1 Bell system D1 24-channel system

This, sometimes referred to as system T1, was the pioneering p.c.m. system, being the first in extensive service and the first to be used in short-haul (junction) networks.[3] Synchronisation, or, more accurately, frame alignment, is based on the use of an extra bit (193 instead of 192 bits) per frame, it not being particularly significant whether the extra bit is the 1st or the 193rd of a frame. 24-channel working would normally result in a 1·536 Mbit/s system line bit rate, but the additional frame delineation bit gives a total line bit rate of 1·544 Mbit/s.

Bit 8 of the 8 bits in each time slot is used for in-slot signalling, all time slots in all frames being treated in the same way from the signalling aspect. As the signalling is the equivalent of outband, the bits available for speech encoding are reduced from 8 to 7. Signalling is at the 8 kbit/s rate (one signalling bit per speech channel every 125 μs), the signalling capacity being one binary 'channel' due to the single signalling bit per speech channel. The term 'signalling channel' should not be confused with the term 'speech channel'. Signalling channels in p.c.m. signalling relate to the number of discrete signalling bits per speech channel, and, as 4-wire working is implicit in p.c.m. transmission, independent forward and backward signalling paths apply and the signalling channel(s) operate in each direction without mutual interference. All built-in p.c.m. signalling is 'continuous', the signal meaning being given by the bit pattern of the signalling bit(s). Respective codes may be quickly identified, starting at any point in the code sequence, such codes being called 'comma free'. Owing to the single signalling channel, two sequential codes 0000 . . . and 1111 . . . are available in the D1 system, which conforms to the Bell system philosophy of 2-state (on-hook, off-hook) continuous signalling in each direction and also conforms to the Bell system E- and M-lead control. The same signalling state may be used to indicate more than one signal by taking advantage of the fixed order of occurrence of specific signals. The signal code for the D1 system is the same as that described for the D2 system. The Bell D1 24-channel p.c.m. system is not specified by the CCITT.

6.2.2 UK Post Office first-generation 24-channel system (short-haul)

This has some similarity to the Bell D1 system, but the signalling bit, bit 1 of the 8 bits in each time slot, is not used solely for signalling; its use is organised on a 4-frame multiframe basis to provide frame and multiframe alignment synchronisation and signalling.[4] Signalling is thus in-slot, is the equivalent of outband, and the bits available for speech encoding are reduced from 8 to 7.

The UK network requires more signals than those given by the Bell system; typically, metering over junction circuits is required, meter pulses being passed back over the junction to the end exchange when, with nonitemised call charging being used, the s.t.d. bulk billing charge rate is determined at the trunk exchange. 2-state signalling with a single signalling bit is not adequate for the signalling requirement, two signalling bits per speech channel being necessary to expand the signalling bit coding possibility. Using more than one bit of each 8-bit channel time slot would unduly degrade speech transmission, and, to avoid this, a system of multiframing is adopted. Multiframing consists of identifying frames in a consecutive sequence by count, the number of frames to be identified depending on the size of the multiframe adopted, e.g. 4 frame, 12 frame, 16 frame. As the frames in a multiframe are time displaced, and can be time identified, a signalling bit per speech channel in each constituent frame is time displaced and a discrete number of signalling channels can be obtained per speech channel. Thus multiframing facilitates the attainment of more than one signalling channel per speech channel for built-in p.c.m. signalling, enabling the signalling bits to be time assigned to the appropriate speech channel. It is not essential to use a signalling bit in each constituent frame when multiframing; the multiframe concept is flexible and different detailed arrangements may be adopted to meet the requirements of different situations.

The UK 24-channel p.c.m. system adopts a 4-frame multiframe (Fig. 6.1). Bit 1 of each 8-bit time slot in frames 1 and 3 contains the signalling information for the individual speech channels. Bit 1 of each 8-bit time slot in frame 4 is used for frame and multiframe alignment synchronisation with a pattern of 16 bits (bit 1 of 16 time slots) in the form 1101010101010101, the remaining bits 1 in frame 4 not being used. Bit 1 in frame 2 is, in general, not used, but clearly could be used to extend the signalling capacity should this ever be a future requirement. With frame alignment based on some of the bit 1 time slot capacity, the Bell D1 system additional frame alignment bit per frame is not needed, the frame consists of 192 bits and not 193, and the line bit rate is 24×64 kbit/s or $1 \cdot 536$ Mbit/s. As two frames in the multiframe provide a signalling bit each, two signalling channels per speech channel are available. These may be used independently or the two bits may be coded in combination, as desired. The signalling capacity per speech channel is two binary channels each at 2 kbit/s, or, as a signalling bit is available every $250 \, \mu s$ in every speech channel, one channel at 4 kbit/s. With the two signalling channels, three sequential codes 0000 ..., 1111 ..., 1010 ..., are available in the UK system, and, here again,

Table 6.1   *Typical signal code for UK 24-channel p.c.m.*

	Signal	Typical codes only
Forward:	Idle	0000 . . .
	Seizure	1111 . . .
	Decadic address	0000 . . . (dial breaks)
		1111 . . . (dial makes)
Clear forward		0000 . . .
Trunk offer		1010 . . .
Backward:	Idle	0000 . . .
	Called subscriber free	1111 . . .
	Called subscriber busy	0000 . . .
	Answer	1010 . . .
	Clear back	0000 . . .
	Manual hold	1010 . . .
	Meter pulses	1111 . . .

the same signalling coding may be used to indicate more than one signal by taking advantage of the signal sequence. The three sequential codes may be used separately in each direction. On this basis, a typical signal code could be as in Table 6.1. Typical examples only are stressed, the example being given merely to illustrate how the signalling bit codings could be used in short-haul application.[5]

It should be noted that multiframing is not required in the Bell D1 system, as, here, one in-slot signalling channel only is required per speech channel, and, as all the time slots in all frames are treated in the same way from the signalling bit aspect, it is not necessary to identify particular frames. Multiframing is not required, of course, when common channel signalling is applied, as here, unlike in built-in p.c.m. signalling, the signals are circuit label addressed and not time assigned to a particular speech channel. The UK 24-channel p.c.m. system is not specified by the CCITT. It is now superseded in the UK network by the CEPT 30-channel system,[10] but the description is included to demonstrate the principle of the multiframing technique.

6.2.3 Bell system D2 24-channel system

8-bit speech encoding is preferred to reduce quantising noise and thus to secure better speech transmission performance on the several conversions on multilink (long-haul) connections.[2] This is a feature of the CCITT(CEPT) 30-channel p.c.m. system. The Bell system, for similar reasons, has introduced a long-haul-version 24-channel p.c.m. system, known as D2. As in the D1 system, the line bit rate is 1·544 Mbit/s (193 bits per frame) and the signalling is in-slot, the equivalent of outband.[6]

While the Bell system 2-state continuous signalling in each direction is met by p.c.m. signalling with a single signalling channel, two signalling channels are given by the D2 system to facilitate the wider application of the system, and multiframing is used. A 12-frame multiframe is adopted for D2 and the additional alignment bit per frame performs, in respective alternate frames of the multiframe, the respective functions of frame alignment and multiframe alignment. Frames 6 and 12 are designated as signalling frames, giving the two signalling channels, and bit 8 of the 8 bits in each channel time slot is used in every signalling frame (frames 6, 12 etc.) to carry the signalling associated with that speech channel. Bit 8 of each time slot is used for speech encoding in the intervening frames 1–5, 7–11 etc. This 'bit stealing' technique, which allows 7 bits for speech encoding in the signalling frames (every sixth frame), allows all 8 bits in all the other frames. This improves the speech transmission relative to the D1 system, but the full gain of 8-bit speech encoding is not quite achieved. Table 6.2 shows the multiframe structure of the D2 system, the sequence shown being repetitive.

The signalling capacity per speech channel is two signalling channels A and B, each at 0·65 kbit/s, or, as a signalling bit is available every sixth frame, one channel at 1·3 kbit/s. This is significantly less than the 8 kbit/s of D1, but is still generous. As multiframing is not required with common channel signalling, the multiframe alignment S-bit is used as a 4 kbit/s bearer (or a submultiple of this) for common channel signalling, and all time slots are 8-bit speech encoding.

As has been mentioned, the Bell system 2-state continuous signal-

Table 6.2   *Multiframe structure of D2 system*

| Frame number | 193rd bit | | Bit number(s) in each channel time slot | | P.C.M. signalling channel |
	Frame alignment	Multiframe alignment (S-bit)	For speech	For signalling	
1	1		1–8		
2		0	1–8		
3	0		1–8		
4		0	1–8		
5	1		1–8		
6		1	1–7	8	A
7	0		1–8		
8		1	1–8		
9	1		1–8		
10		1	1–8		
11	0		1–8		
12		0	1–7	8	B

Table 6.3   *Bell line signalling D2 system*

| Signal | Transmitted sending duration | Signalling bit transmitted. State: 0 = on-hook, 1 = off-hook | |
		Originating end	Terminating end
Idle	Continuous	0	0
Forward			
Seizure (off-hook)	Continuous	1	0
Decadic address, when applied (sequential on-hook, off-hook)	Pulsing	0 (dial break) 1 (dial make)	0
Clear forward (on-hook)	Continuous	0	0 or 1*
Forward transfer (on-hook)	65–135 ms pulse	0	0 or 1†
Backward			
Delay dialling (off-hook)	Continuous¹	1	1
Proceed-to-send start dialling (on-hook)	Continuous¹	1	0
Answer (off-hook)	Continuous	1	1
Clear back (on-hook)	Continuous	1	0
Blocking (off-hook)	Continuous	0	1

* Depending on whether clear back prior to clear forward or not.

† Depending on whether prior answer received or not.

Notes:
(1) The duration of these signals is variable and depends on when the succeeding signals occur. To ensure proper registration of these signals,
the transmitted signal duration should not be less than 140 ms.
(2) A backward release guard signal is not given in Bell system practice.

ling in each direction does not require two signalling channels, and, in
this application of the D2 system, the same signalling information is
sent on both the signalling channels A and B. Table 6.3 shows the D2
system codings for the Bell system standard line signalling.

It is stressed that the above is a typical signal code application
of two signalling channel p.c.m. signalling on the D2 system, the same
signalling information being transmitted on both signalling channels.
The concept is flexible and allows other signal code possibilities,
as required. For example, should decadic address signalling be required,
this could be signalled by sequential 0 or 1 signalling bit states in the
forward direction, as shown in Table 6.3.

The Bell D2 24-channel p.c.m. system, with the two signalling channels arrangement, is specified by the CCITT.[6]

6.3 P.C.M. signalling 30-channel p.c.m. system

This p.c.m. system adopts 8-bit encoding for all speech channels and 32 time slots instead of 24. Framing is controlled by sacrificing one slot (time slot 0), in which a frame identification word (for frame alignment) and a status word (concerning urgent and nonurgent alarms) are transmitted alternately. Another time slot (time slot 16) is sacrificed for signalling. The speech channels are thus 30, but unrestricted 8-bit encoding is available on each, as the signalling is out-slot. The line bit rate is $32 \times 8 \times 8$ kbit/s = 2·048 Mbit/s. Speech channels 1–15 and 16–30 are transmitted in time slots 1–15 and 17–31, respectively, the time slots of the system being numbered 0–31.[7,11,12]

The channel time slot 16 is used for signalling for the 30 speech channels of the system, providing 1 64 kbit/s digital path which is submultiplexed into lower rate paths by using the multiframe alignment signal as a reference. The out-slot signalling is a form of separate channel signalling, the signalling bits being associated with the relevant speech channel on a time assigned basis by virtue of the 16-frame multiframe adopted, the frames being numbered 0–15 (Fig. 6.2). This signalling is often referred to as 'bunched'.[8,9,11,12]

The first four bits in frame 0 time slot 16 are used to define the multiframe; the remaining four bits are spare, but may be used as desired, typically to indicate loss of multiframe alignment. The 8 bits of frames 1–15 time slot 16 are divided into two groups of four bits each for the two separate speech channels. The speech channels to which the signalling bits in a particular frame are allocated are separated by 16 time slots (Fig. 6.2). Thus, in every 2 ms, the 16-frame multiframe time, four bits (channels A, B, C, D) are available for each speech channel. These may be regarded as being four independent signalling channels, each of 500 bit/s, or as an overall signalling bit rate of 2 kbit/s per speech channel. All four bits may not always be required in particular applications, and, if not, bits B, C or D are given the values B = 1, C = 0, D = 1 according to the bit signalling channel(s) not used.

The multiframe alignment word is 0000 (bit positions 1–4 frame 0 time slot 16). To avoid confusion with this, the coding 0000 is not used as a combination of signalling bits A, B, C and D. This restricts to 15 the maximum number of signalling conditions of speech channels 1–15, and, for uniformity, speech channels 16–30, in both the forward and backward signalling directions, but this restriction is of no consequence as the signalling capacity is ample.

Channel time slot 16 of the 30-channel p.c.m. system can be used to provide a 64 kbit/s output and a 64 kbit/s input port, to allow a 64 kbit/s signalling capability, which could be used for common channel signalling at the 64 kbit/s signalling bit rate. The multiframing would not be required when common channel signalling is used. It should be noted that, as at present specified,[6] the 24-channel system does not allow a 64 kbit/s bearer capability for common channel signalling, which is a weakness, as it is at present preferred for common channel signalling in digital networks to be optimised at the 64 kbit/s signalling bit rate. The S-bit in the present specified 24-channel system permits common channel signalling at a maximum rate of 4 kbit/s. It will be understood, of course, that all channels in p.c.m. systems have a 64 kbit/s capability, and, with a change of the present specified arrangements of the 24-channel system, a 64 kbit/s bearer could be isolated and made available for common channel signalling, if desired.

This 30-channel p.c.m. system, with channel time slot 16 bunched signalling, is specified by the CCITT and standardised by the CEPT, and is often referred to as the CEPT 30-channel p.c.m. system.[7]

6.4 Comments on built-in p.c.m. signalling

Built-in p.c.m. signalling on a point-to-point application of the CCITT specified p.c.m. systems in analogue networks makes adequate provision for line (supervisory) signalling with or without decadic address information signalling. At present, interregister m.f. signalling is provided by using the analogue m.f. signalling system, the m.f. signals being coded in the same manner as speech, and decoded and recorded in the register as in the analogue case. As a possibility with integrated digital transmission and switching, the speech channel bits could be used for interregister signalling. During call set up, all 8 bits could be available for signalling purposes and some or all of these could be manipulated in the registers to provide fast coded interregister signalling. This is mentioned as a possibility, but is not used at present.

Built-in time assigned p.c.m. signalling gives signalling capability on a per-p.c.m. system basis. Such signalling is lightly loaded in the telephony service, telephone signals being relatively infrequent, particularly the line supervisory signals. Time assigned channel associated signalling is distinct from circuit label addressed signalling as applies in common channel signalling, which signalling does not require multiframing. A digital common channel signalling system can serve the signalling requirements of hundreds of speech circuits and thus of many p.c.m. systems, particularly when a 64 kbit/s capability is available for common channel signalling. The built-in time assigned p.c.m. signalling principles could be applied in integrated digital networks, but there is little merit, and considerable complication, in this, and digital common channel signalling is the preferred approach.

With built-in p.c.m. signalling, while the total signalling is t.d.m., each traffic circuit has exclusive use of its own signalling means and hence is not subjected to the queuing delays that arise in common channel signalling. Further, unlike common channel signalling, built-in p.c.m. signalling does not require special means to ensure speech path continuity.

References

1 CATTERMOLE, K. W.: 'Principles of pulse code modulation' (Iliffe, 1969)
2 BYLANSKI, P., and INGRAM, D. G. W.: 'Digital transmission systems' (Peter Peregrinus, London, 1976)
3 FULTZ, K. E., and PENICK, D. B.: 'The T1 carrier system', *Bell Syst. Tech. J.*, 1962, 44, p. 25
4 BOLTON, L. J., and BENNETT, G. H.: 'Design features and application of the British Post Office 24-channel pulse-code-modulation system', *Post Off. Electr. Eng. J.*, 1968, 61, p. 95
5 CREW, G. L.: 'A line signalling scheme for PCM telephone systems', *Telecommun. J. Australia*, 1968, 18, (1), pp. 59–62
6 CCITT: Green Book, 6, Pt. 1, Recommendation Q47 ITU, Geneva, 1973
7 CCITT: Green Book, 6, Pt. 1, Recommendation Q46 ITU, Geneva, 1973
8 POSPISCHIL, R., and SCHWEIZER, L.: '30- kanel PCM system fur den nahverkehr', *Tech. Rundsch.* No. 9, 1971
9 CHRISTIANSEN, H M., and SENFT, R.: 'PCM 30/32, a time division multiplex transmission system for local and short-haul networks', *Siemens Rev.*, 1971, 38 pp. 190–195; Special issue on 'Communications engineering'
10 BOAG, J. F.: 'The end of the first pulse code modulation era in the UK', *Post. Off. Electr. Eng. J.*, 1978, 71, Pt. 1, pp. 2–4
11 VOGEL, E. C., and McLINTOCK, R. W.: '30-channel pulse code modulation system – Pt. 1 multiplex equipment', *ibid.*, 1978, 71, Pt. 1, pp. 5–11
12 WHETTER, J., and RICHMAN, N. J.: '30-channel pulse code modulation system – Pt. 2 2·048 Mbit/s digital line system', *ibid.*, 1978, 71, Pt. 2, pp. 82–89

Chapter 7

Influence of transmission on signalling

7.1 General

In addition to the influence of the basic type of transmission system on the signalling method, d.c., v.f., outband, m.f. and p.c.m. digital, as discussed in Section 1.3.3, various features and conditions of the application of transmission plant influence the details of signalling systems. In general, such influences concern long-distance circuits in national and international networks, and it is appropriate to examine these influences before discussing signalling systems for long-distance circuits.

7.2 4-wire circuits

Long-distance analogue speech circuits are usually, and digital circuits always, 4-wire, and, with 4-wire switching, the 4-wire speech circuit must, of course, be extended to the switching equipment.[1] This is also preferred practice with 2-wire switching, the 4-wire/2-wire terminating set being located at the switching equipment.[2] E and M control of signalling by leads separate from the speech leads (Section 7.10) permits the terminating set to be located remotely from the switching equipment when using 2-wire switching, but this does not negate the signalling merit of 4-wire circuits. 4-wire circuits permit separate and independent forward and backward a.c. (v.f., outband or m.f.) signalling paths, which greatly eases the signalling arrangements, as simultaneous signalling in the two directions may occur without mutual interference, and, in particular, signal discrimination by direction is easily achieved. This is of particular significance when continuous signalling in the two directions is desired, as, without the independent signalling paths, pulse a.c. signalling would normally be necessary. The duplex condition requirement is given by the two a.c. simplex signalling paths.

When the line signalling system terminals are colocated with the switching equipment, duplex d.c. signalling on the speech leads obtains at the switching equipment, and this signalling inputs and outputs the a.c. line signalling at the switching equipment. When adopted, the interposed E and M d.c. signalling stage inputs and outputs the a.c. line signalling system, the duplex condition being given by the E and M simplex d.c. signalling paths. Thus, with 2-wire switching and E and M control, the 4-wire/2-wire line termination and the a.c./d.c., d.c./a.c. conversions may be remote from the switching equipment, but the signalling merit of independent a.c. signalling paths in each direction is not disturbed.

7.3 Echo suppressors

Echo suppressors are not normally required in many national networks, but may be required in large countries and on international circuits.[3-5] The preferred type of echo suppressor is a terminal, differential, half-echo suppressor operated from the far end. When required, echo suppressors, which are usually located at the transmission terminals, may be permanently associated with the circuit, or may be provided in a pool common to a number of circuits, and arrangements are made to associate a suppressor with any circuit requiring one.[6]

When fitted, echo suppressors would normally disturb simultaneous forward and backward signalling on the speech path unless precautions are taken, such as:

(*a*) Locating the line signalling system terminals on the line side of echo suppressors, the suppressor being located between the signalling terminal and the switching equipment.

(*b*) Disabling the action of echo suppressors should these be located on the line side of the line signalling system terminals by means of an appropriate condition extended from the signalling equipment to the echo suppressor while signalling is in progress.

Method (*a*) is preferred, and, when separate leads E and M control of signalling not adopted, the requirement is facilitated by extending the 4-wire circuit to the switching equipment. With E and M control, the line signalling system terminals are located somewhat remotely from the switching equipment (and possibly at the transmission terminal), which simplifies the arrangement to locate the signalling terminals on the line side of the suppressors without extending the 4-wire speech circuit to the 2-wire switching equipment.

Interregister m.f. signalling terminals are usually located with the registers (senders) and thus with the switching equipment. As, in this case, the signalling terminals would not be located on the line side of the echo suppressors, local arrangements would be necessary to disable the echo suppressor while interregister signalling is in progress, should the interregister signalling require simultaneous signal transmission in each direction, and should the echo suppressor be line associated during the interregister signalling. This disabling is not necessary when the interregister signalling is forward signalling only.

When echo suppressors are provided from pools, signals in the signalling system are necessary to give information to relevant points as to the echo suppressor requirement at such points; e.g. an incoming half-echo suppressor is required. This philosophy requires an outgoing exchange to take a decision by analysing its echo suppressor requirements at the time the outgoing circuit is selected, and to pass an appropriate signal forward to guide subsequent exchanges on the necessary action to be taken on the possible association of an incoming echo suppressor, the control actions being concerned with 'inserting' or 'not inserting'. The relevant signalling is usually carried by the interregister m.f. signalling system. The concept allows for the possibility of inserting echo suppressors after the interregister signalling is completed and so to avoid local arrangements to prevent echo suppressor disturbance to simultaneous forward and backward interregister signalling.

This control of echo suppressor provision from pools by means of signals is presently limited to certain CCITT international signalling systems.[7] The CCITT international signalling systems 4, 5bis and R2 pass signals to ensure association of an incoming half-echo suppressor when required. Systems 5 and R1 do not include echo suppressor control signals. In system 5, the normal field of application would usually indicate the presence of echo suppressors. In system R1, the regional control procedures applied do not require echo suppressor signals. Arrangements are necessary in the common channel signalling systems to prevent echo suppressor action disturbing the procedure for making a continuity check of the speech path should a per-call continuity check be performed.

When echo suppressors are inserted from pools, there is a small probability that an echo suppressor will not be available when needed. In this event, it is recommended that an equipment congestion tone be returned to the caller.

7.4 Long propagation time

This is of concern with any form of acknowledged signalling, the signalling being slowed and the holding time of relevant equipment and the postdialling delay being increased. Also, with any form of line signalling, there is increased likelihood of double seizure on bothway working of circuits owing to the increased unguard time. Although long international circuits are the main concern, the problems also arise in large national networks. The maximum round-trip signalling time for a satellite circuit is assumed to be 1300 ms for planning purposes, which takes account of the terrestrial circuits between the earth stations and the relevant exchanges, and some 500 ms for wholly terrestrial circuits, and, although terrestrial circuits could introduce problems, the satellite circuit is clearly the main concern.

The acknowledged continuous line signalling of the CCITT international signalling systems 5 and 5bis is dictated by the requirements of time assignment speech interpolation (t.a.s.i.) equipment, the consequent penalties due to long propagation times being accepted as the t.a.s.i. requirements were considered to be of greater importance and were thus overriding. The serious problems arise with per-digit acknowledged address information signalling, as the delays have a serious impact on the postdialling delay. The pulse per-digit acknowledged CCITT 4 system involves two propagation times per digit signalling sequence, and while this system could be applied to satellite circuits, this is inadvisable, owing to the increased postdialling delay. Continuous compelled signalling of the address information, as instanced by the CCITT R2 system, would introduce serious postdialling delay problems in view of the four propagation times (two tone-on, two tone-off) and the four signal recognition times (two tone-on, two tone-off) per digit signalling sequence, and such an address signalling technique is inadmissible on satellite circuits. Thus the nature of the transmission circuit has a direct influence on compelled mode address signalling, as alternative arrangements to fully continuous compelled signalling must be adopted for long propagation time circuits. The alternative arrangement must be based on speeding up the digit signalling sequence, and it will be seen later that a semicompelled philosophy is proposed when system R2 is applied to long propagation time circuits such as those occurring in satellite systems, a pulse (as distinct from a continuous) signal acknowledging the continuous primary signal. This involves two propagation times and two signal recognition times per digit signalling sequence instead of the four of each when signalling is fully continuous compelled. Even with this alternative arrangement, the address signalling is relatively slow, but per-digit acknowledgment is an inherent feature of system R2 (together with the primary signal being continuous) and any alternative arrangement of address information transfer must be based on this philosophy for this system.

In common channel signalling with error control involving error correction by retransmission, short propagation times admit the possibility of a so-called compelled error control method (one message only at a time in the error control loop delay), whereas long propagation times necessitate a noncompelled method (a number of messages at a time in the error control loop delay). In the latter case, the sequence numbering range of signal units (or message) and the size of the retransmission store are related, for a given signalling bit rate, to the maximum error control loop delay (and thus to the maximum propagation time of the signalling circuits) to which the system is to be applied.

7.5 Satellite circuits

In addition to the delays to signalling, the long propagation time of satellite circuits delays the speech transfer. The mean 1-way propagation times between earth stations for two illustrative single-hop satellite systems are

satellite at 14 000 km altitude: 110 ms
satellite at 36 000 km altitude: 260 ms.

Further delays arise on the circuits linking the earth stations to the relevant exchanges.[8] For planning purposes, it is considered advisable to adopt the extreme case of a double-hop satellite link, which accounts for the 1300 ms loop delay stated in Section 7.4. The magnitude of the satellite delays makes it desirable to impose some routing restriction on their use to guard against the inclusion of two or more satellite links in a connection where this can be avoided. In very exceptional circumstances, a connection with more than one satellite link could be used; for example, where no other reliable means of communication is available, or where the connection is required for special purposes. To achieve the necessary safeguard, some signalling systems incorporate a history of routing indicator in the signal repertoire to inform subsequent switching points whether or not prior usage of a satellite link has been taken. This enables switching points to impose a routing restriction to avoid, when possible, the inclusion of a further satellite link in the forward routing of the connection. International circuits are clearly the main concern and some CCITT signalling systems, e.g. 5bis, 6, 7, include such satellite indicators, and most of the others have spare signal capacity to enable the indicators to be added.[9] The specification of the CCITT system R2 is in process of being modified for optional semicompelled, instead of fully continuous compelled, working to facilitate the application of the system to satellite circuits, and the inclusion of a satellite indicator in the R2 interregister signalling will be necessary as part of the modification.

7.6 Time assignment speech interpolation equipment

The normal speech activity of individual subscribers on a call is some 35% and the separate speech transmission channels of 4-wire circuits are thus normally less than half utilised. Time assignment speech interpolation (t.a.s.i.) equipment on the transmission facility increases the utilisation, and thus the traffic carrying capacity, of transmission circuits by assigning a channel to a trunk when the channel is required to transfer speech, and, on cessation of the speech burst, disconnecting the channel from the trunk and assigning it to another trunk requiring speech transfer. Channels are thus interpolated and assigned to trunks when speech transfer is required, and a given number of circuits can be accessed by a greater number (somewhat more than double) of trunks at each end.[10-12]

The trunks at each end of, say, a transoceanic cable, are each equipped with a speech detector, and, when the presence of speech is detected at one end, the t.a.s.i. system arranges for a channel in the transoceanic cable to be connected to the corresponding trunks at the two ends to enable the speech to be transferred between the trunks. The speech burst is thus clipped for the period required to perform the trunk–channel association. The speech detector has a short hangover time to maintain the trunk–channel association during short gaps in speech to reduce the extent of interpolation, the trunk–channel association being disconnected when a speech cessation duration exceeds the detector hangover time.

As t.a.s.i. is relatively expensive, its application to date is limited to expensive transmission circuits, such as those in transoceanic cables (Atlantic, Pacific etc.) to increase the effective capacity of a cable to somewhat more than double its original capacity. The relevant signalling systems concerned when t.a.s.i. is applied are those of the international type. CCITT signalling systems, 5, 5bis, 6, 7 are suitable for t.a.s.i. application – the others are not – and systems 5 and 5bis are specially designed for the purpose.

As with speech bursts, t.a.s.i. clips a signal, owing to the time required to perform trunk–channel association to transfer the signal the clip being some 17 ms when a free channel is available. The clip duration increases under busy traffic conditions when a free channel may not be immediately available, and the possibility of a partial, or even a complete, 'freeze-out' of a signal arises. To avoid freeze-out, signals must be of sufficient length to assure trunk–channel association and to permit signal recognition, if the association is not arranged prior to signal transmission. With pulse signalling, it has been determined that some 500 ms would be required for the extreme trunk–channel association condition, to make a transmitted pulse some 850 ± 200 ms when allowing for the signal recognition time. Pulse signals of such lengths would slow the signalling and result in unnecessary t.a.s.i. signalling activity when the t.a.s.i. system is lightly loaded, as here the trunk–channel association time would always be less than 500 ms. Fixed-length pulses cannot take advantage of this lightly loaded condition, and for this reason the line signalling of signalling systems for t.a.s.i. application (noncommon channel) should be continuous compelled, trunk–channel association always being assured in the actual time required for this function. The requirement for an acknowledgment signal in continuous compelled signalling increases the t.a.s.i. signalling time and activity, but this is not a penalty should a line signal require an acknowledgment in the normal signalling function. The adoption of continuous compelled signalling instead of a single pulse for signals not normally requiring a return signal could incurr signalling time and activity penalties when t.a.s.i. is heavily loaded, but could incur advantages when lightly loaded, and, on balance, continuous compelled signalling is preferred and adopted for the line signalling of the conventional signalling systems.

Conventional system continuous compelled interregister signalling would be far too slow on long propagation time t.a.s.i. equipped circuits; the t.a.s.i. signalling activity penalty would be excessive, and pulse signalling is preferred for the transfer of address information. Clipping cannot be tolerated for fast, pulse, address signals and it is essential that a t.a.s.i. channel be prior associated for pulse address signalling and remain associated during the gaps between successive address pulse signals, and thus signal clip does not occur. Assuming prior trunk–channel association, the association may be maintained during the address signalling by (*a*) complete *en bloc* of the address information at the outgoing register (or sender), the gaps in the outpulsed address signals being controlled and of less duration than the t.a.s.i. speech detector hang-over time

or

(*b*) a lock tone when overlap operation of the address information at the registers is adopted, as, in this mode, the gaps between consecutive pulses are not always machine controlled and are thus likely to exceed the speech detector hangover time.

When parts of t.a.s.i. are occupied with signalling, these parts cannot be utilised for speech. As the purpose of t.a.s.i. is to increase the traffic carrying capacity of expensive circuits, and as the greater the t.a.s.i. signalling activity the less the permissible t.a.s.i. speech activity, ideally, the signalling activity should be a minimum, the signalling being as rapid as possible in the two directions over the t.a.s.i. link, and the signals as few as possible. On the other hand, the number of signals should be consistent with the facilities to be given. Of the two CCITT signalling systems 5 and 5bis specially designed for t.a.s.i. application:

(*a*) System 5 adopts continuous compelled line signalling (with the exception of the forward transfer signal, which is pulse). The inter-register m.f. signalling is pulse forward signalling only and complete *en bloc*. This approach was deliberately adopted to minimise the interregister signalling t.a.s.i. signalling activity, this being considered to be of main importance. The *en bloc* of the address information also facilitates a measure of validity checking by digit count to avoid expensive intercontinental circuits being taken ineffectively on incomplete dialling. On the other hand, the *en bloc* approach increases the postdialling delay, and the absence of backward interregister signalling limits the facilities.

(*b*) System 5bis adopts the same line signalling as system 5, but the interregister m.f. signalling is both forward and backward pulse signalling, with overlap operation at the registers. A lock tone in each direction maintains t.a.s.i. trunk–channel association during the gaps between consecutive address pulses. Relative to system 5, system 5bis reduces the postdialling delay by virtue of the overlap operation, permits more facilities, owing to the inclusion of backward interregister signalling, but increases the t.a.s.i. signalling activity.

The choice between the two systems depends on the significance given to the t.a.s.i. signalling activity in the main, and, in the result, system 5 is the more widely used.

When interpolating, t.a.s.i. clips speech to a degree depending on the traffic load. Although, with a multiplicity of t.a.s.i. systems on a connection, the total clipping would be on a random basis, it is clear that, in certain circumstances, the clipping could well be assessed as being unacceptable. At present, the relevant international signalling systems do not include history of routing indicator signals for the specific purpose of controlling the inclusion of further t.a.s.i. systems in the forward routing. Should the requirement for such a restriction arise in future, spare capacity is available in the signalling systems to include such indicators.

7.7 Compandors

Should compandors be equipped on transmission systems, these affect short-pulse compound signalling, e.g. interregister m.f. pulse signalling, owing to distortion and the production of intermodulation frequencies.[13-15] Detection of low-level signals is an onerous condition and, as low received level signals may arise with end-to-end interregister m.f. signalling, link-by-link signalling eases the signalling problem in the presence of compandors. Usually, taking account of the permitted transmit signal level, end-to-end signalling with a modest number of constituent links is satisfactory in the presence of compandors, providing the pulse duration is adequate to take account of the distortion.

Should speech be required in the presence of a signalling tone, as arises in the Bell SF signalling system on noncharged calls and the backward signalling tone not being ceased, the presence of the tone may reduce the compandor crosstalk and noise advantage, but this is of little significance.

7.8 Signal power

A high transmitted signal level greatly eases signal detection, particularly in the face of noise and other interference, and simplifies the signalling system. On the other hand, transmission systems place restrictions on the permitted magnitude of the transmitted signal level to limit crosstalk and to avoid overload of the common amplifiers of transmission systems.

For reasons of crosstalk, the maximum permissible absolute power level of a signal pulse is limited (Section 2.3), the limitation being a CCITT recommendation.[16] For reasons of overload, the CCITT recommends that the maximum energy which may be transmitted by all signals and tones during the busy hour should not exceed

$36\,000\,\mu$Ws per channel, i.e. for one direction of transmission
$72\,000\,\mu$Ws per circuit, i.e. for both directions of transmission.

The permissible transmitted signal level will clearly depend on the number and duration of signals per call, and could thus vary, depending on the features of particular signalling systems. The recommendation is of particular significance for a.c. signalling systems (v.f., outband and interregister m.f.) employing continuous signals. With outband signalling it normally precludes continuous signalling during the speech period, although this may be permissible, but not preferred, if the number of such signalling circuits are relatively few in a transmission system. The recommendation is particularly onerous for tone-on during idle line signalling, v.f. or outband, as the transmitted signal level is required to be particularly low (typically -20 dBm0), which introduces problems with this type of v.f. signalling in particular, as discussed for the Bell SF v.f. signalling system. A much higher level (typically -6 dBm0) is permitted for pulse signalling, which eases the signal detection arrangements.

The usual forms of interregister m.f. signalling (continuous and pulse) allow significant periods in the signalling sequences when no signal energy is present, and the permissible transmitted signal level may be reasonably high and still meet the requirements of the recommendation when the interregister and line signalling loads are combined. The pulse level (typically -8 dBm0 per frequency) may often be higher than that of the continuous compelled (typically $-11\cdot5$ dBm0 per frequency), owing to the lower time signal energy present. In either case, the permitted transmitted signal level is such as not to introduce severe signal detection problems when end-to-end signalling, should this be used.

7.9 Bothway operation of circuits

7.9.1 General
Bothway operation is often adopted for long circuits in national trunk and international networks, more particularly when time-zone differences occur (the busy hours not coinciding at the two ends of the circuits) and/or when there are relatively few circuits in the circuit

group, in order to increase the utilisation of these relatively costly circuits.[17] Additional signalling equipment is necessary to permit bothway operation, and the nature of the transmission plant influences the detail of the signalling systems, and the increase in cost of the signalling equipment resulting from the bothway operation is accepted in view of the often considerable economic advantage derived from this mode of operation. Unidirectional operation of circuits is the usual practice on local exchange area junction circuits, the additional expense of bothway signalling equipment not being justified, junction circuits being shorter and of less costly type than trunk (toll) and international circuits. Unidirectional signalling equipment tends to be simpler and less costly than that for bothway operation, since it can be designed to cater for single-purpose signalling at the respective ends of the circuits, outgoing only at one end and incoming only at the other.

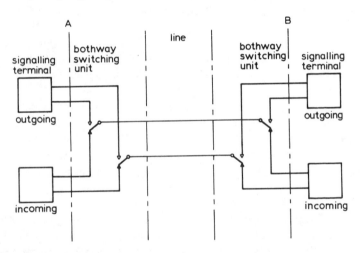

Fig. 7.1   *Bothway operation*

In bothway operation, both outgoing and incoming signalling functions must occur at each end of the circuit, the terms outgoing and incoming denoting the direction of the call. In the basic concept, although the details vary with different signalling systems, each end of the circuit in the idle condition rests on the incoming signalling function and is thus immediately available to accept an incoming call seizure signal (Fig. 7.1). A bothway switching function at each end switches an end from the incoming to the outgoing function on call origination at that end.

The additional signalling complexity consequent on bothway

operation varies for particular signalling systems. In Bell system practice, for example, the additional signalling complexity is absorbed in the trunk circuit at the switching equipment at each end, and the signalling system itself is kept simple by designing for symmetrical operation at each terminal, as instanced by the Bell CX, DX and SF signalling systems. With symmetrical operation, each of the input/output functions is applied, conveyed and delivered in an identical manner for each direction of operation. It is claimed that this arrangement reduces the additional signalling cost of bothway operation, but this is perhaps valid only because of the Bell system simple 2-state continuous signalling philosophy in each direction. The same economic advantage would not result when pulse signalling, owing to the additional logic and memory functions required, nor when the signalling is not symmetrical in the two directions. Other approaches adopt specially designed bothway signalling terminals and yet others equip the unidirectional signalling system outgoing and incoming terminals at each end of the circuit together with a bothway switching unit.

In view of the additional cost and complexity of bothway signalling, the practice is often adopted of dividing the circuit group into three parts, one part unidirectional A–B, one part unidirectional B–A and one part bothway. This method has also the merit of giving each end undisputed access to some circuits as, should all the circuits in the circuit group be worked bothway, the possibility could arise of all the circuits being taken by one end to prevent the other from originating a call set-up.

7.9.2 Double seizure
When outgoing equipment at one end is seized on call origination, there is an unguarded delay before the distant-end equipment is seized and its associated outgoing equipment is busied. During this delay, the latter outgoing equipment may be seized by a caller attempting to set up a call in the reverse direction to the original call. Double seizure (or 'glare') conditions exist when both sets of outgoing equipment are simultaneously in a seized condition, and is detected when an end receives an incoming seizure when in the outgoing seizure condition. The occurrence of double seizures will clearly increase with increasing circuit propagation time and with t.a.s.i. equipment delays to signalling, owing to the increased unguarded interval.

Consideration of two basic techniques, (*a*) recovery and (*b*) prevention, arises for the double seizure problem:

(*a*) *The recovery technique:* This allows double seizures to occur, but they are detected automatically and appropriate recovery action is taken. Automatic detection is achieved when a terminal receives an incoming seizure when it is in the outgoing seizure condition to that circuit. To minimise the probability of double seizure, there is clear merit in the circuit selection at the two ends being such that, as far as possible, double seizure can only occur when a single circuit of a group remains free; for example, by selection of circuits in opposite order at the respective ends of the circuit group.

The recovery action taken may take different forms, depending on an administration's policy and on the capability of particular signalling system. Thus:

(i) An appropriate tone (busy, plant congestion) is sent to the two callers, and the circuit is released on subscriber on-hook clear (or when an outgoing register is forced to release on time-out).

(ii) The circuit is released automatically and an appropriate tone is sent to the two callers. Both (i) and (ii) may well prompt a repeat attempt by the callers, but, of course, this would not be an automatic repeat attempt.

(iii) The circuit is released automatically and automatic repeat attempts are made to set up the two calls. The repeat attempts may or may not be limited to the circuit used on the first attempt.

(iv) The call originating at one end is allowed to proceed on the circuit. The call originating at the other end backs off and either an automatic repeat attempt is made to set up the call by using a different speech circuit, or an appropriate tone is sent to that caller.

(v) As in (iv), but in conjunction with a controlling end procedure, the end from which the call is allowed to proceed being pre-assigned. A controlling end could control, from the bothway working aspect, all the bothway circuits of a speech group, or the circuits could be divided between the two ends for bothway control purposes by either subdivision or by one end controlling odd-numbered speech circuits and the other the even-numbered circuits.

There are other possibilities:

(b) *The prevention technique:* This eliminates double seizures and is a possibility for common channel signalling only. Controlling end allocation of any desired arrangement is used and speech circuits are allocated at the controlling end of a circuit group for calls originating at that end. The noncontrolling end requests a speech circuit allocation from the other (controlling) end before sending the initial address message, which is the first signal sent, as a discrete seizure signal is not given in common channel signalling. This request would include a 'call tag', and the controlling end returns a signal including the call tag and the circuit label of the speech circuit allocated at that end for the call originating at the other end. The noncontrolling end then sends the initial address message, which carries the allocated speech circuit label. The prevention technique is possible with common channel signalling as, owing to the fast signalling, it would not result in any significant increase in the postdialling delay. Adequate capacity is available in common channel signalling for the additional signalling requirement, but, on the other hand, the additional signalling requirement is a slight penalty.

It could be reasoned that, with the fast signalling speed of common channel signalling, the probability of double seizure is reasonably remote, particularly with not-so-long propagation times. In view of this, combined with the recovery technique being somewhat simpler to implement, the prevention technique is not at present favoured and the recovery technique is adopted for common channel signalling as well as for noncommon channel signalling systems.

7.10 E- and M-lead control of signalling

Functionally, signalling system terminals may be regarded as being (*a*) part of, or (*b*) separate from, the switching equipment. Typically, loop d.c. signalling in local junction areas usually conforms to (*a*). Other types of signalling system, such as l.d.d.c., v.f. and outband, may conform to (*b*), and, in this event, the signalling equipment is normally located between the relevant part of the switching equipment (the part being termed the trunk circuit in Bell system practice) and the line. Signalling connections between the trunk circuit equipment and the signalling system terminals are obtained via a d.c. interface consisting of a uniform system of leads designated E and M in Bell practice, separate from the speech leads, which provide the capability of always delivering and accepting uniform signal conditions. When adopted, the E- and M-lead control of signalling in technique (*b*) is usually in association with amplified transmission plant, and in this context it is convenient to assume that the transmission influences the details of the signalling possibilities.

E- and M-lead control of signalling, pioneered by the Bell system, is adopted by Bell for toll network signalling in the main.[18,19] It is also adopted for a few other national networks, but most adopt technique (*a*) above, except for outband signalling.

The Bell system adopts three types of E and M d.c. interface: Types I, II, and III.

Bell Type-I E and M interface: This, the original interface and produced for electromechanical switching equipment, has one lead only
for each direction of transmission (Fig. 7.2). The M-lead carries d.c.
signals from the trunk circuit (colocated with the switching equipment)
to operate the outgoing part of the signalling system terminal, and

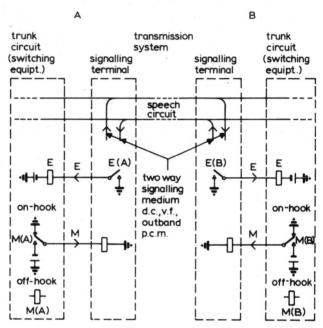

Fig. 7.2   *Bell Type-I interface*
E- and M-lead control of signalling

corresponding signals are transmitted over the line. The E-lead carries
d.c. signals from the incoming part of the signalling terminal to the
trunk circuit equipment. Thus d.c. signals from exchange A on signalling to exchange B leave on the M-lead at A, and corresponding d.c.
signals are received on the E-lead at B, the type of signal on the signal
medium being according to the type of signalling system equipped, e.g.
v.f. Signals from B leave on the M-lead and are received on the E-lead
at A (Fig. 7.2). The M-lead transmits the near-end condition of the
trunk, the d.c. signalling from the switching equipment (trunk circuit)
to the signalling system terminal being earth and resistance battery
signals over the M-lead from contact M(A) [or M(B)], battery for
off-hook and earth for on-hook. Thus, when a trunk is seized (caller
off-hook), or when the call is answered (called party off-hook), the
battery substituted for the earth on the appropriate M-lead causes a

signal to be transmitted to the other end of the line. The E-lead is the signal receiving condition and reflects the far-end condition of the line. Signalling from the signalling system terminal to the trunk circuit is by open and earth conditions on the appropriate E-lead from contact E(A) [or E(B)], earth for off-hook and open (disconnect) for on-hook.

The basic signalling conditions on the E- and M-leads reflect the Bell system signalling philosophy of simple 2-state continuous signalling in the two directions, the basic conditions on the leads being as Table 7.1. Table 7.2 shows the E- and M-lead conditions for various signals sent over interexchange circuits for call A to B.

Table 7.1 *Bell basic E- and M-lead conditions*

Signal A to B	Signal B to A	Condition at A		Condition at B	
		M-lead	E-lead	M-lead	E-lead
On-hook	On-hook	Earth	Open	Earth	Open
Off-hook	On-hook	Battery	Open	Earth	Earth
On-hook	Off-hook	Earth	Earth	Battery	Open
Off-hook	Off-hook	Battery	Earth	Battery	Earth

It will be understood that the 2-way signalling medium carries signals of a type appropriate to the signalling system used (d.c., v.f., outband or p.c.m.). The signalling medium may be single conductor (Bell CX and DX systems), or separate forward and backward signalling paths on 4-wire circuits (v.f., outband and p.c.m.). As the E- and M-leads are separate from the speech leads, and as they interconnect the trunk circuit equipment and the signalling equipment, in theory the 2-way signalling medium could be separate from the speech circuit used for the call. It is clearly preferable, however, that the signalling medium be the call speech circuit, as here the signalling would assure speech path continuity for that call. It should be noted that, when the signalling medium is the call speech circuit, owing to the E- and M-leads being separate from the speech leads, the signalling does not give assurance of continuity of this relevant part of the speech circuit, but this is of little significance.

The Type-I E and M interface has only one lead for each direction of transmission, the d.c. flowing between the switching and signalling equipment being returned over a common earth path. Further, the signalling leads have a greater noise influence than if they were balanced pairs. The Type-I interface is satisfactory for electromechanical systems, but not for electronic systems.

Table 7.2   *Bell Type-I interface E- and M-lead conditions for inter-exchange signalling*

Exchange A			Direction of transmitted signal	Exchange B		
Calling subscriber state	E-lead	M-lead		E-lead	M-lead	Called subscriber state
Trunk idle on-hook	Open	Earth	None	Open	Earth	On-hook
Seizure off-hook	Open	Battery	→	Earth	Earth	On-hook
Off-hook	Earth	Battery	←	Earth	Battery	Delay dialling off-hook (when applicable)
Off-hook	Open	Battery	←	Earth	Earth	Proceed-to-send on-hook (when applicable)
Decadic address sequential on-hook off-hook (when applicable)	Open	Pulse break on-hook earth Pulse make off-hook battery	→	Pulse break open Pulse make earth	Earth	On-hook
Speech off-hook	Earth	Battery	←	Earth	Battery	Answer, speech off-hook
Off-hook	Open	Battery	←	Earth	Earth	Clear back on-hook
Clear forward on-hook (after clear back)	Open	Earth	→	Open	Earth	On-hook
Clear forward on-hook (before clear back)	Earth	Earth	→	Open	Battery	Off-hook
Trunk idle on-hook	Open	Earth	None	Open	Earth	On-hook

Bell Type-II E and M interface: This is the preferred interface in the electronic environment, although not all Bell electronic switching systems can be used with the Type-II system, since this is a 4-wire, fully looped arrangement using open and closure signals in each direction (Fig. 7.3). Signalling from the trunk circuit to the signalling equipment is over the M and SB pair. The signalling equipment must supply the resistance battery on the SB lead and sense for the open and closure signals on the M-lead. Signalling from the signalling equipment to the trunk circuit is by means of opens and closures on the E- and SG leads. The trunk circuit must apply earth to the SG lead and a sensing device to the E-lead. On-hook signals are openings, and off-hook signals closures, in both directions. The Type-II interface has the advantage that trunk circuit equipment can be interconnected directly without convertor circuits. Neither the Type-I nor the Type-III offer this advantage.

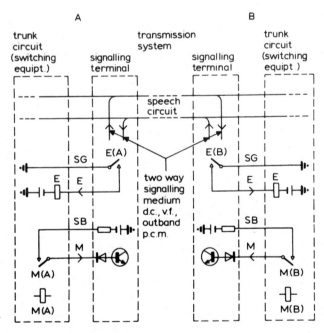

Fig. 7.3   *Bell Type-II interface*
E- and M-lead control of signalling

Bell Type-III E and M interface: This interface is a compromise partially looped, 4-wire arrangement for use in electronic switching applications where the fully looped Type-II interface cannot be used.

The loop portion uses three leads (SB, M and SG) for signalling from the trunk circuit to the signalling equipment. The signalling over the M-lead is the same as in Type I, except that the trunk circuit obtains signalling battery and earth from the signalling equipment over the leads SB and SG, respectively.

Bell E and M signalling limits: The signalling equipment determines the M-lead d.c. range, while the trunk circuits determine the E-lead d.c. range. The objective range for the E- and M-leads in new Type-I and Type-III interfaces is $100\,\Omega$ (older types have ranges as low as $25\,\Omega$). The objective range for the Type-II interface is $300\,\Omega$ loop ($150\,\Omega$ 1-way).

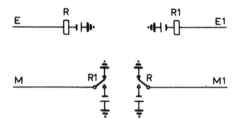

Fig. 7.4   *Bell pulse link repeater between two E and M interfaces of same type*

E and M repeaters and convertors: The E and M capability of delivering and accepting uniform signal conditions can be exploited when a traffic circuit is made up of a number of patched through, but not switched, sections in tandem, each using different types of signalling system, owing, typically, to the different types of transmission system on relevant sections. With E and M control, the signals can be transferred on a d.c. basis between adjacent sections via the E- and M-leads. If two adjacent sections have the same type of E and M signalling interface, an auxiliary pulse link repeater may be included to repeat the signals (Fig. 7.4). During on-hook (idle) conditions, the E- and M-leads are open, relays R and R1 are released, and the M- and M1-leads are earthed. Information received on each E-lead (open or earth) from one signalling system is converted into M-lead information (earth or battery) for transmission over the other signalling system. Should the adjacent sections have incompatible E and M signalling, a trunk link repeater of the type shown in Fig. 7.5 (simplified) may be used. Should a section not be equipped with E and M, e.g. a section having loop d.c. signalling, a convertor is used to perform the change from loop (non E and M) to E and M signalling, and vice versa.

The realisation of this possibility for E and M control would be costly, each section of the traffic circuit being equipped with a signalling system with both outgoing and incoming signalling terminals, to achieve the signalling facility on the whole traffic circuit. It should be noted that v.f. inband signalling may be applied to any type of transmission plant affording speech transmission, and that the one signalling system may be applied to the complete circuit no matter how sectionalised. This would be in line with the practice adopted in the same circumstances of sectionalised traffic circuits when E- and M-lead control is not adopted.

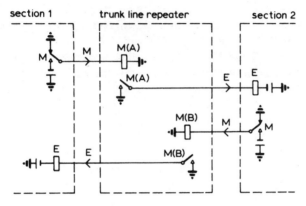

Fig. 7.5   *Typical Bell trunk line repeater interconnecting two sections having incompatible E and M signalling interfaces*

Address information signalling: When adopted, E and M control is applied to per-circuit provided signalling systems. It cannot be applied to interregister m.f. signalling systems, as such systems are functional parts of registers (senders) and thus of the switching equipment. The function of E and M in this situation is to establish the speech path between the two exchanges over which the m.f. signals are transferred, the E- and M-leads having no function on the actual m.f. signal transmission. Should the interregister signalling be decadic d.c. address pulsing (and not m.f.) and the E and M equipped line signalling system be of a type requiring the loop-disconnect d.c. address pulse signals to be converted to a form compatible with the type of signalling system, e.g. converted to v.f. address pulse signals, then sequential on-hook/off-hook E and M d.c. signals may be used for the decadic address signalling, the conversion being performed at the signalling equipment.

Comments on E- and M-lead control: E and M control has the capability of always delivering and accepting uniform signal conditions, which has the potential merit that the d.c. signalling arrangements of the switching machine itself can be standard, regardless of the type of switching system and of the various types of signalling system. This is instanced by the Bell system toll conventional line signalling, as here all the E- and M-lead d.c. signalling is uniform 2-state continuous in each direction, and, similarly, all the toll line signalling systems are 2-state continuous, regardless of type. This significant merit, however, perhaps gives rise to some penalty in other respects; for example, in the evolution of signalling, some signal(s) may be desirable in particular future system(s), but not be necessary in others. Should the E and M control be required to cater for such signal(s), this would be incompatible with the concept of uniform E and M d.c. signalling.

While, with E and M, the d.c. signalling of the switching machines can be uniform, it will be understood that the various different types of conventional line signalling systems will still be installed, but with the possibility that these be located separately from the switching equipment, e.g. at the transmission terminal.

E and M control has particular merit with outband signalling on f.d.m. transmission systems and is universally adopted in this application. Outband signalling equipment is integrated, and located with, the transmission equipment, being controlled by d.c. signals from the switching equipment. E and M control using leads separate from the speech leads has disadvantages when the signalling terminals are located remotely from the switching equipment, as arises with outband signalling and a remotely located transmission terminal. The separate E- and M-leads would then need to be extended on tie cables between the switching and transmission terminals, which is not always convenient. No problem arises, of course, when the transmission terminal is colocated with the switching centre, as here the E- and M-leads would be internal wiring. As an alternative to separate leads, many administrations adopt the principle of E and M d.c. signalling over the phantoms of the side circuits of the 4-wire speech circuit between exchanges and remote transmission terminals for outband signalling.

The designations E and M are historical in the Bell system documentation. While E and M apply functionally to incoming and outgoing d.c. conditions, respectively, relative to the switching machine, the designations do not relate to the words 'incoming' or 'receive' and 'outgoing' or 'transmit' directly. Other countries adopting the E and M philosophy often adopt different designations, SZ1, S*a*, SR for E, and SZ2, S*b*, SS for M, being known.

The d.c. signalling utilisation of the E and M philosophy is flexible, and other countries adopting the E and M principle have the choice to depart from the Bell system utilisation:

Type-I interface: E – open (on-hook), earth (off-hook)
                    M – earth (on-hook), battery (off-hook)
Type-II interface: E/SG – open (on-hook), closed (off-hook)
                    M/SB – open (on-hook), closed (off-hook)

to suit their own conditions. Typically, Japan adopts

SR (equivalent to E) – open (on-hook), earth (off-hook)
SS (equivalent to M) – open (on-hook), earth (off-hook)

While the utilisation is flexible, the simplest arrangements are achieved with 2-state continuous signalling on the E- and M-leads, the signalling on the signalling systems being correspondingly 2-state continuous.

References

1 FLOOD, J. E. (Ed.): 'Telecommunication networks' (Peter Peregrinus, London, 1975), pp. 38–40, 66–67
2 HILLS, M. T., and EVANS, B. C.: 'Transmission systems' (Allen & Unwin, London, 1973), pp. 39–48
3 SHANKS, P. H.: 'A new echo suppressor for long-distance communications', *Post Office Elec. Eng. J.,* 1968, 60, p. 288
4 FLOOD, J. E. (Ed.): 'Telecommunication networks' (Peter Peregrinus, London, 1975), pp. 40–45, 91, 190
5 HILLS, M. T., and EVANS, B. C.: 'Transmission systems' (Allen & Unwin, London, 1973), pp. 49–54
6 CCITT: Green Book, 6, Pt. 1, Recommendation Q42, pp. 85–88, ITU, Geneva, 1973
7 CCITT: Green Book, 6, Pt. 2, Recommendation Q115, pp. 250–253, ITU, Geneva, 1973
8 CCITT: Green Book, 6, Pt. 1, Recommendation Q41 pp. 83–84, Recommendation Q48 pp. 109–112, ITU, Geneva, 1973
9 CCITT: Green Book, 6, Pt. 2 Recommendation Q216 pp. 388–389, Pt. 3 Recommendation Q261 pp. 463–464, ITU, Geneva, 1973
10 BULLINGTON, K., and FRASER, J. M.: 'Engineering aspects of t.a.s.i., *Bell Syst. Tech. J.,* 1959, 38, p. 353
11 LEOPOLD, G. R.: 'T.A.S.I.-B system for restoration and expansion of overseas circuits', *Bell Lab. Rec.,* 1970, 48, p. 299
12 CLINCH, C. E. E.: 'Time assignment speech interpolation (TASI)', *Post. Off. Electr. Eng. J.,* 1960, 53, pp. 197–200
13 FLOOD, J. E. (Ed.): 'Telecommunication networks' (Peter Peregrinus, 1975), p. 49
14 CARTER, R. O.: 'Theory of syllabic compandors', *Proc. IEE,* 1964, 111, (3), pp. 503–513

15 HILL, M. T., and EVANS, B. C.: 'Transmission systems' (Allen & Unwin, 1973), pp. 55–56

16 CCITT: Green Book, **6**, Pt. 1, Recommendation Q15 pp. 54–55, ITU, Geneva, 1973

17 CCITT: Green Book, **6**, Pt. 2, Recommendation Q108 pp. 242–243, ITU, Geneva, 1973

18 Blue Book 'Notes on distance dialling', Section 5 (AT & T Co., 1975), pp. 17–21

19 BREEN, C., and DAHLBOM, C. A.: 'Signalling systems for control of telephone switching', *Bell Syst. Tech. J.*, 1960, **39**, pp. 1381–1444, and Bell System Monograph 3736, 1960

Chapter 8

Voice frequency signalling

8.1 General

Voice frequency (v.f.) signalling is a.c. signalling, the signalling frequency, or frequencies, being within the voice range of the speech channel, the speech and signals sharing the 300–3400 Hz transmission channel. The signalling is thus inband, is performed on, and is associated with, the speech channel and is completely flexible in application as, being inband, v.f. line-signalling systems can operate over any channel, or combination of channels, which afford speech transmission. The signalling frequency is amplified by the transmission-system amplifiers and thus the signalling range is greater than that of low-frequency a.c. signalling. Compared with d.c. signalling, v.f. signalling has longer delays in signalling time.

While 2-wire v.f. signalling systems exist, the method is usually used on 4-wire circuits, the forward and backward signalling paths being separate, and the duplex signalling condition being achieved by the two simplex signalling paths (Fig. 8.1). The v.f. receivers are permanently associated with the speech path so that they are ready to respond to incoming signals. As speech may contain the signalling frequency, the signalling system is subject to speech interference which could cause false line splits, or, at worst, simulate signals. Thus v.f. signalling systems must incorporate features to minimise false operation by speech (signal imitation) or any other interference, which complicates the system. A buffer amplifier of unity gain in the forward direction and some 60 dB loss in the reverse, on each speech channel protects the signal receiver from near-end interference from the switching equipment. While on multilink connections end-to-end v.f. signalling could clearly be used, the signalling is usually link-by-link and an arrangement of line splits adopted to isolate the v.f. signal to the particular channel. Signalling during the speech period is inadmissible with v.f. signalling.

8.2. Speech immunity

Protection against signal imitation (false line splitting or false signalling) by speech or other interference relies on exploiting differences between speech and signal currents,[1,2] e.g.

(*a*) adoption of a signal frequency which is less liable to exist, and persist, and has less energy than other frequencies, in speech
(*b*) speech currents containing the signal frequency usually have other frequencies
(*c*) signals made longer than the normal persistence of the signalling frequency occuring in speech
(*d*) signals of two frequencies compounded are less liable to occur, and persist, in speech than one frequency
(*e*) signals of other forms which the voice finds difficult to imitate, e.g. a number of short pulses, pulses of amplitude or frequency modulated carrier, etc.

There are other possibilities.

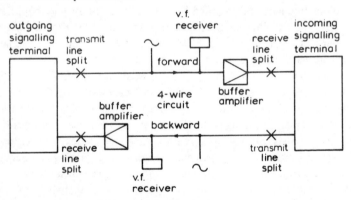

Fig. 8.1   *General arrangement v.f. signalling system*

Modern v.f. signalling systems exploit mainly (*a*), (*b*) and (*c*). Factor (*d*) is used in some cases, but the desire to simplify and cheapen per-line provided v.f. signalling systems indicates 1v.f. rather that 2v.f. for 4-wire circuits in national networks in particular. Further, interregister signalling systems, when used, allow simple line-signalling systems in that relatively few signals are required, and the modest line signal repertoire also points to 1 v.f. line signalling rather than 2 v.f. On the other hand, 2 v.f. conveniently allows extended signal repertoire, aids signal discrimination, and, in the compound mode, improves signal imitation, and is adopted for certain signalling systems (e.g. CCITT systems 4, 5 and 5 bis) where the availability of the two frequencies has merit in the particular application conditions.

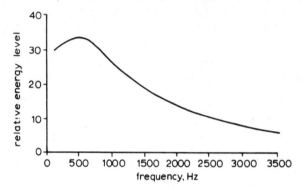

Fig. 8.2   *Spectra of normal speech*

In regard to:

(*a*) *Choice of signalling frequency:* Fig. 8.2 shows the relative energy levels of normal speech. This characteristic is modified by the characteristics of telephone microphones, but the general characteristic holds and indicates maximum energy at about 500 Hz, thereafter decreasing with increased frequency. In theory, this indicates that in the choice of signalling frequency, as high a frequency the transmission system will allow should be adopted as the low signal energy in speech would be less prone to operate the signal receiver and thus reduce signal imitation. Low frequencies should be avoided for signalling as these have relatively high energy level in speech. In practice the following apply:

(i) because of crosstalk considerations, the higher the frequency the less the permissible transmitted level, the receiver would require to be more sensitive to receive the low level, and thus more liable to signal imitation by speech
(ii) there is a sharp rise in the attenuation of the speech channel filter characteristic at the extreme upper range of the speech band, and slight deviations in a high signal frequency would result in significant level variations

(iii) old existing line plant in national networks may cut-off lower than 3400 Hz.

For the above reasons, the signalling frequency should not exceed a value of the order 3000 Hz. The actual choice for a particular network depends on the conditions of that network and typical signalling frequencies adopted are 2040, 2280, 2400, 2600, 2800 and 3000 Hz, each occurring in the higher range of the speech band. Some old v.f. signalling systems exist having lower signalling frequencies (e.g. 600/ 750 Hz), but these date back to the pioneer approaches to v.f. signalling.

(*b*) *Receiver guard circuit:* The v.f. signal receiver includes two elements:

(i) a signal circuit, tuned to the signal frequency, and tending to cause receiver operation
(ii) a guard circuit tending to prevent receiver operation.

Genuine signals strongly excite the signal circuit, but excite the guard circuit only weakly, if at all, and the receiver is able to operate. Frequencies other than the signal frequency, as may arise in speech, excite the guard circuit more strongly than the signal circuit, tending to prevent receiver operation. A difference guard is adopted, the outputs from the signal and guard circuits being compared and the receiver operates, or does not operate, to the difference (Fig. 8.3).

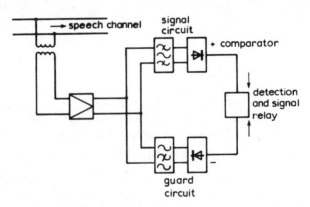

Fig. 8.3   *Elements of v.f. signalling receiver*

In this way signal imitation by speech is decreased. The guard circuit output does not completely negate the effect of the signal circuit as this would prevent receiver operation to a genuine signal accompanied by low-level interference such as line noise (signal interference). A highly-sensitive guard, while giving good protection against signal imitation, would give excessive signal interference. The guard-circuit sensitivity therefore tends to be a compromise between signal imitation and signal interference, and while being a powerful factor, the guard itself is unable to effect complete protection against signal imitation; it must be used in conjunction with signal recognition delay times and other factors to increase the protection.

The term 'guard coefficient' has been used to define the sensitivity of the guard circuit. It is defined as the ratio of two voltages at the

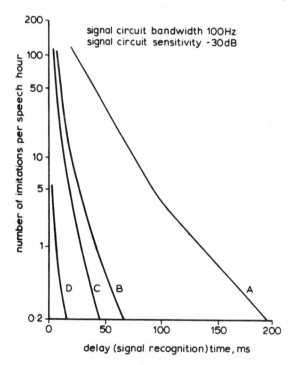

Fig. 8.4 *Effect of guard sensitivity on signal imitation*
signal circuit bandwidth 100 Hz
signal circuit sensitivity — 30 dB
A guard coefficient 0·5 ⎫
B guard coefficient 2 ⎭ simple frequency 2200 Hz
C guard coefficient 0·5 ⎫
D guard coefficient 2 ⎭ compound frequency 2000 + 2400 Hz

input to the receiver; the first voltage having the frequency of maximum sensitivity of the signal circuit and magnitude within the range of received signals, and the second voltage having the frequency of maximum sensitivity of the guard circuit outside the signal frequency band. The two voltages are such that their combined effect just inhibits operation of the receiver. Clearly, the higher the guard coefficient the less the signal imitation, and vice versa. Fig. 8.4 shows indication for various typical conditions.[1]

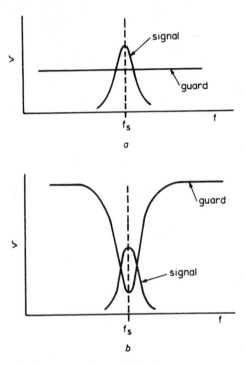

Fig. 8.5   *Typical guard-circuit characteristics*
a aperiodic
b tuned

The guard characteristic may be 'aperiodic' (Fig. 8.5*a*) or 'tuned' (Fig. 8.5*b*). The tuned-type guard is more usually adopted. Fig. 8.5 shows the 1 v.f. condition; the 2 v.f. is the same in principle, there being two signal responses and two dips in the tuned-guard characteristic. The aperiodic guard response is substantially constant throughout the received frequency band. This limits the guard sensitivity as the signal frequency excites the guard as well as the signal circuit, and with high guard coefficient the guard voltage could exceed the signal voltage, which is inadmissible. The aperiodic guard coefficient should not exceed about 0·7. The tuned guard can be more sensitive than the aperiodic as its maximum response is outside the signal frequency band

and a guard coefficient of the order six is the probable maximum for 1 v.f. systems; less for 2 v.f.

(*c*) *Signal duration:* Signal imitation can result in false line splits, and, at worst, false signals. Thus two signal recognition delays arise, one to minimise false line splitting and the other to minimise imitation of a signal, and the latter delay recognition is longer than the former. Usually, a possible signal is not recognised as such until it has persisted for some 40 ms but the complete signal recognition time can be much longer than this, particular for important signals such as the clear forward. The line-splitting recognition delay may be less than the 40 ms delay mentioned above; the shorter it is the greater the line splitting during speech, and vice versa. Some systems have a short splitting delay prior to speech and a longer one during the speech period, the answer signal being used to bring about the change.

In continuous v.f. signalling systems the cessation of a signal frequency has signal meaning, and signal cessation recognition time arises to guard against false interruptions simulating a signal condition.

(*d*) *2 v.f. signalling:* A signal of two frequencies compounded is less liable to be imitated by speech than a single frequency signal (Fig. 8.4). This is sometimes exploited in 2 v.f. signalling systems, typically in a signal consisting of a short compound prefix and a longer single frequency suffix (e.g. CCITT system 4). Both conditions are required to be received before the signal is effective. The prefix is the initial portion of the signal and has the function of preparing the equipment for the receipt of the significant signal, which is the suffix transmitted immediately after the prefix. This technique is a very powerful safeguard against signal imitation and is superior to 1 v.f. for equal signal recognition times (Fig. 8.4). Recognition of the compound prefix controls the line split, a delay that can be shorter than that of 1 v.f. systems for equal false line splitting during speech. In some 2 v.f. systems (e.g. CCITT systems 5 and 5bis) the whole signal consists of compound. When in a 2 v.f. system a single frequency has signal meaning, this is termed 'simple' operation as distinct from 'compound'.

The 2 v.f. receiver guard may be aperiodic or tuned. In either case there is a certain amount of guard energisation at each signalling frequency (guard leak). In the aperiodic guard, this is deliberately designed to be so; in the tuned guard, the condition arises owing to practical limitations of design. Because of this, difficulties arise when two signalling frequencies are required to be received as the guard circuit operates on either signal-output stage independently. The guard leak increases compared with single frequency reception and too much guard may be produced which would reduce the receiver operate margins and at worse prevent response to genuine signals. The margins are still further reduced when the two frequencies of a compound signal are received at different levels (typically 6 dB). This increase in the guard action when genuine signals are received puts a limit to the guard sensitivity of compound receivers, and the maximum permissible guard coefficient is generally much less than that which can be realised with simple frequency receivers.

The 2 v.f. technique, with simple (single) frequency working in the respective directions, is sometimes adopted when the signalling equipment is connected at a 2-wire point. The different single frequencies in the respective directions give the necessary signalling direction discrimination. Compound operation does not apply in such systems.

8.3 Signal information transfer on multilink connections

8.3.1 General
With v.f. signalling, signal information may be transferred end-to-end or link-by-link on multilink connections, the end-to-end mode being possible owing to v.f. signalling being inband. Unlike v.f. signalling, link-by-link signalling is implicit in d.c., outband and common channel signalling.

8.3.2 End-to-end signalling
The v.f. signals pass straight through the connection, whole or part, not being converted to d.c. at transit switching points. Thus certain signals (e.g. answer, clear forward) can be quickly transferred. Depending upon the number of links in the connection, the signal detection may be difficult owing to the cumulative deviations in signal level and frequency, and onerous signal/noise. End-to-end signalling does not conveniently allow the network to evolve with alternative types of signalling systems.

8.3.3 Link-by-link signalling
The v.f. signals are converted to d.c. at each transit switching point. The mode has the merits of:

(*a*) the various deviations (signal level, frequency, etc.) are limited to the single link condition, which simplifies receiver design as the signal detection and signal/noise problems are eased (*b*) allowing the network to evolve with alternative types of signalling system as these may be equipped on individual links to interwork with existing signalling systems on other links.

On the other hand, owing to the v.f.-d.c. and d.c.-v.f. conversions at each point, the signal repetition at transit points slows the signalling compared with end-to-end signalling. The slower signalling of the supervisory line signals increases the holding time of equipment, but does not react on the postdialling delay. Should the v.f. line-signalling system incorporate decadic address signalling, the slower speed of transfer of the address information increases the postdialling delay, and a 2-link v.f. signalling connection in the trunk network is the probable maximum allowing for further postdialling delay in the local exchange network parts of the connection.

Link-by-link line v.f. signalling is preferred, and usually adopted, to achieve (*a*) and (*b*) above, with, perhaps, greater emphasis on (*b*). Link-by-link signalling requires line splits.

8.3.4 V.F. signalling line splits

With link-by-link v.f. line signalling, signals are required to be confined to the particular link concerned. This is achieved by means of a receive line split on each signalling channel (Fig. 8.1), initiated by receiver operation and persisting with received signal. Without the split, the v.f. signal frequency would pass to subsequent link(s) to cause false signalling on those link(s). The receive line split is delayed for a short time (typically 20 ms) from the start of v.f. signal receipt, and thus some 20 ms of the signal will spill over to the subsequent link(s). This spillover is of shorter duration than the time required to recognise the shortest signal and thus false signalling does not occur on the subsequent link(s) due to the spillover. The splitting delay, which controls the spillover duration, is determined by the signal imitation requirement in the main as false line splitting occurs owing to signal imitation by speech. Short recognition delays (e.g. 10 ms) increase, and longer delays (e.g. 30 ms) decrease, the number of false line splits during speech (Fig. 8.4), but the longer delays would slow detection of genuine signals in certain situations. A splitting delay of some 20 ms is a reasonable compromise, which could be increased during the speech period if desired. The line split is maintained for the duration of the received signal and ceases within a short time (typically 200–300 ms) of signal termination.

The receive line split may be by a physical line disconnection, or by the suppression of the passage of all frequencies by the buffer amplifier, and usually either of these is adopted for the more usual pulse v.f. line signalling; or by a bandstop filter attenuating the signal frequency, a method that is necessary in tone-on idle continuous v.f. signalling (e.g. Bell SF system).

A transmit line split is also included on each channel to prevent transients from the switching equipment at the transmit signal end from interfering with the transmitted v.f. signal. This split is applied a short time (typically 20 ms) before a v.f. signal is transmitted. In pulse v.f. signalling, the split is maintained for the duration of the transmitted v.f. signal and stopped within a short time (typically 200–300 ms) of signal termination. In tone-on idle v.f. signalling (e.g. Bell SF system), the split does not always persist with a transmitted signal, and when it does not, is maintained for a short time only (typically 350–750 ms).

The line splits preclude signalling during the speech period and thus, typically, meter-pulse signalling during speech is inadmissible. This arises as subscribers would notice the interruptions to speech and hear the spillover v.f. tone.

Problems occur with the line splits if a short verbal answer follows immediately on the electrical answer signal, as the normal line splitting associated with the transfer of the answer signal over the connection may clip the verbal answer, partially or completely. If the verbal answer is not repeated, this could leave both subscribers waiting for verbal response, with the possibility of an abandoned call. The multi-link link-by-link signalling connection is clearly an onerous condition in this regard, as a measure of additive line split durations results. It is of interest to comment that it was recognition of this problem in the international service which first gave rise to the initial consideration of common channel signalling by the CCITT.

8.4 V.F. receiver signal-guard

There are many different designs of v.f. receivers, but the main functions (Fig. 8.6) are reasonably common. Most receivers include a limiter to bring the received voltages in the working range to a constant level before connection to the tuned circuit, thus simplifying the design of the comparator. The limiter, however, (i) produces harmonics of f_s which must later be attenuated so as not to produce a false guard signal, and (ii) introduces complications in its property to suppress low-level components of a signal in the presence of a high-level component, which requires selective attentuation of the f_s signal prior to the limiter and hence a second frequency-selective circuit is required.

The main interest is in the signal-guard arrangement (Fig. 8.3), and, in the evolution of v.f. signalling system design, considerable development effort has been spent on simplifying this feature. There are many different signal-guard arrangements; space precludes a comprehensive survey, but it is of interest to describe a selection and to trace the evolution of the UK development trend as a typical case in the 1 v.f. 2280 Hz environment.

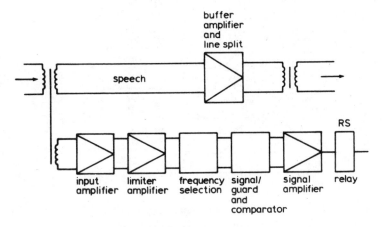

Fig. 8.6   *Block functions of v.f. receiver*

(a) *Early electromechanical system valve receiver:*[3] The tuned guard is achieved using a frequency-selective bridge (Fig. 8.7). $C1$ and $T1$ are tuned to approximately 1400 Hz to obtain sufficient pre-emphasis of the guard frequencies compared with the signal frequency f_s. The centre-tapped transformer $T1$ secondary forms two arms of the bridge, the tuned primary $T2$ and $R1$ form the other two arms. $C2$ tunes $T2$ to f_s, the impedance being high at f_s, and decreasing at other frequencies. When f_s only is received, tuned $C2$, $T2$ near balances $R1$, the potential difference across points X and Y is a minimum and the voltage across $T2$ a maximum. Complete balance of the tuned circuit at resonance by $R1$ is not attempted as it is necessary to develop some guard output when f_s is received so that the operation bandwidth of the receiver may be controlled. The tuned circuit has low impedance at frequencies other than f_s, and the bridge is unbalanced. Virtually no voltage is developed across $T2$, and a higher potential is developed across X and Y as guard volts. Maximum voltage is developed across X and Y at approximately 1400 Hz. The guard and signal voltages from the bridge feed two separate voltage-doubler networks. $R3$ and $R4$

form a signal-guard comparator and the difference voltage between V_s and V_g applied as signal or guard to the final d.c. stage of the receiver.

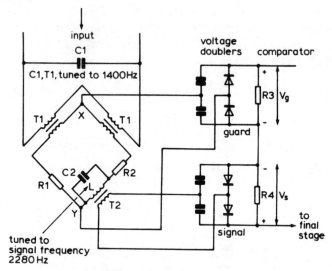

Fig. 8.7   *Signal-guard circuit UK 1 v.f. valve receiver*

(b) *Early transistorised receiver:*[4] This employed a simplified and improved signal-guard feature (Fig. 8.8) to achieve the tuned guard. TR1, a current amplifier, feeds series $L1$, $C1$ tuned to f_s in parallel with $R1$. At f_s, the signal voltage V_s across the secondary of T1 will be high and the guard voltage V_g across $R1$ low. At guard frequencies f_g, V_s will be low and V_g high. TR3 turns on when V_s exceeds the $R2$ bias. TR2, with V_g at base, acts as a rectifier and charges $C2$ to peak V_g to apply bias to oppose the turn on of TR3 by V_s at base. TR3 is thus a signal-guard comparator and the driving transistor to the final d.c. stage of the receiver.

(c) *Later transistorised receiver:*[5] This further simplified the tuned guard (Fig. 8.9). A signal voltage V_s develops across the inductor L with a peak at f_s, and a guard voltage across the series LC circuit tuned to f_s. The voltage across the tuned circuit is a minimum at f_s, a maximum well removed from f_s, and after amplification is taken as V_g. V_s and V_g are applied to a comparator.

(d) *Latest integrated-circuit receiver:*[5] This latest design of v.f. receiver and buffer amplifier for the UK miniaturised wholly electronic 1 v.f. signalling system is based on the use of integrated-circuit

operational amplifiers, which makes possible the elimination of transformers, inductors and large capacitors. The receiver design (Fig. 8.10)
follows the same philosophy as Fig. 8.9, but replaces the series-tuned

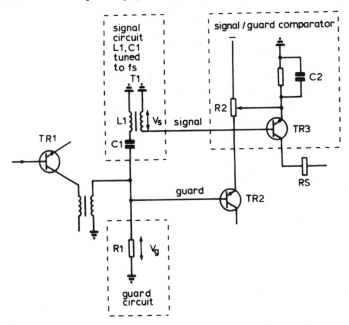

Fig. 8.8   *Signal-guard circuit UK 1 v.f. early transistorised receiver*

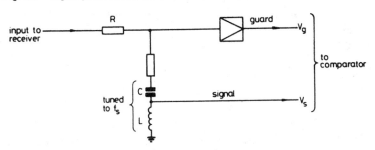

Fig. 8.9   *Signal-guard circuit UK 1 v.f. later transistorised receiver*

circuit with an active *RC* circuit using a gyrator.[6] The use of a sensitive
comparator permitted by the new design eliminates the limiter and de-
emphasis networks. Integrated-circuit operational amplifiers are, in
effect, high-gain broad-band amplifiers provided with inverting and
noninverting high-impedance inputs. External feedback components
determine the amplifier characteristics.

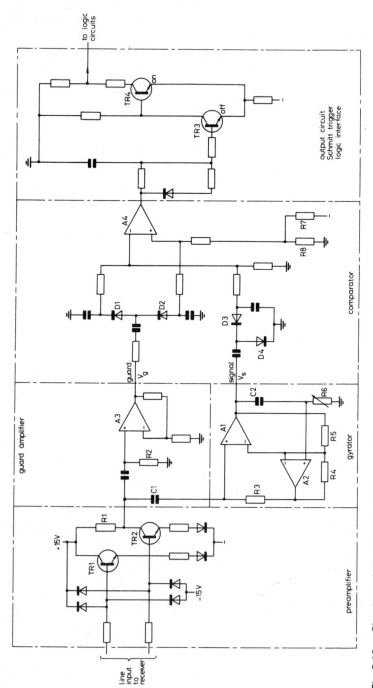

Fig. 8.10   *Signal-guard circuit UK 1 v.f. integrated-circuit receiver*

The receiver preamplifier, transistors TR1 and TR2, d.c. connected to the line via a balanced high-impedance tap:

(i) isolates the receiver from the speech path
(ii) provides the correct driving resistance for the tuned circuit
(iii) adjusts for optimum level of received signals for the subsequent circuits.

It is protected against high-voltage line transients by diodes, which clamp the transistor bases to ± 15 V. The high input impedance (some $500 \, k\Omega$) ensures negligible receiver connecting loss across the line.

The parallel combination of $R1$ and $R2$ constitutes the driving resistance for the tuned circuit consisting of $C1$ in series with the simulated inductance based on a gyrator circuit[6] that consists of operational amplifiers A1 and A2, resistors $R3$–$R6$ and capacitor $C2$. The guard voltage is developed across the tuned circuit, and is amplified by the operational amplifier A3 to produce V_g, and then rectified by diodes D1 and D2. Signal voltage is developed across the simulated inductance, and V_s at the output of the operational amplifier A1 is rectified by diodes D3 and D4. V_s and V_g are applied to the comparator, operational amplifier A4, whose output changes from $+$ ve to $-$ ve on reception of a valid 2280 Hz signal. To ensure that the comparator remains off when the line is silent, or when a signal of less than -28 dBm is present, a bias voltage from $R7$, $R8$ is connected to the inverting input terminal.

Transistors TR3 and TR4 form a Schmitt trigger circuit to interface between the operational amplifiers of the receiver and the low-power t.t.l. of the signalling terminals. This type of circuit is adopted for its fast operation time and wide voltage differential between its stable states to minimise transients from causing false operation of the logic. The network driving the trigger circuit optimises the delay and minimises distortion of the output pulses. At the expense of delay, it is possible to produce zero distortion.

Bell SF system receiver signal-guard[7,8,10]
This design (Fig. 8.11) to achieve a tuned guard circuit has been stable for many years. The output from a limiter amplifier enters a load consisting of a parallel tuned circuit and a series tuned circuit, connected in series, both tuned to the signal frequency f_s 2600 Hz. At f_s, the impedance of the parallel tuned (signal) circuit is high, and that of the series tuned (guard) circuit low, a relatively large voltage V_s is produced across the parallel circuit and a small voltage V_g across the series circuit. These relative voltage magnitudes are reversed at inputs other than f_s.

The voltage across each tuned circuit is separately rectified and stored on capacitors $C1$ and $C2$, respectively. V_s tends to operate, and V_g to inhibit operation of the receiver, the difference, determined by the comparator, being applied to the final stages of the receiver. The sensitivity of the guard may be varied by $R3$ which varies the effective Q of the series tuned circuit.

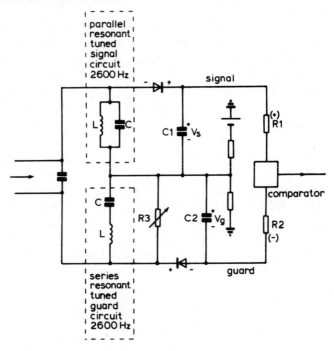

Fig. 8.11   *Tuned guard circuit: Bell SF system*

8.5 General arrangement of v.f. line signalling

Fig. 8.12, an expansion of Fig. 8.1, shows the general arrangement, 2-wire switching and the more usual 4-wire circuit and link-by-link signalling application being assumed. The same basic arrangement applies for 4-wire switching, in which case the 4-wire/2-wire terminations would not be provided. With either 2-wire or 4-wire switching, the v.f. signalling is 4-wire, with separate forward and backward signalling paths.

As v.f. is the most widely used line signalling system in analogue long-distance networks, considerable effort has been made to simplify,

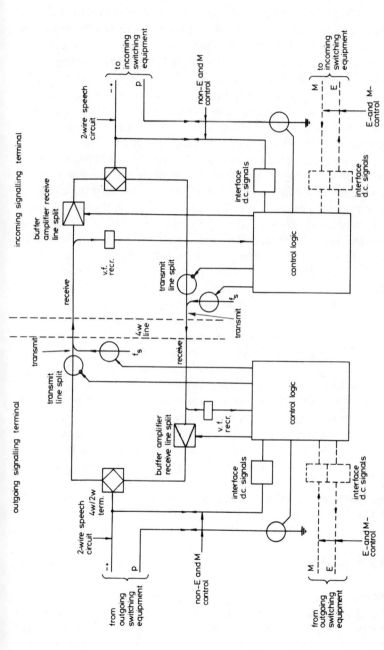

Fig. 8.12 *Schematic 1 v.f. electronic line signalling system*

miniaturise, and reduce the power consumption and maintenance requirement of this signalling technique, and most modern v.f. line-signalling systems are electronic. However, considerable quantities of electromechanical system equipment (with valve or transistorised-type receivers) exist in networks, the signalling terminals using relays for the logic and timing functions. The concept behind Fig. 8.12 applies to both the electromechanical and electronic designs, and the outgoing and incoming signalling terminals may be broken down into four basic parts, i.e. interface, logic, 4-wire/2-wire termination (assuming 2-wire switching) and a buffer amplifier and receiver.

The outgoing signalling terminal converts the d.c. signals from the exchange side into v.f. signals for transmission over the line. The v.f. signals are recognised on the 4-wire side of the incoming signalling terminal and converted to d.c. signals by the receiver for forward transmission to the incoming exchange equipment. Each signalling terminal has a buffer amplifier which may be switched on or off to split the 4-wire receive path at a point following the v.f. receiver to avoid signals being passed to the next link(s). The receive split may be by suppression of all frequencies by the buffer amplifier, or suppression of the signalling frequency only, depending upon the particular system. Each signalling terminal also includes a transmit line split to avoid near-end interference to a transmitted v.f. signal.

The d.c. signalling between the switching equipment and the signalling terminals may be performed over the speech wires (the negative and positive wires in the 2-wire switching case, Fig. 8.12), or over separate E- and M-leads, the d.c. signalling being of appropriate type in either case. Typically, in the UK philosophy (E and M control not applying), loop-disconnect d.c. signalling applies on the speech wires and earth-disconnect d.c. signalling over the p-wire. In the Bell system philosophy (E and M control applying) the d.c. signalling on the E- and M-leads may be, typically, as shown in Fig. 8.15, i.e. earth-battery on the M-lead and earth-disconnect on the E-lead (but see Section 7.10).[10]

8.6 Type of v.f. signal

Signal discrimination in v.f. signalling may be given by direction, signal sequence, frequency content of signal or pulse length, depending on the particular system employed. Both the basic types of v.f. signal, continuous and pulse, exploit some or all of the above discrimination possibilities in the formulation of the required signal repertoire, and clearly, compared with 1 v.f., 2 v.f. systems have more discrimination possibilities, particularly if pulse.

8.6.1 Continuous v.f. signalling

This employs uninterrupted (nonpulse) signals to indicate the signal condition and may be:

(*a*) continuous compelled (e.g. CCITT systems 5 and 5bis) or
(*b*) 2-state continuous non-compelled (e.g. Bell SF system).

In mode (*a*), the signal meaning is in the content of the primary signal itself together with the other signal discrimination possibilities and an acknowledgment signal (which may contain other signal information) is required to cease the primary signal. When the continuous primary signal ceases, this is not because a new signal is sent, but because it has been detected and interpreted correctly. The signals may be 1 v.f. or 2 v.f. depending on the system. In mode (*b*), the signal information is contained in the change of state, i.e. signal tone-on or -off, and this mode conveniently transfers the simple 2-state on-hook, off-hook continuous signal conditions in each direction, as in the Bell SF system. Acknowledgment signalling is not involved as the same switching condition is maintained with signal.

Compared with pulse, continuous signalling has merit in that on tone-on detection, the signal persistence recognition is not negated by signal interference to the initial part of the signal, as the system can wait upon receipt of a valid signal condition. On the other hand, after detection, the mode is more affected by transient interruptions and interference bursts relative to pulse as the transmission media is occupied for a longer time for a given amount of information transferred. This normally requires safeguarding logic elements in the signalling terminals. It is considered that the continuous signalling mode (*a*) is marginally more reliable than pulse and that pulse is marginally more reliable than the continuous mode (*b*). The mode (*a*), being circuit propagation time-dependent, is slower than mode (*b*), which is not so dependent.

Mode (*a*) is not usually adopted for v.f. line signalling except for special requirements such as the t.a.s.i. application CCITT systems 5 and 5bis. The signalling is relatively slow, but this is acceptable in the special application, and with 2 v.f. signalling, functioning in both the compound and simple frequency modes, an adequate line signal repertoire is obtained.

As, in mode (*a*), the signal meaning is given by signal content, and not by change of state (on/off) of the signal; memory logic is required in the signalling terminals to hold the memory of the signal condition, and in this regard, this mode of continuous signalling is little different from the pulse mode.

In mode (*b*), the signal meaning is given by 1 v.f. tone-on or -off, which enables the basic d.c. signalling condition, on-hook/off-hook, in each direction, to be simulated. Basically, this gives two signals in each direction, the same switching condition being maintained for the duration of a signal. Additional signal(s) can be obtained by sequence discrimination within the basic 2-state concept (e.g. the proceed-to-send signal in the Bell SF system), but this possibility is limited and yet further signal(s) would require timing and thus pulse (e.g. the forward transfer signal in the Bell SF system). The signal repertoire is somewhat limited even with additional signal(s) exploitation within the basic concept, but is often adequate for line signalling. As the memory of the signal condition is in the signal itself (on/off), little memory logic is required in the signalling terminals, which gives significant potential for simple signalling terminals.

Mode (*b*) signalling is of a more continuous nature than mode (*a*) which has significant silent periods in the signalling sequences, and thus greater care must be taken to avoid overloading the transmission systems. The mode (*b*) transmit signal level is required to be significantly less than that of mode (*a*) to conform to the μWs/busy hour criteria discussed in Section 7.8, which gives rise to v.f. signal detection problems.

The presence of the v.f. signalling frequency during the speech period is inadmissible, and thus, in basic concept, the simplest implementation of 2-state continuous v.f. signalling is signal tone present during idle (on-hook) periods, i.e. tone-on idle. This automatically ensures that the signal tone is ceased in each direction during the off-hook subscriber speech period and permits easy supervision of the on-hook, off-hook conditions. The tone-off idle alternative starts with the transmission of signal frequency in each direction on call origination. This has the virtue, for instance, of being less prone to interference, but requires appropriate arrangements to cease the tone in each direction during speech and thus the presence or absence of signal frequency is not always directly related to the on-hook, off-hook conditions. This is a form of semicontinuous signalling, a system that is somewhat more complex than 2-state continuous. Systems are known in which both techniques are used together, tone-off idle for the forward signals and tone-on idle for the backward. Systems are also known in which both continuous and significant pulse signalling are used. Such system approaches depart from the basic simple concept of tone-on idle 2-state signalling in each direction, but are sometimes used to extend the signal repertoire.

8.6.2 Pulse v.f. signalling

Here, signal condition changes are indicated by timed pulses. Compared with continuous signalling, pulse v.f. signalling:

(*a*) gives potential for more signals and thus facilities

(*b*) allows a higher transmit level which permits a better signal/noise ratio at the receiver, which eases signal detection

(*c*) is less subject to interference as the total signalling time is less

(*d*) complicates the d.c. and a.c. conversions owing to the pulse-timing requirements

(*e*) as, unlike 2-state continuous, the signalling does not simulate the basic d.c. on-hook/off-hook condition, the signalling terminals are complicated by the requirement for appropriate signal memory logic elements.

In modern common control switched networks the interregister m.f. forward and backward signalling permits facility enhancement and in this event it could be reasoned that the modest signal repertoire then required of the line signalling system could be achieved by 2-state continuous signalling in the two directions and thus, from the signal repertoire aspect pulse v.f. signalling is not essential. Should interregister m.f. signalling not apply, the v.f. line signalling system carrying decadic address information signals, and if a measure of facility enhancement be required, this would point to the adoption of a pulse-type v.f. line signalling system to achieve the required number of signals and the necessary signal discrimination by convenient means. It is a question of degree.

The main advantage claimed for 2-state continuous signalling is the potential for simple signalling terminals. Accepting the limited signal repertoire, simplicity is undoubtedly achieved in 2-state continuous d.c. and outband signalling, and the continuous mode is preferred for these systems. Different conditions arise, however, with 2-state continuous v.f. signalling, and here the same relative simplicity compared with pulse v.f. signalling is not achieved in system realisation, and this factor must also be taken into account in the consideration of choice.

8.7 Bell SF 1 v.f. line signalling system

The Bell SF line v.f. signalling analogue and digital versions in combination with the Bell interregister m.f. signalling system are collectively known as signalling system R1 (Regional system R1) in the CCITT series.[11]

8.7.1 Analogue version line signalling

This is tone-on idle, 2-state continuous inband v.f. signalling, employing the same signalling frequency 2600 Hz in both signalling directions on 4-wire circuits.[7-10] Due to the application flexibility of v.f. signalling, the SF system is by far the most extensively used line signalling system in the Bell system N. American toll analogue telephone network. Frequencies 2600 Hz and 2400 Hz, respectively, are used in the respective directions should 2-wire circuits be used. Two analogue versions of SF, E-type and a later F-type are available, the principle of both being the same, the F-type employing greater sophistication in the electronic design. The system delivers and accepts d.c. signals to and from the switching trunk equipment in accordance with the Bell system

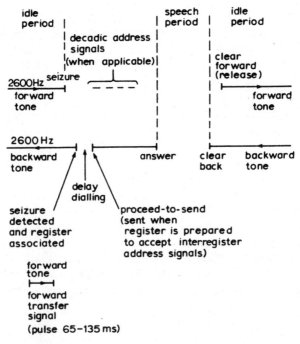

Fig. 8.13   *Tone-on idle continuous v.f. signalling: Bell SF system*

standardised E- and M-lead control philosophy (Section 7.10), the d.c. signals being converted to v.f. signals on the line side and vice versa. Table 8.1 shows the corresponding v.f. signals (Fig. 8.13) and the

Table 8.1   *Bell SF system v.f. signalling conditions*

Signal	Corresponding v.f. signal condition	
	Forward	Backward
Idle (on-hook)	On	On
Forward		
Seizure (off-hook)	Off	On
Decadic address – when required (sequential on-hook, off-hook)	On (dial break) Off (dial make)	On
Clear forward – release (on-hook)	On	Off or on (called party off-hook or on-hook)
Forward transfer (on-hook)	65–135 ms pulse	Off or on
Backward		
Delay dialling (off-hook)	Off	Off
Proceed-to-send (on-hook)	Off	On
Answer (off-hook)	Off	Off
Clear back (on-hook)	Off	On
Blocking (off-hook)	On	Off

signal repertoire which is standard for all Bell system toll line signal-ling systems.[9]

Signal discrimination is by direction and sequence except for the on-hook decadic-address pulse breaks and the on-hook forward transfer, discrimination between these and the on-hook clear forward being by signal recognition time, that of the clear forward being longer.

The connection release is normally on the clear-forward signal on calling party clear, this signal being overriding to initiate release and thus whether or not the clear back backward tone-on applies. If the called party has not cleared, the backward tone-on on-hook condition is returned when the circuit is released to the clear forward. The clear back stops the charge and measurement of call duration and can release the connection by automatically initiating a clear-forward signal if the calling party has not cleared within 10–120 s after recognition of the clear-back signal.

The on-hook forward transfer signal is initiated by an outgoing operator to recall an operator at a point further ahead in the connection. The signal may be sent before, or after, the called subscriber has answered and thus the backward signal tone may be on or off.

The off-hook delay-dialling signal is sent in the backward direction to delay the outpulsing of the outgoing register (or sender). The subsequent on-hook proceed-to-send signal, sent when the incoming register is prepared to accept address signals, terminates the delay-dialling signal to indicate that outpulsing may commence. The delay-dialling signal, sent as a consequence of the correct receipt of the seizure signal, also performs a signalling integrity check in that its receipt indicates that both the forward and backward paths are capable of transferring signals satisfactorily.

The delay dialling, proceed-to-send, is normally required to be a line-signalling sequence if controlled outpulsing decadic address dial-pulsing is transferred between registers. In Bell system practice, it is also required to be a line-signalling sequence if controlled outpulsing interregister m.f. signalling is used as the Bell R1 interregister signalling does not incorporate backward signalling, which could otherwise perform the proceed-to-send (and thus delay dialling) and integrity functions, and normally does. The SF system thus adopts a uniform line-signalling practice for delay dialling regardless of the type of register outpulsing, and the same philosophy is adopted for all Bell system toll line-signalling systems. The delay-dialling signal is not essential should the control of outpulsing be unnecessary, typically due to liberal register provision, but even here, however, its inclusion is advantageous for signalling integrity check purposes.

The delay-dialling feature may take either of the following forms:

(*a*) Delay dialling sent on, and acknowledging, receipt of the seizure signal. The delay-dialling signal duration is variable (but a minimum of 140 ms is required) depending on the time taken to associate an incoming register and the register prepared to accept address signals, at which time the tone-on proceed-to-send signal is sent. The duration would depend upon the magnitude of register provision and the traffic.
(*b*) On receipt of the seizure signal, the incoming exchange requests register association, but does not immediately return a tone-off delay-dialling signal. The idle tone-on to the outgoing exchange is maintained until an incoming register is associated, at which time the tone-on is changed to tone-off, which is maintained until the incoming register

is prepared to accept address signals. The tone-on proceed-to-send signal is then sent. The duration of the tone-off delay-dialling signal is in the range of 140–290 ms and being more controlled, is less variable than that in (*a*). The transitions of the backward tone from tone-on idle to tone-off delay dialling to tone-on proceed-to-send, with the duration of the tone-off delay dialling as stated, is known as 'wink operation', this being the latest practice (Fig. 8.13).

As the tone-off period in (*b*) is more controlled than that in (*a*), wink operation has potential for faster detection of double seizure on bothway operation should it be desired that double seizure detection be earlier than the time out for nonreceipt of the tone-on proceed-to-send signal.

The blocking signal busies the outgoing end during maintenance attention at the incoming end.

The backward signal tone is on during speech on noncharged calls as an electrical answer signal is not given for this condition in Bell system practice, and appropriate arrangements are incorporated in the system to overcome this.

Operation

Fig. 8.14 shows the main features of the SF system in greatly simplified form. The same arrangement appears at each end of the 4-wire circuit and the duplex condition is obtained by the two simplex signalling paths. It is convenient to adopt relay circuitry logic for explanatory purposes, but the same basic features apply in the solid-state electronic design of the system.

On 2-wire circuits, different frequencies, 2600 Hz and 2400 Hz, respectively, are used in the respective directions.

The various transmit and receive line split arrangements during the various signalling sequences are described later.

Assume a call in direction A to B. In the idle condition, both signalling paths are tone-on, both receiver RG relays are operated, both G relays are released, both the E- and M-leads at each end are in the on-hook (no d.c.) condition, both M relays are released, and the transmit and receive line splits are operative at each end of the circuit.

The idle signal tones are at low level (12 dB pad in circuit) to avoid overloading speech transmission systems. The keyer relay M, controlled by the on-hook/off-hook condition d.c. conditions on the M-lead, alternately applies and removes a 2600 Hz signal tone on the transmit path. When continuous tone-on on-hook signals are to be transmitted (clear forward, etc.), the release of M to the M-lead condition releases

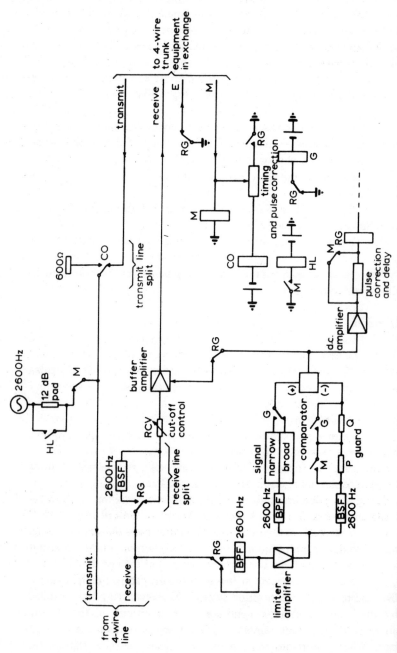

high-level relay HL and an augmented signal level is transmitted during the slow release HL (300–550 ms) by removal of the 12 dB pad. This secures an improved signal/noise receiver operating environment. Thereafter, the transmitted signal level is low on release of HL to hold the distant receiver operated.

Seizure originating at A results in a change of state on the A end M-lead from earth to battery. Relay M at A operates to remove the signal tone from the transmit path. CO terminates the transmit line split a short interval after application of the tone-off seizure. The cessation of tone is detected at B and receiver relay RG released changes the E-lead at B from the open to an earth condition, which results in the extension of the seizure to exchange B and request for incoming register connection. The trunk circuit equipment at exchange B changes the M-lead state at B from earth to battery resulting, on relay M operation, in tone-off B to A. The resulting release of RG applies earth on the E-lead at A to give the delay-dialling (or wink operation) signal to the switching equipment at A.

When an incoming register is connected at B and prepared to accept address signals, the trunk circuit equipment at B restores earth to the M-lead and the tone-on proceed-to-send signal sent B to A. This tone-on detection at A opens the E-lead on RG operation to indicate that outpulsing of the address information from the outgoing register (or sender) can begin.

If decadic dial-pulsing is used for addressing, the keyer relay M at A is operated and released in accordance with the dial pulses being transmitted, resulting in pulses of tone-on, one for each dial break, at the augmented signal level as HL is slow release. The transmit line split is maintained during pulsing transmission. Interdigital pauses of 0·6 s are inserted between dial pulse trains when the outpulsing is from a register (or sender). The receiver at B responds to the address signals which are repeated to exchange B by the opening of the E-lead during the pulse breaks. The SF system accepts and transmits decadic address pulses 8–12 p.p.s. 46–76% break. A pulse corrector in the receiver corrects for the received pulse ratio, and a pulse corrector may be included on the M-lead to correct for the transmitted pulse ratio.

At the end of pulsing, tone-off applies in the forward direction, the E-lead at B is earthed to extend the forward off-hook condition to exchange B. Tone-on applies in the backward direction and the E-lead at A is open to extend the backward on-hook condition to exchange A. At this stage, any supervisory tones are extended to the caller as, in the backward direction, the transmit line split will not be operative in this condition, and the receive line split is by means of a bandstop filter suppressing the 2600 Hz signal frequency band only, RG being in the operated state.

If the register outpulsing is m.f. and not decadic, forward tone-off applies during the outpulsing and tone-on applies in the backward direction.

On called-subscriber answer, the resulting off-hook condition is extended on the M-lead at B from exchange B, to operate M, which results in backward tone-off. Detection of this at A releases the receiver and RG released extends the off-hook answer condition to exchange A by earthing the E-lead at A. No supervisory tones are transmitted during the speech condition and no band-restricting components are present on the speech path, both receivers and both RG relays are released to the tone-off.

On called-party on-hook clear, the clear-back condition is received on the M-lead at B and relay M releases to apply the backward tone-on B to A. The resulting receiver operation at A operates RG to open the E-lead to extend the clear back to exchange A.

On calling party on-hook clear, the clear-forward condition is received on the M-lead at A and relay M releases to apply the forward tone-on A to B, where the resulting receiver relay RG operation opens the E-lead to extend the clear forward to exchange B. The clear-forward signal initiates clear down of the connection (link-by-link) after appropriate signal recognition, the equipment at A being busied by a time delay to cover the release time of the equipment at B, a release guard signal B to A not being given in Bell system practice. The tone-on clear-forward signal is overriding, causing connection clear down in either backward tone-on or tone-off conditions. If the clear back has not been received, the called party being off-hook, receipt of the tone-on clear forward is utilised to cause the return of tone-on on the backward path.

On noncharged calls (service calls, etc.) the backward signal tone is not ceased, but as the receiver and RG remain operated, the bandstop filter receive line split prevents the signal tone reaching the caller. On transmission systems equipped with compandors, the presence of the backward signal tone may reduce the compandor crosstalk and noise advantage. The tone-on clear forward-clear down on this type of call is always in the condition of backward tone-on as a tone-off answer signal is not given.

Table 8.2 shows the various E- and M-lead d.c. conditions for the Bell Type I interface (but see Section 7.10) and the corresponding v.f. signal conditions for the SF system in relation to Fig. 8.15.

Table 8.2 *Bell SF system: E- and M-leads (Type I interface) and v.f. signal conditions*

Signal	Outgoing exchange A		Incoming exchange B	
	E- and M-leads d.c. condition	Corresponding v.f. tone condition (forward)	Corresponding v.f. tone condition (backward)	E- and M-leads d.c. condition
Idle (on-hook)	Earth M-lead Open E-lead	On	On	Earth M-lead Open E-lead
Forward				
Seizure (off-hook)	Battery M-lead Open E-lead	Off	On	Earth M-lead Earth E-lead
Decadic address pulsing (sequential on-hook, off-hook)	Earth M-lead (pulse break) Battery M-lead (pulse make) Open E-lead	On (pulse break) Off (pulse make)	On	Earth M-lead Open E-lead (pulse break) Earth E-lead (pulse make)
Clear forward (on-hook)	Earth M-lead Earth or open E-lead	On	Off or on (called party off-hook or on-hook)	Battery or Earth M-lead (called party off-hook or on-hook) Open E-lead

Table 8.2 *(Continued)*

Signal	Outgoing exchange A		Incoming exchange B	
	E- and M-leads d.c. condition	Corresponding v.f. tone condition (forward)	Corresponding v.f. tone condition (backward)	E- and M-leads d.c. condition
Forward transfer (on-hook)	Earth M-lead (pulse 65–135 ms)	On (pulse 65–135 ms)	Off or on	Battery or Earth M-lead Open E-lead (pulse 65–135 ms)
Backward				
Delay dialling (off-hook)	Battery M-lead Earth E-lead	Off	Off	Battery M-lead Earth E-lead
Proceed-to-send (on-hook)	Battery M-lead Open E-lead	Off	On	Earth M-lead Earth E-lead
Answer (off-hook)	Battery M-lead Earth E-lead	Off	Off	Battery M-lead Earth E-lead
Clear back (on-hook)	Battery M-lead Open E-lead	Off	On	Earth M-lead Earth E-lead
Blocking (off-hook)	Earth M-lead Earth E-lead	On	Off	Battery M-lead Open E-lead

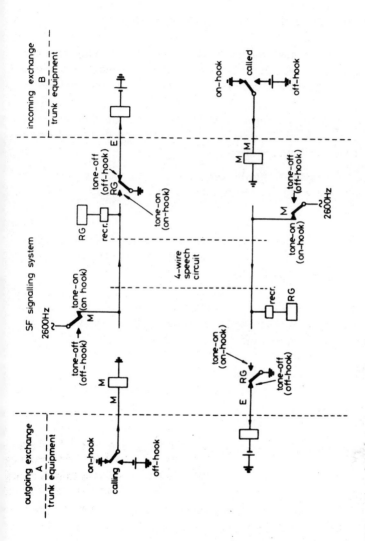

Fig. 8.15 *Bell SF system E- and M-leads tone-off tone-on conditions*

Transmit line split

The nature of the transmit line split depends on the type of signal to be sent and the condition of the receive signal path, the line being terminated with 600 Ω on the split. The transmit split arrangements of the E- and F-type SF systems differ in minor detail, those for the R1 version of the SF system being typical, as follows:

The receive path detecting tone-off: the transmit line split occurs within a period of 20 ms before to 5 ms after application of tone-on and remains split for 350–750 ms.

The receive path detecting tone-on: the transmit line split occurs within a period of 20 ms before to 15 ms after application of tone-on and remains split until either:

(*a*) a tone-off signal is to be sent, in which case the transmit line split is terminated within 75–160 ms after the tone is removed

or

(*b*) a tone-off condition is detected on the receive path, in which case the transmit line split is terminated within 350–750 ms of the receipt of the received tone-off condition.

Sending tone-off signals: under all receive-path conditions the transmit line split occurs within a period of 20 ms before to 5 ms after application of the tone-off condition and remains split for 75–160 ms.

Receipt of tone-on signals: if during the sending of a tone-on signal, a tone-on condition is detected on the receive path, the transmit line split is introduced within 250 ms of receipt of the incoming tone-on.
The above arrangements:

(i) establish a transmit path split at both ends of the circuit during the idle condition
(ii) ensure that a caller can receive supervisory tones prior to the answer condition
(iii) ensure that speech in the backward direction may take place if the backward signal tone has not ceased, typically on noncharged service calls when an electrical answer signal is not given.

In (i), the transmit line split persists with transmitted signal tone-on. (ii) and (iii) preclude a transmit line split persisting with transmitted signal, the split being terminated on cessation of receive path tone-on.

Receive line split

To prevent tone-on signals on a link from causing disturbances to signalling system(s) on subsequent link(s), the receive path is split when the signal frequency is received. The use of a bandstop filter suppressing the signal frequency band is necessary to give the split for the reasons discussed as follows.

The receive line split arrangements for the E- and F-type versions of the SF system differ in minor detail. The receive path of the early E-type is split immediately the signal frequency is received. This is achieved by means of a no-delay output from the comparator (whose output, if maintained for an appropriate time delay, would have the capability of eventually operating the receiver relay RG) operating on the buffer amplifier to suppress the onward transmission of all frequencies. On receiver relay RG operation, this total-frequency buffer amplifier split is terminated and the line split taken over the insertion of the bandstop filter (Fig. 8.14). Spillover of the signal frequency to subsequent link(s) is virtually nonexistent, but owing to the very short delay before splitting, the possibility arises of numerous short duration signal-imitation false line splits occurring during speech.

Delayed line splitting increases the spill over, but reduces the extent of false line splitting during speech. This is recognised in the more modern F-type system as here the bandstop filter split is applied more rapidly than in the E-type, and the initial buffer amplifier split is dispensed with. The R1 version of the SF system (the latest requirement) specifies that the spillover duration should not exceed 20 ms, and as this is less than the tone-on signal recognition time of minimum 30 ms, no problems arise.

With the tone-on idle, a continuous tone-on backward signal is present (*a*) prior to answer on all calls, and (*b*) during speech on non-charged calls. As a backward transmission path is required in condition (*a*) to enable the caller to receive supervisory tones, and in (*b*) to enable the caller to receive the called party's speech, the receive line split cannot be by means of a physical line disconnection (or by any means of suppressing the onward transmission of all frequencies) persisting with signal. The split is achieved in the SF system by a 2600 Hz bandstop filter, which is inserted for the duration of any received on-hook tone-on condition. This prevents the backward signal tone from reaching the caller, but permits the reception of supervisory tones. The signal frequency band, approximately 100 Hz centred on 2600 Hz, is suppressed from the called-to-calling party speech on noncharged calls, whose speech is degraded, but this is acceptable in the condition postulated. Speech transmission, of course, is not degraded on charged calls, as here the receiver is released during the tone-off speech condition and the bandstop filter is not inserted in the speech path.

The level of the signal leak transmitted to a subsequent link when the bandstop filter is inserted is at least 35 dB below the received signal level. The bandstop filter receive line split is maintained for the duration of the received tone-on signal, ceasing within 150–300 ms of received tone cessation.

Receiver signal-guard arrangements

The receive side includes a buffer amplifier, bandstop networks BSF and the signal receiver (Fig. 8.14). In addition to blocking the near-end interference from the exchange side to prevent interference to signal reception, the buffer amplifier makes up in the forward direction for the insertion loss of the v.f. equipment in the receive speech path. The signal receiver includes a limiter amplifier, the signal-guard network, a d.c. amplifier and a pulse corrector, the output of which operates the final stage relay RG to repeat the signals on the E-lead to the trunk-circuit equipment at the exchange. The signal-guard arrangement (Fig. 8.11) is described in Section 8.4. The guard feature itself is insufficient to assure that signal imitation will not cause false operation of the receiver. Additional protection is provided by a time delay which ensures that spurious voltages and bursts of noise within the signal frequency band of less than approximately 30 ms duration do not cause operation of receiver relay RG.

As signalling is by change of state on-hook to off-hook and off-hook to on-hook the SF system must be protected under all conditions. When in the on-hook condition tone-on, interference to the signal tone would tend to change the signal to the off-hook condition, even though the signal frequency is present. When in the off-hook tone-off speech condition, interference by frequencies at or near the signal frequency 2600 Hz would tend to change the signal condition to on-hook. These idle and speech condition requirements are conflicting, and necessitate the receiver signal frequency-band response and the guard circuit efficiency to be varied, depending upon the signalling condition. This is achieved by adoption of the following receiver signal-guard logic features (Fig. 8.14):

(*a*) In the idle condition (tone-on in both directions), receiver relay RG is operated, M is released as there is no off-hook d.c. condition on the M lead. G is released to RG operated to make the receiver signal circuit responsive to a wider range of frequencies relative to the speech tone-off condition. M and G released greatly decrease the sensitivity of the guard circuit by switching in resistors P and Q. These two actions at each end of the circuit greatly minimise false release of the respective receivers to minimise false seizures on interference to the idle tone-on. This is further aided by the insertion of a 2600 Hz bandpass filter BPF in the input to the receiver when RG is operated to the receive tone-on. (*b*) In the speech condition (tone-off in both directions), receiver relay RG is released, M is operated to the M lead off-hook d.c. condition. G operated to RG released results in a narrow bandwidth receiver signal circuit. M and G operated greatly increases the guard sensitivity by switching out both P and Q. In this condition of each receiver an almost pure 2600 Hz tone is required for operation, which tone is required to persist for the appropriate signal recognition time (at least 30 ms) for signal, or for an imitation.

(*c*) On noncharged call speech (no electrical answer signal), a condition that also applies on interception and recorded announcements, tone-off will apply in the forward direction and tone-on in the backward. The incoming end receiver is released to the tone-off, RG is released and G operated, and M will be released owing to the on-hook condition on the M-lead. This receiver signal bandwidth is narrow, and, owing to G operated switching out Q, the guard sensitivity increased relative to that in the idle condition, but decreased relative to that in the electrical signal answered condition when both P and Q are switched out. This can be tolerated as the signal circuit bandwidth is narrow.

The equipment terminating the backward path at the outgoing end is in the condition M operated (due to the off-hook condition on the M-lead at that end), receiver relay RG operated to the tone-on, and G released. This receiver signal bandwidth is thus wide, and, due to M operated switching out P, the guard sensitivity increased relative to that in the idle condition, but decreased relative to that in the electrical signal answered speech condition when both P and Q are switched out. Thus, except for the guard sensitivity, the receiver on the backward path is in much the same condition as when idle tone-on, but substantially the reverse to that when the backward speech is with tone-off. This, however, is the required condition as protection is required against false release of the receiver when speech is in the face of backward tone-on as distinct from false operation of the receiver when backward tone-off in the speech condition.

(*d*) If decadic address signalling applies:

(i) The receiver on the backward path at the outgoing end is in much the same condition as the noncharged call speech condition, i.e. wide signal circuit bandwidth and a low order of guard sensitivity.

(ii) Prior to pulsing, and during the interdigital pauses, the receiver on the forward path at the incoming end is in the normal tone-off seized condition of narrow signal circuit bandwidth and guard sensitivity due to P in circuit (M released as M-lead is on-hook) and Q switched out (G operated to RG released). Slow release of G maintains operation during the digit to maintain this same receiver condition. The guard sensitivity is less than that in the answered speech condition, but this is of no consequence since speech is not present during address pulsing.

Thus it is seen that, to secure optimum overall operation, the characteristics of the signal and guard features are shifted during the various signalling sequences. Additional safeguards, related to the system and not to the receiver, are:

(*a*) False interruptions to tone-on signals are protected against by time delay, interruptions of 30 ms or less (40 ms or less if the previous tone-on was 350 ms or longer) are not recognised as signal conditions.

(*b*) To protect against momentary interruptions to an idle tone-on causing a continuous succession of false seizure and clear-forward signals, response to the second of two closely-spaced seizure signals is delayed. The delay, which is a function of the round trip signalling time (terrestrial circuits 500 ± 100 ms maximum, satellite circuits 1300 ± 100 ms maximum) is started at the end of the initial seizure signal, or on recognition of the clear forward, and a second seizure persisting beyond the delay is assumed to be a valid seizure.

Bothway operation
The SF system is symmetrical, the outgoing and incoming signalling terminals being identical and the same signalling terminals apply for either unidirectional or bothway operation. This is standard Bell system toll signalling system philosophy and eases the bothway equipment requirement as far as the signalling system itself is concerned, the additional signalling and switching complexity consequent on bothway operation being included in the trunk circuit equipment at each exchange.

On double seizure (glare), the outgoing seizure at each end is forward tone-off and thus the incoming seizure at each end is tone-off. Double seizure detection at each end occurs when both transmitted and received conditions are tone-off, but as the received condition for the delay-dialling is also tone-off, a delay safeguard is necessary before the assumption is made that a double seizure exists. The delay philosophy may be:

(*a*) if a tone-on proceed-to-send signal, which would normally terminate a tone-off delay dialling signal, is not received within the time-out interval (typically 5 s) for receipt of the proceed-to-send, double seizure is assumed (*b*) in appropriate applications, where it is assumed that in normal operation a proceed-to-send would always be received in much less time than the proceed-to-send time-out, double seizure may be assumed after the normal expectancy of the proceed-to-send receipt.

It will be clear that the double seizure detection would be reasonably immediate should the delay-dialling feature not be incorporated in any particular application of the SF system.

Either of the following may apply on double seizure detection depending upon the arrangements adopted in particular applications:

(i)   an automatic repeat attempt made to set up the call or
(ii) a reorder (tone) indication given to the callers and no automatic repeat attempt made.

With either method, the double seized circuit is automatically released and the exchange which first assumed (based on timing) that the double seizure had occurred transmits a forward tone-on signal (100–200 ms) followed by a tone-off signal before the final forward tone-on clear-forward signal is sent. The tone-off signal is recognised as an 'unexpected tone-off signal' at the distant end, and this recognition, subsequent to the recognition of a proceed-to-send tone-on condition but prior to the completion of register outpulsing, releases the distant register.

Should the automatic repeat attempt apply, the postdialling delay would be increased for that call should the proceed-to-send time-out philosophy be assumed for double seizure detection. Double seizure detection in a time much less than the proceed-to-send time-out would greatly ease this postdialling delay problem.

The difficulty giving rise to the delay requirement for double seizure detection arises from both seizure and delay-dialling being similar state signals (tone-off) and thus the lack of precise interpretation in the double seizure condition as discrimination by sequence cannot apply. Systems with seizure and proceed-to-send conditions of unique signalling states, as occurs, typically, when the seizure is a line signal and the proceed-to-send an interregister, and not a line, signal, do not have the difficulty of delayed detection of double seizure.

Relevant data

Transmitted signal level: augmented -8 ± 1 dBm0, low -20 ± 1 dBm0
Receiver sensitivity: -29 ± 1 dBm0

Transmitted frequency: 2600 ± 5 Hz
Receiver signal circuit bandwidth in speech condition: 2600 ± 50 Hz

A tone-on forward signal 300 ms or longer is recognised as a valid clear-forward signal. The clear forward achieves release from any condition and prior to incoming register association, a 30 ms clear-forward signal recognition applies, this being faster as the connection is not in the speech condition.

Interregister m.f. signalling (forward signalling only in the Bell system) is not affected by the SF system transmit line split. The SF signalling delay plus the time required to associate an incoming register exceeds the line split disconnect period. The receive line split is not operative, the m.f. signalling frequencies clearing the SF signal frequency band.

8.7.2 Comments on Bell SF system

In accordance with Bell system practice for line signalling, in the SF system:

(*a*) the proceed-to-send/delay dialling sequence is given by line signalling
(*b*) a release guard signal, signalling to the outgoing end that the incoming equipment has released to the clear-forward, is not given.

Usually, and preferably when m.f. signalling applies, the proceed-to-send is given by a backward interregister m.f. signal. Backward signalling, however, is not included in the Bell interregister signalling system, and in this circumstance the proceed-to-send must be a line signal, which is consistent with the requirement for the proceed-to-send being a line signal when decadic dial-pulse address signalling applies.

Without the release guard signal, the outgoing end guard is given by time delay with the consequential problems of catering for different release times without maintaining the guard unnecessarily. The R1 SF system specifies a delay guard 750–1250 ms (1050–1250 ms for satellite circuits) after initiation of the clear-forward signal, which delay could be less for particular applications as appropriate.

A release guard signal would have merit in that:

(i) The outgoing end busy is maintained for the actual time required and automatically caters for the different propagation times and different incoming equipment release times.
(ii) As it functions as an acknowledgment for the important clear-forward signal, its nonreceipt within a certain time owing to malfunction in the release can alarm the condition. Without the release

guard, failure of incoming equipment to release for any reasons results in complication to recover from the condition.

Note: In the SF system, receipt of the backward continuous tone-on while the tone-on clear forward is transmitted does not give, nor is it intended to give, assurance of incoming equipment release.

In pulse systems, inclusion of the release guard signal permits repeat attempts to clear down the circuit if it is not received within a certain time, and to ultimately raise an alarm. With the continuous tone-on clear forward of the SF system, it may be reasoned that this repeat attempt feature would not be necessary.

It is considered that the inclusion of a release guard signal has merit for any long-distance network line-signalling system. It is adopted by most administrations and its noninclusion in the SF system (and all Bell toll line signalling systems) could be considered as being somewhat of a penalty. The signal could be catered for within the basic 2-state continuous signalling concept of the backward signal tone, but at the expense of some complication to the system design.

Signalling simplicity reduces cost and contributes to reliability. The SF line-signalling concept based on simulating the basic d.c. signalling condition aims at signalling simplicity, but the v.f. environment is different from that of d.c. (and of outband) and different problems arise, the resolution of which tends to depart from the basic simplicity desired, e.g.:

(*a*) various timing requirements arise even though the relevant signal conditions are within the basic concept of 2-state continuous signalling, as discussed in Section 8.7.1

(*b*) the tone-on idle concept, while satisfying the requirement of tone-off during speech by simple means introduces signal level and signal-guard switching problems

(*c*) the concept of tone-off (and receiver release) being a signal condition is a negative-type signal requiring appropriate safeguards against transient interruptions, particularly should the interruption simulate a signal

(*d*) the speech transmission in the face of backward tone-on on non-electrical signal answer calls raises speech transmission (signal band suppressed from speech band) and line splitting problems

(*e*) the transmit line split is of variable type (being either momentary or persisting with signal) as distinct from being a simple split persisting with signal (e.g. pulse signal)

(*f*) the receive line split is by signal frequency band suppression instead of being a simple total band suppression

(g) the continuous tone concept complicates various conditions (e.g. double seizure detection, repeated interruptions simulating false seizures and clear forwards)
etc.

While resolution of the individual problems is accomplished by reasonably simple means, the totality tends to a signalling system design which erodes the basic simplicity aimed at. This comment is relative to v.f. signalling, as assuming the same signal repertoire, the desired signalling simplicity of simulating the basic d.c. signalling condition is achieved with d.c. and outband line signalling.

Pulse v.f. signalling:

(i) has potential to give more signals than 2-state continuous
(ii) gives positive indication, as all changes in signal condition are by tone-on pulses
(iii) avoids signal level and signal-guard circuit switching, and speech transmission in the face of continuous tone
(iv) allows simple line splits persisting with signal
(v) as all signal conditions have unique meaning, permits simple detection of such conditions as double seizure, etc.

A number of the above points ease the design relative to 2-state continuous signalling. On the other hand, the signalling terminals are complicated by the send and receive timing requirements and by the additional logic functions required to retain memory of signal conditions.

Assuming the same signal repertoire of 2-state continuous signalling, 1 v.f. pulse signalling would have the same signal discrimination capability, i.e. direction and sequence, and require the same timing discriminations, e.g. clear-forward longer than dial-pulse breaks and forward transfer, etc. as 2-state continuous. The various signal recognition times could be the same. The resolution of choice between 2-state continuous v.f. and 1 v.f. pulse for the same signal repertoire would then depend upon relative complexity and cost, and thus largely on the relative weightings given to transmit pulse timing, memory logic functions for the 1 v.f. pulse, and the various arrangements adopted to overcome the problems arising from 2-state continuous v.f. as discussed above. This is a question of degree, being influenced by the emphasis placed on particular points.

2-state continuous signalling in each direction facilitates symmetrical signalling terminals which has equipment merit, particularly in bothway operation (Section 7.9.1). This symmetrical signalling philosophy is standard Bell system practice and valid for the limited signals given, and the 2-state continuous v.f. SF system conforms to the required practice. Pulse v.f. signalling would tend to a somewhat more complex bothway signalling system.

Modern 1 v.f. pulse line signalling systems are relatively simple, and are attractive in view of their ability to incorporate signals additional to the basic on-hook, off-hook conditions by convenient means (e.g. forward transfer, release guard). Such systems, of course, cannot, and do not, aim to simulate the basic d.c. signalling condition and cannot achieve the equivalent simplicity, but as reasoned above, 2-state continuous signalling in the v.f. environment does not achieve the equivalent simplicity. Most administrations adopt pulse v.f. signalling.

8.7.3 Digital version SF line signalling

This is as described in Sections 6.2.1 and 6.2.3.[11] Table 6.3 of Section 6.2.3 gives the signalling bit states for the SF system signal repertoire. As the digital signalling is in-slot, no line splits, transmit or receive, are required. The various time delays for signal recognition of the changes of signalling bit state from 0 to 1 (corresponding to change from tone-on to tone-off) and from 1 to 0 (corresponding to change from tone-off to tone-on) are as for the SF analogue system.

As for all interregister m.f. signalling at present, the analogue m.f. signalling of the R1 (and thus the Bell) system is applied in the digital transmission environment, the analogue m.f. signals being bit-encoded as for speech.

8.8 Pulse v.f. line signalling

8.8.1 General

As stated, most national networks adopt the pulse mode for v.f. inband line signalling (see Table 8.5 and Reference 14). Unlike the 2-state continuous mode, pulse v.f. systems cause discrete signals of v.f. pulses of different duration to be generated to give the signal meaning as they do not simulate the basic loop d.c. signalling condition and do not give corresponding changes of state of the signal tone for on-hook, off-hook. As a result, pulse v.f. signalling requires the inclusion of memory logic elements in the signalling terminals.

Most modern national network pulse v.f. line-signalling systems are 1 v.f., with the same signal frequency in the respective directions on 4-wire circuits. 2 v.f. signalling, with simple frequency signalling in the respective directions, is sometimes adopted, more particularly when v.f. signalling on 2-wire circuits is required. Cases also exist where a national network, desiring v.f. signalling rationalisation, adopts an international 2 v.f. pulse compound signalling system for national application (e.g. Italy — see Table 8.5).

Space precludes treatment of the many variants in the detail of 1 v.f. pulse signalling, but as the basic philosophy is much the same for all, it is perhaps sufficient to describe the 1 v.f. line-signalling systems (AC9 and AC11) applied in the UK telephone network,[12] the basic concept being typical of many national network pulse v.f. line-signalling systems. This approach would at least demonstrate the basic differences between the pulse and 2-state continuous philosophies.

Early UK v.f. line signalling was 2 v.f. pulse (AC1 system 600 and 750 Hz) designed for 2-wire connection, now superseded by 1 v.f., which is much less expensive, more compact and simpler, achieved largely through the use of a single frequency and 4-wire connection. The 1 v.f. AC9 system incorporates decadic address signalling, and is used in the trunk network in situations where the switching equipment is decadic with the addition of register-translator equipment to give the trunk facility; the earlier version (AC9) being electromechanical and the later (AC9M) solid-state electronic. The AC11 system, at present available with electromechanical relay sets, is applied for line signalling in the trunk network when common control switching and interregister m.f. signalling, applies, and this system does not incorporate address signalling. All three systems AC9, AC9M and AC11 are link-by-link, 4-wire, using the same signal frequency 2280 Hz in the respective signalling directions, the duplex requirement being given by the two simplex signalling paths.

8.8.2 UK/AC9 and AC11 1 v.f. pulse line signalling systems

AC9 system[3]
Fig. 8.16 shows the main features in simplified form, the relay logic being adopted for convenience of explanation. E- and M-lead control of signalling does not apply, the loop-disconnect d.c. signalling at each end being on the speech path. The v.f. receiver design evolved as discussed in Section 8.4 (*a*), (*b*) and (*c*).

In the signal code Table 8.3, signal discrimination is by direction, sequence and pulse duration. The clear forward signal achieves release in any situation of the connection. The answer and clear-back signals have the same pulse duration and recognition, discrimination being on sequence as when an answer signal is given it naturally always precedes a clear back. No v.f. signal less than 20 ms is recognised. A

Table 8.3   *UK AC9 1 v.f. system signal code*

Signal	Transmitted pulse duration	Recognition time
Forward	ms	ms
Seizure	50–95	20–45
Decadic address pulses (breaks)	55–65	20–40
Clear forward (release)	700–1000	400–600
Backward		
Answer	200–300	100–150
Clear back	200–300	100–150
Release guard	650 minimum	400–600
Blocking (a) unidirectional	Continuous	45
(b) bothway	Continuous (1·7 s min.)	250

forward transfer signal is not given. A delay-dialling/proceed-to-send backward line-signalling sequence is not given as the type of incoming switching equipment applying does not give, and is not required to give, this indication.

Fig. 8.16 shows the simplified features for the forward signalling in the main, the backward signalling being much the same in principle except that there is no pulse correction. Relays A and D respond to the on-hook, off-hook conditions of the calling and called ends, respectively. Relays AA and B, controlled by the receiver relay RS at the incoming end, extend the forward signals on the speech path to the incoming switching equipment, and relay DA, controlled by the receiver relay RS at the outgoing end, extends the backward signals on the speech path to the outgoing switching equipment. The transmission bridge at each end is incorporated in the 4-wire/2-wire line termination in the signalling terminals, the 4-wire transmission line being extended to the switching centres.

The buffer amplifier is unity-gain forward in the nonreceive line split condition, and some 70 dB loss in the backward in both the normal and line split conditions. The receive line split is performed by effectively switching off the buffer amplifier (on RS and BS operated), the forward loss being increased by some 70 dB. The receive line split

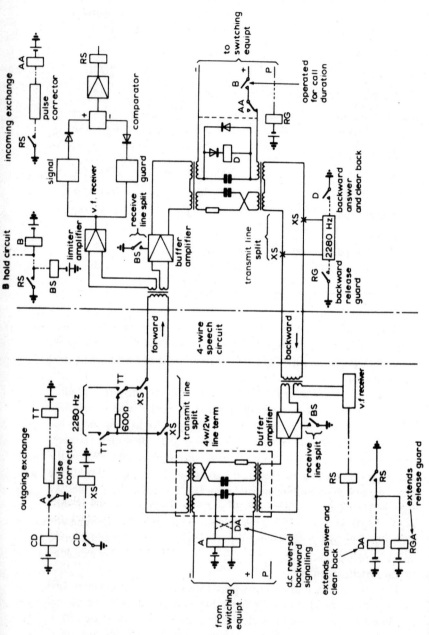

Fig 8.16   UK AC9.1 vf signalling system

delay is 10 ms in the unanswered state and 20–35 ms in the answered, the latter reducing false line splitting during speech. The split persists with received v.f. signal and is terminated 20 ms after cessation of signal.

The transmit line split is by line disconnection (XS released) and applies in the idle condition to terminate the transmission line with the 600 Ω characteristic line impedance at each end. The split is thus present prior to the v.f. seizure signal transmission. When signalling after seizure, the split is applied 15–35 ms prior to v.f. signal transmission (the so called 'silent prefix' to allow time for the distant receiver to recover from the effects of any preceding surges), is maintained with signal and terminated 100–200 ms after cessation of transmitted signal. The transmit line split is maintained during the transmission of a v.f. address digit pulse train and is terminated at the end of each digit train to allow transmission conditions on the speech path. Arrangements are adopted for an end to give the transmit line split whenever a receive line split occurs on signal received on the other path.

Fixed output break pulse correctors are equipped at each end for the forward signalling. The outgoing corrector controls TT which transmits the v.f. seizure pulse signal (A operated) at a time when the idle transmit line split is operative, the split being terminated after the seizure signal transmission. On decadic address signalling, the A relay release on pulse break operates CD which releases XS to give the transmit line split. TT operates after the silent prefix and transmits the digit train of break pulses 60 ± 5 ms. The silent prefix applies at the beginning of each digit train. The distant receiver relay RS operates B to seize the incoming switching equipment and RS, responding to the decadic address pulsing, operates AA via the incoming corrector and AA repeats fixed-break d.c. pulses of desired duration forward.

The UK network adopts the release guard signal philosophy on connection clear down, as distinct from outgoing end fixed time busying. The engaged condition of the incoming switching equipment is indicated by RG operated on the P-wire, which returns the release-guard signal after clear-forward signal recognition to maintain the outgoing end busy condition. RG releases on incoming equipment released ceases the release-guard signal to unbusy the outgoing end. Nonreceipt of a release-guard signal at the outgoing end within a certain time indicates a fault, and to automatically return a circuit to service following a short-term fault, particularly necessary where circuits are controlled at unattended exchanges, automatic repeat clearing is adopted. The seizure and clear-forward signals are transmitted as a repeated test sequence every 50 s, each clear forward inviting the return of a release guard. The circuit is released in the normal way when the release-guard signal is eventually received correctly, but the condition is alarmed should the release-guard signal not be received after, typically, 6–12 min. The repeat seizure signal is sent, in addition to the repeat clear-forward, to cover the case of non-receipt of the release guard due to backward channel line fault, the incoming equipment having released correctly, it thus being necessary to reseize the incoming equipment to enable the clear forward to be effective in returning a release-guard signal. This seizure/clear-forward repeated test sequence cannot be given when fixed time busying is adopted. On a bothway circuit, the repeat seizure maintains the circuit busy at the outgoing function at the distant exchange.

As the release-guard signal is an effective acknowledgment for the clear-forward signal, the outgoing end busy condition is maintained for the actual time required, and thus accommodates for different circuit propagation times and for different incoming equipment release times. It also safeguards against the possibility of an outgoing equipment being unbusied whilst an incoming equipment is still busy, or, alternatively, the outgoing end busy being maintained unnecessarily long. These significant merits cannot be achieved with outgoing end fixed time busying and there is little doubt that the release-guard signal technique is much superior.

In any v.f. line-signalling system, it is vitally important that signal imitation should not cause false release of an established connection. For this reason, the important clear-forward signal is safeguarded by the relatively long recognition time of 400–600 ms in the UK system (which may be compared with, say, 300 ms of the Bell SF 2600 Hz system, although this is accounted for in part by the lower signal frequency 2280 Hz). As a further safeguard, a check is made on the clear-forward signal in the UK 1 v.f. philosophy. Advantage is taken of the condition that after address information, the next forward signal expected in normal cause is the clear-forward, sent if the caller is on-hook for at least 250 ms. An incoming terminal, detecting an incoming signal (200–300 ms recognition), returns a 2280 Hz check tone to verify that the signal is being originated by the outgoing end terminal, and not by some source preceding it such as speech. Receipt of the check tone at the outgoing end operates the receive line split, which, in turn, operates the transmit line split. Spurious signals from preceding sources are disconnected at the transmit line split before the incoming terminal interprets the spurious signal as a clear-forward signal. A genuine signal is not disconnected as it is applied to the transmit path at a point where the transmit line split is operative in normal course for genuine v.f.

signal transmission. Cessation of a spurious signal due to the outgoing end transmit line split, operative on the receive line split on check tone received, causes removal of the check tone at the incoming end 30–80 ms after cessation of the incoming end receive line split. The incoming end transmit line split is terminated on cessation of check tone transmission and thus both the receive and transmit paths of the 4-wire circuit are restored to the speech condition. Should the spurious signal return, the above sequences between the incoming and outgoing equipments continue as a repeated pattern until the spurious signal ceases. The pattern is so timed that the check tone duration is insufficient to imitate answer or clear-back, and the duration of the signal being checked is insufficient to imitate a clear-forward. Should the signal being checked be a genuine clear-forward it would continue to be recognised as such, the transmission of the check tone being maintained for the remaining duration of the clear-forward signal. On cessation of a recognised clear-forward signal, the transmitted check tone continues as a release-guard signal (if desired, the release guard could be a newly commenced signal).

Calling party release applies in the UK network. The purpose of the clear-back signal is supervision, but it is also used to cease the timed charge after a short delay when it occurs before the clear-forward. Release is always in direction forward from the originating exchange, and on multilink connections the links release successively link-by-link, a clear-forward signal being transmitted on a link after a clear-forward signal recognition on the previous link, not waiting for the cessation of the release-guard signal on the previous link.

The backward blocking signal busies the outgoing end equipment of a circuit when the incoming end equipment is under maintenance attention, or, if for any other reason, it is desired at the incoming end that the outgoing end should not accept traffic. In the UK v.f. systems, the blocking signal is a continuous tone-on for both unidirectional and bothway circuits, the backward busying being maintained by the blocking signal. On bothway, the signal is required to persist for at least 1·7 s, being initially recognised as a seizure signal at the other end, but after 250 ms recognition, which is significantly longer than a genuine seizure recognition, the seizure condition is cancelled, but the outgoing equipment function remains busied to the continuous blocking signal.

When it is required to backward-busy a group of circuits (e.g. when a number of circuits are terminated on one incoming switching equipment such as a crossbar switch), the transmission of a number of continuous tone blocking signals may overload transmission systems.

In this situation, blocking signals are not transmitted on the group of circuits concerned should the incoming switch be faulty. The circuits will eventually be busied-out owing to nonreceipt of the release guard after being seized, and subsequently the release attempted on caller on-hook clear on follow on calls, which calls will fail.

Bothway operation: In the bothway equipment arrangement of the UK AC9 system, the associated incoming and outgoing equipment functions at each end share a common v.f. receiver, v.f. sending device, and buffer amplifier, the circuit resting on the incoming function at each end in the idle condition.

On double seizure, detected when simultaneous seizure conditions apply at both the outgoing and incoming functions at a terminal, the incoming function is rendered inactive, and, to prevent further calls being originated at the distant end, a blocking signal is sent. Equipment engaged tone is returned to the caller and release of the circuit is by the subscriber on-hook clear. A similar sequence occurs at the other terminal. The blocking signals in each direction busy the respective outgoing functions until they have both been released. The continuous blocking signals are transmitted for at least 1·7 s. A blocking signal is ceased (after 1·7–2·3 s) when an outgoing equipment function is released, but the blocking received from the distant end maintains the outgoing function busy until this received signal ceases, which indicates release of the distant outgoing function. The first outgoing equipment function to be disassociated from the seizure conditions restores its associated incoming equipment function to idle within 100 ms after cessation of the received blocking signal. The last outgoing equipment function disassociated from seizure conditions ceases the transmission of the blocking signal (after 1·7–2·3 s) and removes the busy condition 500 ms later.

AC9M system

In view of the large provision of v.f. line signalling equipment in most long-distance networks, it is a logical development evolution that such systems be solid-state electronic to achieve savings in space, maintenance and power. In line with this, the latest version (AC9M)[5] of the UK AC9 type system is miniaturised solid-state electronic based on low-power transistor-transistor-logic (t.t.l.),[13] the receiver and buffer amplifier using integrated-circuit operational amplifiers.[5,6] The receiver arrangements (Fig. 8.10) are as described in Section 8.4(*d*).

This AC9M electronic version is based on the same functional arrangements as for the AC9 system Fig. 8.16, adopts the same signal code (Table 8.3), and the general arrangement Fig. 8.12 applies. E and M lead control of signalling does not apply.

A significant merit of solid-state electronic relative to electromechanical in any telecommunication equipment is the much reduced fault rate and the consequential reduced maintenance cost. As an indication of this, the mean-time-between-failure of the AC9M signalling units is assessed to be approximately 13 years as compared with the 1·5 years of the electromechanical AC9 system the AC9M system was designed to supersede.

Electronic signalling systems may be applied with various types of switching equipment, electronic or electromechanical, and must meet conditions to be expected on the connections to and from such switching equipments. It is thus a clear requirement that care be taken in the interfacing arrangements, more particularly when E and M control by separate leads does not apply. For this reason, relay switching is used for certain interfacing functions of the AC9M system,[5] e.g. mercury-wetted reed relays are used for the pulsing contacts in the pulsing-out loop of the incoming signalling unit (contacts AA Fig. 8.16) and for the signal tone switching (contacts TT Fig. 8.16) in the outgoing signalling unit, the latter to obtain a minimum signal tone leak to line better than − 80 dBm0. The transmit line split (contacts XS Fig. 8.16) is also relay. The buffer amplifier consists of two operational amplifiers arranged in symmetrical balanced circuit mode that readily provides for noiseless receive line splitting after appropriate delay (10 ms in the unanswered state and 20–35 ms in the answered) by bias to cut-off the operational amplifiers to give a forward loss in excess of 70 dB.

Relevant data UK AC9 and AC9M systems

Transmitted signal level:	− 6 ± 1 dBm0. This signal is applied at a − 4 dBr point of the 4-wire circuit and thus at a nominal level − 10 dBm
Transmitted frequency:	2280 ± 0·3% Hz
Receiver:	operate to 2280 ± 25 Hz in the level range + 3 dBm to − 18 dBm with signal distortion less than 5 ms and delay less than 10 ms.
	not operate to a 2280 Hz signal at a level of − 28 dBm or less or to signals outside the band 2280 ± 75 Hz.
	when operated by a 2280 Hz signal in the

level range + 3 dBm to − 18 dBm, receiver output should be inhibited by the introduction of a signal, at the frequency to which the guard circuit is most sensitive, at a level 12 dB below that of the 2280 Hz signal, but should not be inhibited if the guard signal is 20 dB below the 2280 Hz signal.

ignore interruptions in signal of less than 4 ms, but not ignore interruptions of more than 10 ms.

Speech immunity: one false operation of 20 ms duration in two consecutive speech hours. One of 50 ms in ten speech hours. A speech hour is one hour of speech not including the quiescent periods in normal telephone speech.

AC11 system[4]

This 2280 Hz, 1 v.f., link-by-link, line signalling system does not incorporate address-information signalling, being applied when interregister

Table 8.4   *UK AC11 1 v.f. system signal code*

Signal	Transmitted pulse duration (ms)	Signal recognition time (ms)
Forward		
Seizure	50–80	20–35
Clear forward	650–1000	400–600
Backward		
Answer	200–300	100–150
Clear back	200–300	100–150
Release guard	650 minimum	400–600
Blocking		
(*a*) unidirectional	Continuous	35
(*b*) bothway	Continuous	250
	(1000 ms min.)	(note)

Note: Bothway equipment recognises a signal of duration less than 100 ms as a seizure signal and a signal of 250 ms or more as a blocking signal m.f. signalling is applied, otherwise it is much the same as the AC9 system. The general arrangement Fig. 8.16 applies except that the pulse correctors are not required and the switching may be 2-wire or 4-wire. Fig. 8.6 and the receiver signal-guard circuit Fig. 8.9 apply. The release sequence is as described for the AC9 system and arrangements at the incoming end ensure that incoming register equipment is not periodically taken into use during the repeat seizure and release retest sequences. A delay-dialling/proceed-to-send backward line signalling sequence is not given, the proceed-to-send signal being given by a backward interregister m.f. signal. Table 8.4 shows the signal code.

8.8.3 Other national network v.f. line signalling systems

Table 8.5 gives some general information on a number of national network v.f. signalling systems, these being selected from the more comprehensive list detailed in Reference 14. It will be noted that most administrations adopt the pulse mode. It is of interest to give brief information of one or two of the various systems simply to illustrate broad differences.

(a) Federal Republic of Germany

1 v.f., pulse, link-by-link, 3000 Hz (or 2280 Hz for narrow bandwidth speech circuits).[15-18]

Typical signal code:

Forward	Transmitted pulse (ms)
seizure	40
decadic address	40 (pulse breaks)
clear forward	1600
trunk offering	40
coin box line	40
Backward	
number received	155–350
answer	180
clear back	200 ms pulses continuous
sub. busy and plant congestion	750
release guard	750
blocking	continuous

(b) France

1 v.f., pulse, link-by-link, 2280 Hz line signalling when interregister m.f. signalling is applied.[19,20]

Table 8.5 *Selected national network v.f. signalling systems*

	USA	Canada	UK	Australia	Germany (F.R.)		France	Switzerland	Sweden	Spain	Netherlands	Italy
Type	continuous	continuous	pulse	pulse	pulse		pulse	pulse	pulse	pulse	pulse	pulse
Frequency, Hz	2600	2600 (2400/2600 for 2-wire)	2280	2280	3000	2280*	2280	3000	2400	2500	2400/2500 separate	2040/2400 separate and compound
Transmitted frequency tolerance, Hz	± 5	± 5	± 6	± 6	± 7.5	± 6	± 3	± 3	± 6	± 3	± 2	± 6
Receive line split delay time, ms	35 max	35 max	35 max	35	20	20	35	70	35–40	10	30–55	35
Transmitted level at point of zero relative level, dB	− 8 and after attenuation − 20	− 8 and after attenuation − 20	− 6	− 6	− 8	− 8 * for narrow-band circuits	− 6	− 3·5	− 6	− 6	+ 3·5	− 9

Typical signal code:

Forward	*Transmitted pulse (ms)*
seizure	100
clear forward	500–750
Backward	
answer	100
clear back	pulse 100, gap 233
	(continuous)

8.8.4 Comments on pulse v.f. line signalling
Section 8.7.2 discusses the various problems which arise with 2-state continuous v.f. signalling, and states that relative to continuous, the v.f. pulse technique:

(*a*) has potential to give more signals

(*b*) avoids signal level and receiver signal-guard circuit characteristic switching

(*c*) avoids speech transmission in the face of backward continuous tone-on on nonelectrical answer signal calls

(*d*) allows simple transmit and receive line splits persisting with signal

(*e*) as signal conditions can have more specific meaning, permits simpler detection of such conditions as double seizure, etc.; it should be noted, however, that, in this regard, both the pulse and continuous modes suffer the same penalty on bothway working in that a blocking signal may simulate an incoming seizure, and appropriate safeguarding features are necessary.

The typical pulse v.f. systems described in Section 8.8.2, the basic philosophy of which is common to all pulse v.f. systems, demonstrate the validity of the above points. On the other hand, 2-state continuous signalling minimises the memory logic features required, an advantage not possible with pulse signalling. For example, in 2-state continuous v.f., tone-off is adequate indication to hold the circuit in the speech condition, whereas additional memory logic is necessary with pulse (typically, relays B and DA Fig. 8.16 must be maintained operated during speech by local holding arrangements in the respective signalling terminals).

An advantage of pulse v.f. signalling is the potential of giving a reasonably wide signal repertoire by convenient means. All signals need not be used in a particular network or application, and the one basic signalling arrangement used for a number. To illustrate this point, in addition to the basic on-hook, off-hook signal conditions, a spread of networks (or applications) may require a collective repertoire of proceed-to-send (when required to be a line signal), number received (when required to be a line signal), forward transfer, release guard, trunk offering, etc. signals, an individual application adopting its own requirement. The same result could, of course, be achieved by the addition of significant pulse signals to the basic 2-state continuous signalling, but this would be illogical. This point is perhaps of declining significance with the advent of modern interregister m.f. signalling in networks, but even here the need exists for line signal(s) additional to the basic on-hook, off-hook. On balanced assessment, the pulse technique is usually preferred for v.f. line signalling systems.

References

1 FLOWERS, T. H., and WEIR, D. A.: 'Influence of signal imitation on reception of voice frequency signals', *Proc. IEE*, 1949, **96**, Pt. 3, p. 223

2 WELCH, S.: 'The influence of signal imitation on the design of voice frequency signalling systems'. Institution of Post Office Electrical Engineers, Printed Paper No. 206, 1953

3 MILES, J. V., and KELSON, D.: 'Signalling system AC No. 9', *Post Off. Electr. Eng. J.*, 1962, **55**, Pt. 1, pp. 51–58

4 MILLER, C. B., and MURRAY, W. J.: 'Trunk transit network signalling systems – line signalling systems', *ibid.*, 1970, **63**, Pt. 3, pp. 159–163

5 HILL, R. A., and GEE, G. J. H.: 'A miniaturised version of signalling system AC No. 9', *ibid*, 1973, **65**, Pt. 4, pp. 216–227

6 ANTONIOU, A.: 'Realisation of gyrators using operational amplifiers, and their use in *RC*-active-network synthesis', *Proc. IEE*, 1969, **116**, (11), pp. 1838–1850

7 NEWELL, N. A., and WEAVER, A.: 'Single frequency signalling system for long telephone trunks', *Trans. AIEE*, 1951, **70**, p. 11, and Bell System Monograph 1841, 1951

8 NEWELL, N. A., and WEAVER, A.: 'Inband single frequency signalling', *Bell Syst. Tech. J.*, 1954, **33**, p. 1309

9 BREEN, C., and DAHLBOM, C. A.: 'Signalling systems for control of telephone switching', *ibid*, 1960, **39**, pp. 1381–1444, and Bell System Monograph 3736, 1960

10 'Notes on distance dialling' Blue Book, Section 5, 'Signalling' pp. 17–21, (AT & T Co., 1975)

11 CCITT: Green Book, **6**, Pt. 3, Recommendations Q310–Q332, 'Specification of signalling system R1', ITU, Geneva, 1973

12 HORSFIELD, B. R.: 'Fast signalling in the U.K. telephone network', *Post Off. Electr. Eng. J.*, 1971, **63**, Pt. 4, pp. 242–252

13 MORTON, W. D.: 'Semiconductor device developments: integrated circuits', *ibid.*, 1967, **60**, p. 20 and 1967, **60**, p. 110

14 CCITT: Green Book, **6**, Pt. 4, ITU, Geneva, 1973, pp. 714–717

15 FUHRER, R.: 'Landesfernwahl – Band 1' (R. Oldenbourg Verlag, Munich, 1966)

16 FUHRER, R.: 'Landesfernwahl – Band 2' (R. Oldenbourg Verlag, Munich, 1962)

17 3000-Hz-Wahl, Unterrichsblatter (B) der Deutschen Bundespost (1955) Nos. 13/14

18 RINGS, F.: 'Probleme der tonwahl im fernsprechverkehr', *Nachrichtentech. Z.*, 1955, 8, pp. 531–536

19 LUCAS, P.: 'Etude des principes de base d'une signalisation rapide', *Ann. Telecommun.*, 1966, 21, pp. 61–87

20 LUCAS, P., LEGARE, R., and DONDOUX, J.: 'Les idees modernes en commutation telephonique', *Commutat. et Electron.*, 1965, 9, pp. 5–37

Chapter 9

Outband signalling

9.1 General

Any on-speech-path signalling technique in which the signalling is performed outside the effective speech frequency band may be regarded as being outband, and that mutual interference between the speech and signal currents is eliminated. Thus, on this logic, such signalling techniques as a.c. signalling below the effective speech band (low-frequency a.c. signalling), d.c., a.c. signalling above the effective speech band, and in-slot p.c.m. signalling, are outband in concept. In practice, however, the term 'outband' is applied to systems based on a.c. signalling within a speech channel spacing (usually 4 kHz) but above the upper limit of the effective speech frequency band (i.e. usually above 3400 Hz), and also to frequency-division multiplex (f.d.m.) transmission systems.

In f.d.m. transmission systems incorporating outband signalling, the channel bandwidth is divided into a speech channel and a signalling channel by filtration. This prevents mutual interference between the speech and signal currents (Fig. 9.1). Usually one signalling frequency only is used, located approximately midway between two adjacent channels, the forward and backward signalling paths being separate on a 4-wire circuit, the duplex signalling requirement being achieved by means of the two simplex signalling paths. With a 4 kHz channel spacing (effective speech transmission 300–3400 Hz) a signalling frequency of 3700, 3825 or 3850 Hz is typical. The CCITT recommends 3825 Hz,[1] which takes account of the CCITT recommended channel filter characteristic for f.d.m. transmission system.[2] Speech transmission is not degraded as the signalling is above 3400 Hz. The separation of the speech and signalling enables these currents to be transmitted simultaneously and independently with little or no mutual interference and as the same channel carries both speech and signalling, a connection cannot be set up on a faulty speech path as the on-speech-channel signalling checks the integrity of the channel.

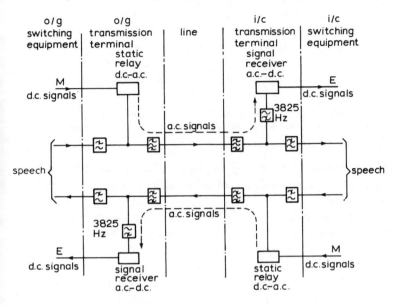

Fig. 9.1 *General arrangement outband signalling*

E- and M-lead control applies for outband signalling. The appropriate d.c. condition on the outgoing M-lead operates the static relay, and the signal tone (typically 3825 Hz) is injected into the main transmission path to the channel modulator at a point after a lowpass filter (Fig. 9.1). The lowpass filter in the speech path removes energy at the signal frequency band which may be present in the incoming speech, or which may be produced by the preceding stages of the equipment. After the demodulation process at the distant end, the signal tone is directed by filtration (nominal 3825 Hz) to the signal receiver which performs the a.c.–d.c. conversion. To prevent the signal tone from being audible during speech, a lowpass filter, permitting the passage of speech in the range 300–3400 Hz on the speech path following the signal-tone extraction point, eliminates the signal frequency. As the signal frequency must not be extended to the switching equipment, outband signalling is implicitly link-by-link, the signals being d.c. at switching points, and end-to-end signalling is precluded. The signal tone, after detection, amplification, limiting and rectification, operates (typically) a relay stage, a contact of which applies the appropriate d.c. condition to the incoming E-lead. The arrangements are the same in the two signalling directions.

Unlike inband v.f. signalling, outband signalling:

(a) does not require protection against signal imitation by speech or other interference, which gives scope for a much simpler system
(b) is free of disturbances owing to echo suppressors, and from disturbances which might arise from connection to other signalling systems
(c) eliminates line split and spillover problems, which complicate v.f. signalling systems
(d) permits signalling during the speech period, which is of importance for example for the transmission of meter pulses during speech when nonitemised bulk billing applies.

On the other hand, inband v.f. signalling can be applied to any type of transmission media, whereas outband signalling can only be applied to carrier circuits.

In outband signalling, the d.c.–a.c. and a.c.–d.c. signalling equipment is an integral part of the f.d.m. transmission equipment, the E- and M-lead d.c. signalling applying between the switching and transmission equipments (Fig. 9.1). The E- and M-leads may be separate from the speech leads, or may utilise the speech leads between the switching and transmission equipments.

As the signalling is independent of speech, there is virtually complete freedom in the choice of signalling mode with reasonably simple signalling arrangements. Signalling may be 2-state continuous tone-on idle (off during speech), tone-off idle (on during speech) a mode, however, that tends to overload the transmission system, 'semicontinuous' tone-off idle and off during speech (not preferred), or pulse. With the 2-state continuous mode in each direction, outband signalling enables the basic loop d.c. signalling condition to be simulated far more easily compared with inband v.f. 2-state continuous signalling, with significant potential for simple signalling terminals. This is because the various factors which tend to complicate continuous v.f. signalling systems (Section 8.7.2) are absent in outband signalling. The pulse mode also has potential for simplicity compared with pulse v.f. The continuous mode outband signal level must be relatively low (CCITT recommendation $-20\,\mathrm{dBm0}$)[1] to avoid overload of transmission systems which requires sensitive signal receivers with consequent signal/noise problems. A higher signal level (CCITT recommendation $-5\,\mathrm{dBm0}$)[1] is permissible with pulse signalling, which greatly eases signal detection.

Should a traffic circuit be made up of a number of patched-through carrier sections, each with outband signalling, signalling on the whole circuit can be achieved by suitable interconnection of the E- and M-leads at the intermediate patched-through points. This is costly since the number of d.c.–a.c. and a.c.–d.c. conversion equipments on the circuit increases with the number of sections making up the circuit. The various signal conversions introduce signal distortion which could be of significance for decadic address signalling, and appropriate precautions may need to be adopted (e.g. pulse correction, restriction on the number of sections permitted). An alternative approach is to eliminate the intermediate static relays and signal receivers. This is possible should the terminal equipments at the intermediate points be capable of transmitting speech and signal currents on a 'through' basis, and thus not difficult when carrier systems are interconnected on a 'group' basis. Difficulties arise when individual channels are interconnected on an audio-frequency basis since it would be necessary to alter the filtration arrangements on the carrier equipment and to ensure that any tie circuits interposed between the carrier systems are capable of transmitting the signal frequency without excessive attenuation.

In the Bell system, a so-called 'through channel unit' is available which may be applied at intermediate points. These units provide demodulation and modulation of the speech channel and the outband signalling frequency together, and, instead of recovering the d.c. signals on the E- and M-leads, the signal frequency is connected to the following carrier system on an a.c. basis.[3]

9.2 Consideration of the signalling mode

A main advantage of outband signalling is its potential for simplicity, particularly when 2-state continuous signalling in the two directions. The pulse mode would not achieve this simplicity owing to the pulse-timing requirements, but nevertheless is sometimes adopted when more signals than can conveniently be given by 2-state continuous signalling are required, the mode also permitting the desired higher signal level. If the line signal repertoire requirement is modest, which is usually the case when interregister m.f. signalling applies, but could also be the case when it does not, continuous signalling is clearly a preferred approach, the issue then resting between tone-off idle (on during speech) and tone-on idle (off during speech). Compared with tone-on idle, it is considered that tone-off idle has the following advantages:

(*a*) Transient bursts of interference in the transmission path do not give rise to false dial pulses or to the premature release of established connections.

(*b*) Short-duration interruptions of the transmission path do not result in seizure of idle incoming circuits. With tone-on idle, such interruptions could cause the simultaneous seizure and release of relatively large numbers of incoming equipment with consequential repercussions (e.g. serious register congestion in the case of circuits having direct access to registers).

(*c*) Tone is connected to the signal receivers immediately on seizure, and, should decadic address pulse signalling be included, is maintained during the interdigital pauses. Thus any a.g.c. features in the signal receiver function effectively for all the dialled pulses in a digit train. With tone-on idle, signal tone is disconnected on seizure and is absent during the interdigital pauses for periods of indefinite duration; during these periods, with some designs of receiver, the sensitivity would drift to its maximum value and cause excessive distortion to the first pulse in each digit train.

(*d*) The presence of the signal tone during speech tends to make the system less prone to signal imitation by harmonics of speech currents. The presence of signal tone cannot be guaranteed, however, during all conversations since on some types of call an electrical answer signal may not be given, and accordingly it is a basic requirement that the voice immunity performance should be satisfactory whether signal tone is present or not.

The above advantages are obtained at the penalty of the following disadvantages:

(i) Short-duration interruptions of the transmission path may cause false dial pulses or premature release of connections.

(ii) Transient bursts of interference may result in seizure of idle incoming circuits. In this regard, the probability of interference bursts occurring simultaneous on many channels in a system is less than that of transient disconnections occurring simultaneously. Further, interference bursts are by their nature of very short duration which reduces the risk of register congestion when circuits have direct access to registers.

(iii) The loading of transmission systems is increased during the busy hour since the engaged channels are required to carry speech and signal currents simultaneously. With tone-on idle, the maximum signal energy is transmitted when the least number of circuits are in use.

(iv) If automatic retest-to/restore-to service circuits that are 'busied out' due to loss of the release-guard signal (e.g. as a result of a fault in the transmission path) is adopted, elements in the outgoing signalling terminal must be incorporated to give this. Such elements may not be essential on unidirectional circuits employing tone-on idle, but would have advantages with either mode on bothway circuits.

Apart from transmission system overload, it is considered that there is little to choose between the two modes tone-on idle and tone-off idle. Should transmission-system overload not be a factor, as may be the case when small-scale application of outband signalling applies, it has been reasoned that tone-off idle (on during speech) offers a slight advantage.[4] If large-scale application of outband signalling occurs in a network, avoidance of transmission system overload would indicate tone-on idle. For the general application case, there is no doubt that tone-on idle (off during speech) has merit and is preferred.

9.3 Application constraints of outband signalling

In the a.c. signalling field on high-frequency (h.f.) line plant, outband line signalling is preferred to v.f. line signalling owing to its relative simplicity and reduced cost, and indeed is the first choice a.c. line-signalling method in analogue networks. Unfortunately, existing conditions in networks often tend to place constraints on outband signalling application. Thus, while f.d.m. is the main transmission facility in analogue long-distance networks, outband signalling provision is modest in many networks, the main signalling facility being v.f.

Clearly, a progress to remove the constraints would increase outband signalling application to the advantage of many networks owing to line signalling simplicity and exploitation. Typically, when nonitemised bulk billing applies for national trunk and international dialled traffic, there is significant economic justification in centralising the meter-pulse charge-rate determination equipment. For national trunk dialling, this equipment is usually located at the lowest-level trunk exchanges, the meter pulses being signalled during the speech period over the local junction line plant to operate the charge meter at the subscriber's exchange. It is clearly preferable that the charge-rate determination for international dialled traffic be performed at the centralised international exchange, the meter pulses being passed back over the trunk (toll) network to the lowest-level trunk exchange and then subsequently to the subscriber's exchange. As v.f. line signalling precludes signalling during speech, this would necessitate outband signalling on the analogue trunk network h.f. transmission plant. Nonavailability of outband signalling on the trunk network would thus require international traffic charge-rate determination to be performed at the more numerous lowest-level trunk exchanges, with consequent economic penalty and complication at these exchanges.

The application constraints of outband signalling are:

(*a*) In cases that arise where traffic circuits are made up in sections, partly in multichannel systems equipped with outband signalling and partly in other transmission media not so equipped, e.g. existing carrier and coaxial systems and amplified audio cables. E- and M-lead d.c. signalling at the intermediate patched-through points would enable different signalling systems on different sections to be interconnected as discussed in Section 7.10. This would require all relevant signalling systems (outband, v.f., d.c.) used in national long-distance networks to be E- and M-equipped, this being the Bell system philosophy. Most administrations, however, while adopting a form of E- and M-control for outband signalling, do not adopt E and M for the other signalling systems. Further, the interconnecting philosophy would require signalling terminals on the sections, which is expensive.

Should E and M not apply on the v.f. and d.c. systems, it is possible to envisage the provision of signal convertors to convert from outband to v.f. or d.c. signalling (including conversion from a continuous to a pulse mode, and vice versa). Here the cost of the convertors plus the cost of the signalling terminals would be at least as great, and probably greater, than the provision of v.f. signalling overall.

Such problems do not arise with v.f. signalling overall on the traffic circuit, v.f. being applicable to any transmission media affording speech transmission. This approach is adopted by most administrations. (*b*) Old existing f.d.m. plant in networks may not be equipped with outband signalling. It would be unrealistic, if at all possible, to modify such line plant to include outband signalling and v.f. tends to be applied.

The above constraints account for v.f. being the main line-signalling facility in most national long-distance networks. It is of interest to note that despite the availability of E- and M-lead control on all relevant signalling systems, the SF v.f. system is by far the main line-signalling facility in the Bell system analogue toll network. It is considered, however, that any new analogue network design should be such as to assure the outband line-signalling facility on the f.d.m. line plant.

9.4 Typical outband line-signalling systems

While generally modest in quantity, most national networks have outband line-signalling provision.[5-12] As all system designs are based

Table 9.1 *UK AC8 system long-distance network outband signal code (tone-off idle version)*

| Signal | Signal tone condition | | Signal recognition time, ms |
	Forward tone	Backward tone	
Idle	Off	Off	
Forward			
Seizure	On	Off	30
Decadic address	Off (pulse breaks)	Off	
Clear forward (release)	Off	On (if clear forward prior to called party clear) Off (if clear back prior to clear forward)	100–150
Backward			
Answer	On	On	100–150
Clear back	On (clear back prior to clear forward)	Off	100–150
Release guard	Off	On	50–75
Blocking	Off	On	30

on the same concept, it is sufficient to discuss a selected few as typical to illustrate realisation of the principle.

Two versions of outband line signalling are equipped in the UK network; the AC8 system which incorporates decadic address signalling and applied when the switching equipment is decadic step-by-step, and the AC12 system, which does not incorporate address signalling and used when interregister m.f. signalling applies in common control switching situations. Both systems are 3825 Hz continuous tone-off idle (on during speech) in present application, this mode being adopted in view of the small application in the network, it being concluded that transmission-system overload owing to the simultaneous presence of speech and signal currents would not be a significant factor. Both system designs allow the option of tone-off idle (on during speech) or tone-on idle (off during speech) as potential for possible future requirement.

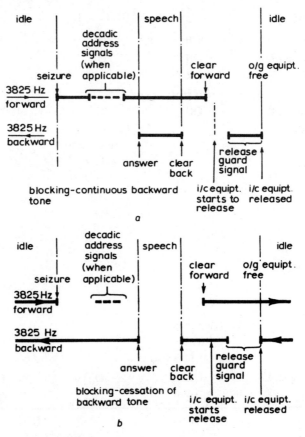

Fig. 9.2   *UK outband basic signal codes*
a Tone-off idle
b Tone-on idle

(a) UK AC8 outband system[5]

Fig. 9.2*a* shows the tone-off idle arrangement in use at present, signal discrimination being by direction and sequence. Fig. 9.2*b* shows the tone-on idle option. The system is applied in both the long-distance

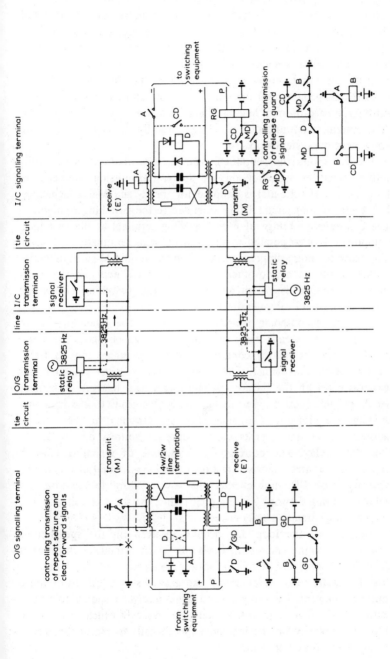

Fig. 9.3 *Basic arrangement UK outband signalling system*

(trunk) and local junction networks in UK. The trunk network application signal repertoire is as Table 9.1. As discussed in Section 3.2.6, the UK junction network application requires additional signals and to avoid pulse complication, the additional signals are limited. Backward meter-pulse signalling can be given in normal course and manual hold by continuous tone-on after receipt of the clear-forward signal, both signals being within the basic 2-state signalling concept. A delay-dialling/proceed-to-send line-signalling sequence is not given by the AC8 system, this indication not being given by the type of switching equipment the system works to.

Fig. 9.3 shows the simplified main features of the system. Earth-return phantom signalling on the speech-transmission side circuits (and on the tie circuits) is adopted for the E- and M-d.c. signalling, the 4-wire speech circuit being extended to the switching equipment. The transmission bridge at each signalling terminal at the switching equipment is incorporated in the 4-wire/2-wire termination. Resistor and capacitor noise-suppression filters on the E- and M-paths minimise inductive interference to other circuits. In the UK network, the transmission terminal is often remote from the switching equipment (typically by some 10 miles or 16 km).

On seizure, outgoing relay A operates to the incoming seizure condition and extends earth over the M-lead which biases the static relay to the conducting condition and seizure signal tone is transmitted over the channel. The distant signal receiver responds and extends earth on the E-lead to operate incoming relay A. A contact of A extends loop d.c. seizure on the speech leads to the incoming switching equipment. Outgoing relay A responds to the decadic address signals and repeats earth-disconnect address d.c. pulses to the static relay. The corresponding response of incoming relay A repeats the address signals forward in the loop-disconnect d.c. mode. Operation of CD short-circuits the 4-wire/2-wire termination and relay D during forward address digit trains, the short circuit removed between digit trains.

On answer, the loop d.c. reversal received on the speech-leads operates incoming end-rectified relay D which extends earth (after at least 75 ms recognition of the answer condition) on the M-lead to bias the static relay which causes the answer signal tone to be transmitted backward. The outgoing end signal receiver responds to extend earth on the E-lead to operate outgoing relay D which reverses the loop d.c. polarity on the incoming speech-leads to repeat the answer to the preceding equipment.

On clear back, incoming relay D releases to the restored loop d.c.

polarity on the outgoing speech leads when the called party clears. The corresponding release of the outgoing-end D relay restores the d.c. loop polarity reversal on the incoming speech-leads to repeat the clear back to the preceding equipment. When the calling party clears, outgoing relay A releases and biases the outgoing static relay to the nonconducting condition to disconnect signal tone on the forward channel to extend the clear forward. On release of the incoming end receiver, release of incoming relay A disconnects the loop d.c. to the incoming switching equipment, which releases. The clear-forward signal achieves clear down of the connection in any condition.

With the optional tone-on idle condition, the operation sequence is the same except that the static relays are conducting when the circuit is idle and are biased to the nonconducting state when earth is extended on the respective M-leads by operation of outgoing relay A or incoming relay D.

Sequenced release: The release is sequenced, a release-guard signal being ensured whatever the operating conditions of the circuit (except for release from double seizure on bothway operation) is at the time the clear-forward signal is sent i.e. seized but unanswered, answered or cleared by the called party. It must also be ensured, should answering or clearing by the called party occur, when release has already commenced at the outgoing end. The release-guard signal is backward tone-on for a certain time when the tone-off idle mode applies, and backward tone-off when tone-on idle. Assuming tone-off idle (Fig. 9.2a):

(i) *Release when the called party clears first:* The backward condition is tone-off, which is the same condition as release prior to answer. Outgoing-end relays A and B release on calling party clear, but GD maintains operation to D released to busy the P-wire. Relay A released transmits the clear-forward condition and on clear-forward signal recognition at the incoming end, incoming relays A, B, CD and MD release in sequence. On release of MD, earth is connected to the incoming end M-lead via RG operated to mark the beginning of the tone-on release-guard signal, which is transmitted at least 650 ms after clear-forward signal recognition. RG maintains operation to the earthed P-wire from the engaged, but releasing, incoming switching equipment. Outgoing-end relay D operates to the received release-guard signal to release GD, but the busy on the incoming P-wire is maintained by D operation. Slow release RG releases when the incoming switching equipment has released and disconnects earth from the M-lead to terminate the release-guard signal. To ensure that the duration of the release-guard signal is sufficient to cover the necessary circuit operations in the outgoing equipment should the incoming switching equipment release quickly, this form of release-guard signal is transmitted for at least 400 ms, an alternative circuit for RG being provided under the control of MD. Outgoing-end relay D releases to the terminated release-guard signal to remove the busying earth from the incoming P-wire, which returns the equipment to the idle condition.

(ii) *Release when the calling party clears first:* The backward tone-on answer condition (incoming-end D relay operated to the called party off-hook) returned from the incoming equipment is maintained and merges with the tone-on release-guard signal. On the calling party clear, outgoing-end relays A, B, and GD release and the busy is maintained on the incoming P-wire until the outgoing-end D relay release on termination of the release-guard signal.

More recent system AC8 tone-off idle equipment adopts the pulse release-guard signal (Fig. 9.4) as described for system AC12. A pulse-type release-guard signal is not a requirement for the tone-on idle option of system AC8 and here Fig. 9.2*b* applies.

Forward retest: If the release-guard signal is not received within a certain specified time (sufficient for the incoming-end equipment to recognise release conditions) due to fault, the seizure and clear-forward signals are transmitted repeatedly in an attempt to clear down the circuit under short-term fault conditions. On nonreceipt of the release-guard within the specified time, outgoing-end relay GD maintains operation which causes outgoing-end circuit arrangements to connect earth to the M-lead to give the reseize. Interruption of this earth after a short time gives the repeat clear-forward signal, an interruption that is repeated every 30 s, a delayed alarm being given after retesting for some 6–12 min. The continuous reseize signal, maintained between successive clear-forward signals during the retest sequence, ensures that a bothway circuit is busied by virtue of the incoming function being in a seized condition.

With tone-on idle, it may be reasoned that the retest sequence is not so necessary since if the backward path be faulty tone-off will apply which busies the outgoing end. The eventual automatic return of tone-on when the path fault clears would then return the circuit to the normal idle condition. If after backward tone-on, the tone-off release-guard signal is not received, the circuit would be blocked and could be delayed-alarmed.

Blocking: An earth connected to the P-wire of the incoming-end equipment during maintenance attention operates RG which connects earth to the M-lead to subsequently operate D at the outgoing-end

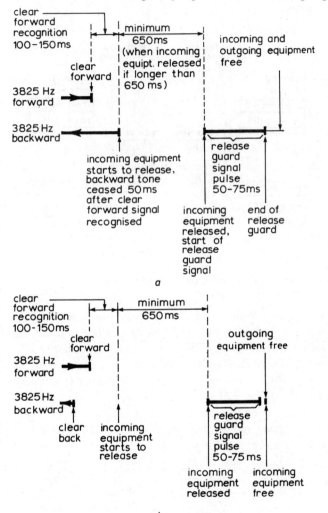

Fig. 9.4   *Release sequence UK tone-off idle outband signalling system*
   a Calling party clears first
   b Called party clears first

equipment to give the backward busying on the incoming P-wire. The blocking signal is tone-on continuous in the tone-off idle mode.

On bothway operation, the blocking signal is received on the E-lead at the incoming equipment function at the distant end, a condition which is the same as that for seizure and will busy the associated outgoing equipment function in normal manner to achieve the backward busying.

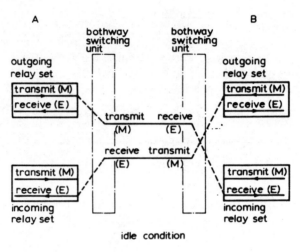

Fig. 9.5 *Bothway arrangement UK outband signalling*

Bothway operation: A special bothway signalling terminal is not designed for the AC8 and AC12 systems since outband signalling application is small in the UK network; also the majority of circuits are operated unidirectionally. Unidirectional outgoing and incoming relay sets, with a bothway switching unit, comprise a bothway signalling terminal (Fig. 9.5) as discussed in Section 7.9.

In the idle condition, each end of the circuit is terminated with both an outgoing and an incoming equipment, both ends being in a condition to accept a seizure signal on the incoming equipment function (Fig. 9.5). On call origination at, say, A, the seizure signal is transmitted to the incoming equipment at B. After a delay, the E-path at A is switched from the incoming to the associated outgoing equipment at A. Recognition of the seizure signal at B causes the outgoing equipment at B to be busied and the transmit M-path of the circuit to be switched from the outgoing to the associated incoming equipment at B. The switched arrangements at A and B are now such that a 4-wire circuit is established between the outgoing equipment at A and the incoming equipment at B, as for unidirectional. A similar sequence applies when a call is originated at B.

A delay occurs between outgoing equipment seizure at one end on call origination and the busying of the outgoing equipment at the other end, and double seizure can occur during this unguarded period. Owing to the delay in transferring the receive E-path from the incoming to the associated outgoing equipment at an end, reception of both seizure signals by the incoming equipments is assured and double seizure is detected when both the incoming and outgoing equipments at an end are in the seized condition. On seizure, each incoming equipment takes control of its associated bothway switching unit and the transmit M-paths are switched from the outgoing to the associated incoming equipment, thereby disconnecting the seizure signals. The incoming equipment at both ends are now connected to the transmit M and receive E-paths of the circuit and each remains in this state until the associated outgoing equipment is freed. The equipment-engaged tone is returned to both callers, both calls fail, and the release of the out-going equipments is under subscriber clear control (or in the AC12 system should an originating register force release on time out). When an outgoing equipment is released by the subscriber clear, the transmit M-path is switched from the incoming to the associated outgoing equipment at that end, the incoming equipment is thus released, and when this occurs at both ends, the circuit is in the idle condition.

If the subscriber at, say, A clears and the other at B does not, the outgoing equipment at A is fully released and available for further calls. The seizure signal for a follow-on call is transmitted to B and as at B both the transmit M and receive E-paths are connected to the incoming equipment, the call set-up will proceed in normal manner and eventually clear down in normal manner, despite the outgoing equipment at B being held to the subscriber not cleared at B, it being released when the retaining subscriber eventually clears.

(b) UK AC12 outband system[6]

The arrangements are much the same as for the AC8 system (Section 9.4a) except that address signalling is not incorporated. No delay-dialling/proceed-to-send line signalling sequence is included as the proceed-to-send is a backward interregister m.f. signal when AC12 line signalling is applied in the network. At present the equipment in use is tone-off idle (Fig. 9.2a) owing to the small-scale provision, but, as with AC8, with the tone-on idle option (Fig. 9.2b).

With the form of sequenced release described in Section 9.4a for the tone-off idle mode, if the forward transmission path faults (e.g. transmission failure) with the possibility of forward tone-on cessation resulting in a false clear-forward signal and premature release, then in the unanswered condition the subsequent false tone-on release-guard signal could be interpreted as a tone-on answer signal to give false call charge. To safeguard against this, the tone-off idle AC12 system, and the more recent equipment of the tone-off idle AC8 system, employ a timed pulse (50–75 ms) tone-on release-guard signal (not shown in Fig. 9.3) which is of shorter duration than the answer signal recognition 100–150 ms. The incoming equipment starts to release when the tone-off clear-forward signal is recognised, and the pulse release-guard signal sent when the incoming equipment has released, but not before 650 ms has elapsed after clear-forward signal recognition. Should the caller clear first on an answered call, the backward tone-on answer condition is changed to tone-off when the clear-forward signal is recognised (Fig. 9.4a), the change to backward tone-off being necessary to enable the release guard to be a timed tone-on pulse. This tone-off prior to the pulse release guard simulates a clear-back tone-off condition, but this is of no consequence. Should the called party clear first (or on release prior to answer), the standing tone-off backward signalling condition facilitates the tone-on release guard pulse (Fig. 9.4b).

The above problem which results in a preference for the pulse release-guard signal does not arise with the optional tone-on idle mode (i.e. tone-off for seizure) as here, failure of the forward transmission path would not give a false tone-on clear forward signal.

On nonreceipt of the release-guard signal due to fault, it is arranged that the receipt of the retest seizure signals by an incoming equipment does not cause association of register or switching equipments.

(c) Bell outband line signalling system[3, 7]
This is tone-on idle (off during speech) outband signalling, the signal frequency being 3700 Hz in each direction, and not 3825 Hz. 2-state continuous signalling applies in the two directions, the signal tone being applied and ceased in accordance with the standard Bell system toll line signalling requirement, the signal code and the various signal conditions being the same as for the Bell inband SF v.f. signalling system (Table 8.1 Section 8.7.1 analogue version, Section 8.7.3 and Table 6.3 Section 6.2.3 digital version) as shown in Fig. 8.13. E and M control with leads separate from the speech leads applies in accordance with Bell system practice for toll line signalling.

(d) Pulse outband signalling
The outband signalling systems discussed so far are of the 2-state continuous type, aiming at system simplicity by simulating the basic loop d.c. signalling condition. The various signal repertoires are modest,

but usually adequate, particularly when interregister m.f. signalling applies. Further signal requirement would tend to remove the basic simplicity. The transmitted signal level is relatively low.

Pulse outband signalling is adopted should it be preferred for any reason such that the signal level is higher than that permitted by continuous signalling, e.g. to ease the signal detection; or when the signal repertoire requirement exceeds that conveniently given by 2-state continuous signalling. Pulse timing is necessary and the signalling system is more complex relative to continuous.

Table 9.2   *Australian T system pulse outband line signal code*

Signal	Transmitted pulse duration (ms)	
	Forward	Backward
Forward		
Seizure	150	
Decadic address*	65	
	(pulse break)	
Clear forward	600	
Forward transfer*	150	
Backward		
Seizure acknowledgment*		150
Answer		150
Reanswer		150
Meter pulses*		150
Clear back		600
Release guard		600
Forced release		600
Blocking		Continuous

* When required.

The Australian T (3825 Hz) system is a typical example of the pulse-outband signalling mode and it is of interest to give brief information to indicate the general nature of the type.[8] The general arrangement of Fig. 9.1 applies and Table 9.2 shows the signal code. Signalling is performed by two signal pulse elements, short (150 ± 30 ms) and long (600 ± 120 ms), in the Australian T system. The decadic address pulses (break) are 65 ms nominal and the blocking signal is continuous tone-on. Signal discrimination is by direction, sequence and pulse timing.

The pulse signal receiving tolerances are for short pulses 80 ± 20 ms, and for long pulses 375 ± 75 ms. Tone-off less than 35 ± 5 ms is ignored by the receiver, except for decadic address pulsing. The system is suitable for application in the trunk and local junction networks when appropriate f.d.m. transmission systems are provided.

(e) Two-frequency outband signalling
While almost all known outband line-signalling systems employ a single frequency in the two directions, and this is preferred, the basic concept of the technique does, of course, allow more than one, and two frequencies have been used in special cases. In, typically, the Lenkurt (USA) type 45 carrier equipment, signal frequencies 3400 and 3550 Hz are used. The 3400 Hz is the trunk idle and on-hook indication, and 3550 Hz the trunk busy and off-hook indication. Supervision and decadic address signals impressed on the M-lead actuate the signalling oscillator causing it to shift from 3400 to 3550 Hz. The resulting signal tone is transmitted in the usual manner to the distant end for detection and subsequent control of the E-lead. The detailed operations generally follow those for 2-state continuous tone-on idle outband signalling.

9.5 CCITT (CEPT) R2 system outband line signalling

9.5.1 General
The CCITT R2 signalling system combines an analogue outband line-signalling system (and a digital version) for 4-wire circuits, and an analogue interregister m.f. signalling system.[13] While the line-signalling system has analogue and digital versions, the R2 interregister m.f. signalling system (Section 10.7) is analogue only, the m.f. signals being encoded in the same manner as for speech encoding in the digital application.

The R2 system was formulated and standardised by the CEPT, and is included in the CCITT series as Regional system 2. To avoid a multiplicity of combinations of the R2 interregister m.f. signalling system and different line-signalling systems, system R2 is specified with its own line signals in an outband system, which reflects the preference for outband line signalling in networks when appropriate transmission plant exists. It is recognised of course that this would be an evolving process in networks, and the relevant CCITT recommendations are in the nature of network planning objectives.

In basic concept the line signalling may be of any type (outband, v.f., d.c., p.c.m.) depending upon the transmission media, when working in combination with the R2 interregister m.f. signalling, providing the required line signals are given. Thus, despite outband line signalling (and the corresponding digital version) being specified for R2, which is perhaps somewhat restrictive, there is no technical reason whatever why various types of line signalling should not be used, depending on administrations' choice. Indeed, a different line-signalling system to that specified for system R2 would be necessary for 2-wire circuits, should this arise.

CCITT signalling specifications do not specify design arrangements of signalling systems. They give information on which design can be based and individual administrations translate the CCITT specification clauses into design realisation and thus the design detail varies according to the administration.

9.5.2 R2 analogue outband line signalling

This is 3825 Hz, tone-on idle (off during speech), 2-state continuous signalling in each direction, signal discrimination being by direction and sequence. Table 9.3 and Fig. 9.6 show the signal code. A delay-dialling/proceed-to-send line signalling sequence is not included as proceed-to-send indication is given by a backward R2 interregister m.f. signal.

Table 9.3 *R2 analogue outband line signal code*

Signal	Signal tone condition	
	Forward	Backward
Idle	On	On
Forward		
Seizure	Off	On
Clear forward (release)	On	Off or on (called party off-hook or on-hook)
Backward		
Answer	Off	Off
Clear back	Off (clear back prior to clear forward)	On
Release guard	On	Off
Blocking	On	Off

The d.c. recognition time for a changed condition, i.e. transition from tone-on to tone-off and vice versa, is 20 ± 7 ms. The total response time of the signal sender and the receiver (typically 30 ms) must be added to this for the a.c. condition.

Consideration is currently being given to the addition of a forward transfer signal in both the analogue and digital versions of R2 line signalling.

In addition to the release by the clear-forward signal under the control of the calling-party clear, and in accordance with standard CCITT recommendation, the system includes a feature to release the connection by a system-generated clear-forward signal, and stop the charge, if the caller has not cleared between 1–2 min after receipt of the clear-back signal; the release preferably being in direction forward from the originating exchange. The time interval stated may be varied in national network application depending on administration requirements.

The release is sequenced, the release-guard signal being ensured whatever the operating state of the circuit at the time the clear-forward signal is sent, i.e. seized but unanswered, answered or cleared, by the called party. It must also be ensured if answering or clearing by the called party occurs when release has already commenced at the outgoing end. The release-guard signal consists of backward tone-off for a certain time.

(a) *Release when calling party clears last, the backward condition being tone-on (which is the same condition as release prior to answer) (Fig. 9.6a):* The backward tone is ceased and the incoming-equipment release initiated on the clear-forward recognition at the incoming end. At the incoming end the sending of an answer signal, should this condition arise when releasing prior to answer, can only be prevented after clear-forward signal recognition. Also, to avoid false operation if answer coincides with release, the transition from tone-on to tone-off backward must not be interpreted as part of the release-guard signal until an interval T1 (250 ± 50 ms) has elapsed after application of the clear forward tone-on. This interval ensures that the clear forward is recognised and the backward tone-off condition established at the incoming end. The outgoing end does not recognise the backward tone-off as the start of the release-guard signal until T1 after sending the clear forward tone-on. The release-guard signal is terminated when the incoming equipment has released, but to avoid false operation in the event of coinciding forward and backward signals or an irregular sequence of signals, the release-guard signal has a duration T2 of

450 ± 90 ms at least. This ensures recognition of the end of the release-guard signal at the outgoing end after T1 has elapsed, which is recognition of the start of the release guard at the outgoing end.

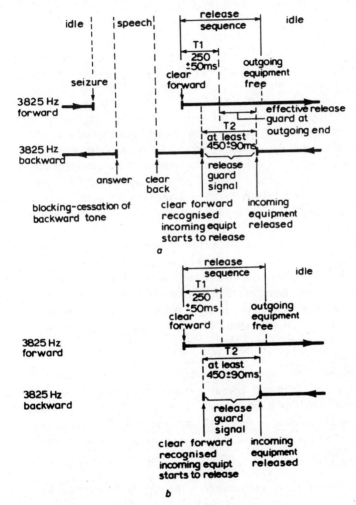

Fig. 9.6   *R2 system analogue outband line-signalling signal code*
a Release sequence: calling party clears last
b Release sequence: calling party clears first

The release-guard sequence is completed and the circuit returned to idle, when the transition from tone-off to tone-on is recognised (after T1) at the outgoing end.

These T1 and T2 intervals assume a maximum 30 ms circuit

propagation time. Longer propagation times (e.g. satellite circuits) would necessitate longer intervals.

(*b*) *Release when calling party clears first, the backward condition being tone-off (Fig. 9.6b):* The seizure tone-off is maintained for at least 100 ms to ensure its recognition at the incoming end to safeguard against an immediate follow-on clear forward tone-on. The same sequence of operation apply as in (*a*) above except that no requirement arises to cease the backward tone at the incoming end on clear forward recognition. The release guard tone-off signal merges with the standing tone-off condition. As in (*a*) above, the outgoing-end recognises the start of the tone-off release guard after T1 has elapsed after the start of sending the clear forward. Sending of a clear-back signal can only be prevented after clear-forward recognition and a coincident condition results in the clear-back tone-on being ceased on clear-forward recognition, and the T1 interval safeguards the situation at the outgoing end.

The retest feature of repeated attempts to clear is not incorporated in R2 line signalling. If, after sending the clear-forward signal the tone-on in the backward direction is not restored, the circuit stays blocked and busied at the outgoing end. The same occurs when, in the idle condition, the backward tone-on is ceased by a fault (the normal blocking signal being continuous tone-off backward) and if the backward tone-on be restored prior to delayed alarm, the circuit returns to normal. If after backward tone-on the release guard tone-off is not received, the circuit is blocked and delayed alarmed.

Interruption control: An interruption-control feature is incorporated to safeguard against unwanted interruptions of signalling channels resulting in false signalling, the control using the group pilots of carrier systems to detect the interruptions. The feature does not apply to interruptions of individual channels, but may be provided if the number of circuits to be used in a given relation is smaller than the capacity of the relevant multiplex.

A receiver at one end monitors the pilot transmitted by the other, and an interruption is assumed when the received pilot level falls from nominal to -33 dBm0. Interruption control then reacts to prevent the unwanted transmission of certain signals on those circuits already seized or to ensure that idle circuits are blocked. The reaction initiated by the control must be faster than the normal tone-on to tone-off transition complete recognition. Various procedures apply depending on the signalling state of the individual circuits.[14] The interruption control reverts to normal when the pilot level rises from

$-$ 33 dBm0 to nominal, the transmission system is re-established and the signalling equipment automatically reverts to normal operating.

Relevant data: In outband signalling, the d.c.–a.c. and a.c.–d.c. conversions are integrated with the f.d.m. transmission system and relevant signalling data is part of the transmission system specification. The following detail is relevant for the R2 analogue outband line-signalling system, and for many national continuous outband line-signalling systems:

Transmitted signal frequency:	3825 ± 4 Hz
Received signal frequency:	3825 ± 6 Hz
Transmitted signal level:	$-$ 20 ± 1 dBm0
Receiver sensitivity:	The signal receiver assumes tone-on condition when the received signal level is $-$ 27 dBm0 or higher

Bothway operation: In the R2 system arrangement, a time factor of 250 ± 50 ms operates at each end of a bothway circuit. When one end sends a seizure tone-off, it verifies whether or not cessation of signal tone in the opposite direction occurs within this 250 ± 50 ms. A tone-off received within this interval would normally be an incoming seizure signal from the other end, and double seizure is detected when both outgoing and incoming signals are tone-off within the 250 ± 50 ms. The other end detects double seizure in similar manner. A tone-off blocking signal cannot be distinguished from a seizure signal at either end and the system R2 specification includes various clauses to safeguard against a received blocking signal coincident with outgoing seizure.[13]

The time interval 250 ± 50 ms is based on a circuit propagation time of maximum 30 ms. Longer propagation times (e.g. satellite circuits) would require a longer interval.

On double-seizure detection, each end of a bothway circuit returns to the idle condition by the release sequence, recognising the incoming tone-on condition after sending the tone-on clear forward. Each end, even if immediately seized for another call, maintains the tone-on condition on the outgoing signalling channel for at least 100 ms to ensure that the end of the double-seizure condition is recognised at the other terminal. Although a double-seizure has been recognised, the backward tone-off condition is passed on backwards to be recognised as an erroneous tone-off answer signal, i.e. an answer signal condition received before the outgoing R2 registers at each end have been dismissed by an interregister signal. This is an abnormal condition leading to automatic release of the connection by the clear forward-release guard sequence. The tone-on clear-forward signal is not sent until the tone-off condition has been maintained for at least 1250 ± 250 ms (as a protection against interruption of a signalling channel corresponding to blocking). After sending the clear forward, each end returns to idle when the idle tone-on condition from the other end is recognised. An appropriate tone may be sent to both callers and both calls lost, or an automatic repeat-attempt made, depending upon administrations' policy.

9.5.3 R2 line signalling digital version

Only two signalling conditions are available in each direction in the R2 analogue line signalling (Section 9.5.2). This simple binary on-off condition is inadequate for the signal repertoire and in the change from 'release' to 'idle' the additional criteria of timing is necessary to ensure a defined sequence corresponding to the transfer of the release-guard sequence. It will be understood of course that even though timing is involved, the release-guard signal is obtained within the basic concept of 2-state continuous signalling. The outband analogue system could be applied to p.c.m. transmission systems by utilising one signalling channel (bit) per speech circuit in each direction. Here again, however, in the simple on-off concept only two signalling conditions would be available in each direction and timing would be necessary for other requirements.

24- and 30-channel p.c.m. systems conveniently allow more than one signalling channel (more than one signalling bit) in each direction for the built-in associated p.c.m. signalling, which introduces greater possibility of achieving a given signal repertoire requirement without timing discrimination and thus simplifying the signalling terminals. The associated signalling of 24-channel p.c.m. systems allows two signalling channels in each direction per speech circuit and that of the 30-channel system allows four (Section 6). Advantage is taken of this and the R2 system digital line signalling uses two signalling channels in each direction per speech circuit and in the case of the 30-channel system these are two of the four available.

The two signalling channels are conveniently referred to as a_f and b_f in the forward direction and a_b and b_b in the backward, their significance being that:

(i) the a_f channel identifies the operating condition of the outgoing equipment and reflects the on-hook, off-hook conditions of the caller

(ii) the b_f channel provides a means for indicating a failure in the forward direction to the incoming equipment

(iii) the a_b channel identifies the operating condition of the incoming equipment and reflects the on-hook, off-hook conditions of the called party

(iv) the b_b channel indicates idle or seized condition of the incoming equipment.

Table 9.4 shows the R2 digital line signal code on this basis.

Table 9.4   *R2 digital line-signal code*

	Forward		Backward	
Signal	a_f	b_f (note)	a_b	b_b
Idle	1	0	1	0
Forward				
Seizure	0	0	1	0
Clear forward (release)	1	0	0	1
			or 1	1
			(depending upon called party off-hook or on-hook)	
Backward				
Seizure acknowledged	0	0	1	1
Answer	0	0	0	1
Clear back	0	0	1	1
Release guard = idle	1	0	1	0
Blocking	1	0	1	1

Note: For all supervisory signals $b_f = 0$. A change to $b_f = 1$ indicates a fault

All conditions are continuous and the codings repetitive (comma free). The recognition time for a transition from state 0 to 1 and vice versa is 20 ± 10 ms, this being defined as the minimum duration that a signal representing 0 or 1 must have at the output of the terminal equipment to be recognised as a valid signalling state by the exchange equipment.

The following are the various operating conditions:

Idle: $b_f = 0$ is established 'permanently'. This state at the incoming end results in $b_b = 0$ in the backward direction providing that the incoming equipment is idle. $a_f = 1$ and $a_b = 1$.

Seizure: This is transmitted only if the condition $a_b = 1$ and $b_b = 0$ is being received. a_f is changed from 1 to 0, which must persist at least until the seizure acknowledgment is received. In this way the outgoing end will only be able to send a clear-forward after the seizure acknowledgment is received. Until then the outgoing end is busied against a new seizure should the caller clear.

Seizure acknowledgment: On seizure signal recognition, the incoming end changes b_b from 0 to 1.

Answer: a_b is changed from 1 to 0. This simple change of one bit only allows the outgoing end to transfer the answer signal condition to the preceding equipment without decoding two signalling channels (bits).

Clear back: a_b is changed from 0 to 1.

Clear forward: a_f is changed from 0 to 1. The outgoing end remains busied until recognition of the release guard signal.

Release guard: This must be ensured whatever the condition of the circuit at the instant of sending the clear-forward signal. Recognition of the clear-forward at the incoming end starts the release of the incoming equipment and initiates the release of the next link, the clear-forward signal on the next link not waiting for the completion of the release-guard sequence. On complete release of the incoming equipment, b_b is changed from 1 to 0. This state, which corresponds to the idle condition, unbusies the outgoing end.

Blocking: The outgoing end is busied for as long as $b_b = 1$ is received from the incoming end.

In the idle condition, $b_f = 0$ results in $b_b = 0$. Should a fault occur during the idle condition, b_f is changed from 0 to 1, and b_b is changed from 0 to 1 as a consequence. This blocks the outgoing end.

Switching and p.c.m. transmission equipment faults cause the transmission of states $a = 1$ and $b = 1$ to the distant end. Thus a fault at the outgoing end results in the sending of the clear forward signal ($a_f = 1$) and the clearing of the connection. b_f is changed from 0 to 1 to busy the outgoing end. A fault at the incoming end results in b_b being changed from 0 to 1, which blocks the outgoing end.

Bothway operation

Double seizure is assumed if an outgoing equipment is in the seized condition and the received condition $a_b = 0$, $b_b = 0$ recognised instead of the seizure acknowledgment $a_b = 1$, $b_b = 1$. The connection is automatically released at both ends by the clear-forward/release-guard sequence. Appropriate tone may be sent to both callers and both calls lost, or alternatively, an automatic repeat attempt made, depending upon administrations' policy. On the automatic release, both ends send the clear-forward $a_f = 1$, $b_f = 0$ and subsequently each end returns to idle on recognition of $a_b = 1$, $b_b = 0$.

References

1 CCITT: Green Book, **6**, Pt. 1, Recommendation Q21, ITU, Geneva, 1973, p. 58
2 CCITT: Green Book, **6**, Pt. 1, Recommendation Q44, ITU, Geneva, 1973, pp. 89–90
3 'Notes on distance dialling', Blue Book, Section 5, 'Signalling', (AT & T Co., 1975), pp. 32–33
4 HORSFIELD, B. R., and GIBSON, R. W.: 'Signalling over carrier channels that provide a built-in out of speech band signalling path', *Post Off. Electr. Eng. J.*, 1957, **50**, Pt. 2, pp. 76–80
5 GIBSON, R. W., and MILLER, C. B.: 'Signalling over carrier channels that provide a built-in out of speech band signalling path', *ibid.*, 1957, **50**, Pt. 3, pp. 165–171
6 MILLER, C. B., and MURRAY, W. J.: 'Trunk transit network signalling system – line signalling systems', *ibid.*, 1970, **63**, Pt. 3, pp. 159–163
7 BREEN, C., and DAHLBOM, C. A.: 'Signalling systems for control of telephone switching', *Bell Syst. Tech. J.*, 1960, **39**, pp. 1381–1444, and Bell System Monograph 3736, 1960
8 CREW, G. L.: 'Line signalling in the Australian Post Office', *Telecommun. J. Aust.*, 1967, **1**, pp. 23–27
9 HEBEL, M.: 'Handbuch für den selbstwahlfernverkehr' (Franckh'sche Verlagshandlung, Stuttgart, 1962)
10 FUHRER, R.: 'Landesfernwahl – Band 1' (R. Oldenbourg Verlag, Munich, 1966)
11 FUHRER, R.: 'Landesfernwahl – Band 2' (R. Oldenbourg Verlag, Munich, 1962)
12 DOHRER, M.: 'Outband signalling for telephone switching signals', Siemens Rev. XXXVIII (1971), Special issue on 'Communications Engineering', pp. 55–57

13 CCITT: Green Book, **6**, Pt. 3, Recommendations Q350–Q368 'Specification of signalling system R2' and Recommendations Q350–Q359 'Line signalling analogue and digital versions', ITU, Geneva, 1973, pp. 589–614
14 CCITT: Green Book, **6**, Pt. 3, Recommendation Q356 'Interrupt control', ITU, Geneva, 1973, pp. 602–607

Chapter 10

Interregister multifrequency signalling

10.1 General

With common control switching, selection is delayed while the address information, or part of it, is processed by the register-translator to give the necessary selection information. This could increase the post-dialling delay, which could be further increased if the interexchange address signalling is decadic, and thus slow. A postdialling delay not exceeding some 5 s is a reasonable design objective for modern networks, and as the common control switching is invariably nondecadic, it is clearly preferable that the interexchange address signalling be non-decadic with high-speed potential to aid the postdialling delay objective. As, by virtue of the storage, the outgoing and incoming registers may function as signal coders and decoders, respectively, the interregister signalling may be coded in a desired way, an address digit value trans-mitted being given by the coding. Coded interregister address signalling could be incorporated in the line signalling system, typically as in the 2 v.f. CCITT 4 system (Section 12.3), but owing to the relatively high cost of analogue signalling in national networks, it is preferred that the per-speech circuit provided line-signalling systems be as simple and as inexpensive as possible. Further, registers facilitate facility enhance-ment in networks, and exploitation of this on a network basis would require interregister signalling additional to address information, to further complicate the line-signalling system. Registers allow the possibility of separating the supervisory (line) and selection-signalling functions and a signalling system, separate from the line-signalling system, has merit for the latter function, which could also include the facility-enhancement signalling. As with the register, such a signal-ling system would be concerned with call connection set-up only, it could conveniently be part of the register function and, with the register, dropped from a call on completion of connection set-up, and prepared to deal with other calls. This common provision nature of interregister signalling would allow a sophisticated system which, while perhaps relatively expensive in itself, would not be unduly so when cost is assessed on a per-speech circuit criteria.

As the address digit value can be coded, fast interregister signalling can be achieved by a single signal condition coded to denote the digit value. In the multifrequency (m.f.) arrangement usually adopted, the available transmission bandwidth is exploited to obtain the fast signalling and to provide a generous signalling capacity. Signals are transmitted in the 2-out-of-N frequencies form. 2-out-of-5 allows ten coding possibilities to signal the digit value, but as usually information additional to address, typically facility-enhancement, is required to be transmitted between registers during connection set up, 2-out-of-6 (2/6 m.f.) is usually adopted, giving 15 coding possibilities. This may apply in the forward signalling direction only, but preferably in both forward and backward; and in the latter case, should one of the directions, usually the backward, not require all the 15 possibilities, 2/5 m.f. or 2/4 m.f. could apply in this direction for economy, the potential of the system would, however, be 2/6 m.f. in each direction.

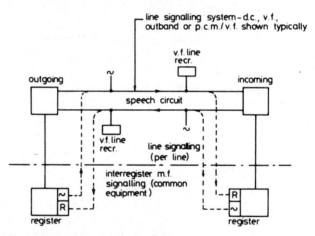

Fig. 10.1   *Line and interregister signalling*

In the general arrangement of Fig. 10.1, the line-signalling system may be of an appropriate type (d.c., v.f., outband, p.c.m.) depending on the type of transmission media. The interregister m.f. signalling system is required to function over any type of transmission system capable of transmitting speech and is thus a.c. within the voice band (inband). Fig. 10.1 shows the typical case of a v.f. line-signalling system and the interregister signalling frequencies must avoid the inband v.f. line-signalling frequency(ies). The registers incorporate the signalling logic, storage, etc. at each end and the interregister-signalling system itself is simply a conveyor of signals between registers, the signalling not subject to signal imitation by speech. When associated with the line, the registers split the speech-transmission path at each end which minimises interference to signalling, which, combined with the elimination of signal imitation by speech, contributes to fast signalling.

Interregister m.f. signalling:

(*a*) is independent of the speed and type of the common control switching equipment

(*b*) can give fast transfer of address (and other relevant information) which can ease postdialling delay problems

(*c*) gives considerable coding possibility with uniform format, the latter contributing to reliability in view of the error-checking possibility

(*d*) is not unduly concerned with signal distortion as the signal information is in the frequency content of the signal and not given by signal length

(*e*) does not permit E- and M-lead control of signalling as the signalling equipment is part of the switching equipment

(*f*) in its basic concept permits many variants: link-by-link, end-to-end, forward signalling only, forward and backward signalling, pulse (acknowledged or unacknowledged), continuous compelled, semi-compelled signalling, etc. the choice being dependent on the application conditions and thus on the characteristics and requirements of particular networks.

10.2 Interregister signalling modes

In both line and interregister signalling systems, interexchange information may be sent as pulse or continuous signals. Usually, when applied in line signalling, the continuous signalling is 2-state reproducing the subscriber on-hook, off-hook conditions, the signalling information being in the change of state and the signal codes having a relatively low information content. In non 2-state continuous signalling (or more simply 'continuous'), which may apply in interregister m.f. signalling and in some v.f. line-signalling systems, the signal codes have a somewhat greater information content. Here, the signal meaning is in the code information content of the signal and not in the change of state on or off, and when the continuous signal ceases it is not because a new signal is sent, but because the original signal has been detected and interpreted correctly. Thus the change of state determines the information content of the signal. Such continuous signalling requires a return signal to acknowledge and cease the information signal, which introduces a signal sequence concept.

Three basic types of signal sequence are identified:

(i) noncompelled
(ii) fully compelled
(iii) semicompelled

A noncompelled sequence may be regarded as one where a signal in one direction has no immediate relation to a signal which may be sent in the other. In theory, it is difficult to visualise a telephony signalling system which truly meets this criteria as a telephone call set-up and clear down procedures follow a reasonably logical repetitive pattern, and each end of a circuit would have some knowledge that a following signal should be a particular one, or one from a limited number of possibilities. System knowledge of signal expectancy owing to the sequence of operations enables some signalling systems to have a limited number of discrete signals, which is of significance when dealing with signal codes of low information content. The noncompelled sequence, however, implies that the duration of a signal in one direction is independent of whether or not a signal is received in the other and that there is no immediate acknowledgment by a specific signal that the information has been correctly received.

For the present purpose it is proposed to adopt an understanding of the compelled sequence as one in which a signal in one direction has a direct relationship to a signal in the other, which implies that each information signal requires an immediate acknowledgment by a specific signal. Thus the compelled sequence requires a signal code reasonably liberal in information content. In the fully-compelled (or usually more simply 'compelled') the signal transfer is fully sequenced and both the information and acknowledgment signals are continuous. In the semicompelled, the signal transfer is not fully sequenced and relevant signals may be continuous or pulse. The duration of a continuous signal in either the fully-or semicompelled sequences is dependent on the receipt of a signal in the opposite direction.

While the above is applicable to both line and interregister signalling, it is convenient to consider the interregister m.f. signalling case to identify points of greater detail.[1] The interregister signalling modes may be:

(*a*) *Pulse:* The signal pulses are of fixed duration, which may not be acknowledged (noncompelled), or may be acknowledged individually (conveniently classified as semicompelled when explicitly acknowledged) or in groups (implicitly or explicitly) by similar signals in the opposite direction (Fig. 10. 2*a*).

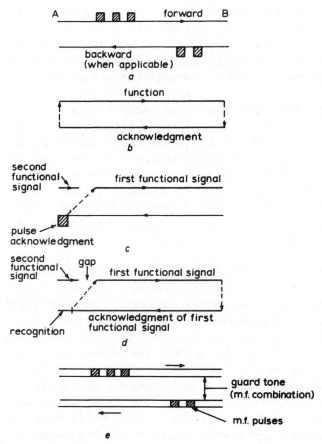

Fig. 10.2   *Types of m.f. signal*
  a Pulse
  b Fully compelled
  c Semicompelled (pulse)
  d Semicompelled (gap)
  e Guarded pulse

(*b*) *Fully-compelled:* Both the information (functional) and acknowledgment signals are continuous. A functional signal persists until the continuous acknowledgment is received. The acknowledgment ceases the functional signal and the detection of this cessation ceases the acknowledgment. The next functional signal may then be transmitted when the acknowledgment cessation is detected. There is a direct relationship between the functional and acknowledgment signals, the duration of either dependent on the signal in the opposite direction, the duration being self-adjusting to the circuit propagation time (Fig. 10.2b). The fully-compelled sequence involves four circuit propagation times (two tone-on, two tone-off) and four signal recognition times (two tone-on, two tone-off) per signal sequence. The speed of signal transfer is slow with long propagation times, which increases the post-dialling delay when address-information transfer is involved.

(c) *Semicompelled (continuous/pulse):* A pulse, instead of a continuous, acknowledgment is utilised to cease the continuous functional signal in the other direction. Both signals are present at the same time as the continuous signal persists until the pulse acknowledgment receipt is recognised, the continuous signal duration being dependent upon the signalling in the other direction. An immediate follow-on functional signal may be transmitted after a delay of approximately two propagation times plus two signal recognition times after the start of transmission of the previous function signal (Fig. 10.2c).

(d) *Semicompelled (continuous/continuous):* Both the functional and acknowledgment signals are continuous, but instead of a follow-on functional signal being transmitted reasonably immediately after the cessation of the acknowledgment as in fully-compelled, it is transmitted after a timed pause after received recognition of the acknowledgment. There is thus a timed gap between two immediately consecutive continuous functional signals (Fig. 10.2d). The duration of either the functional or acknowledgment signal is dependent on the signal in the opposite direction. The first acknowledgment signal overlaps the second, subsequently being replaced by the second. The signal sequence takes place in approximately two propagation times plus two signal recognition times, plus the gap duration.

In all cases, fully- and semicompelled, the acknowledgment may carry additional information, particularly of the signal request nature.

Network conditions and administration policies have influence on the choice from (a)–(d) above for particular networks, typically:

(i) *Network numbering scheme:* In a uniform national or international numbering scheme, the originating register could be aware of complete dialling by address digit count. The originating register may (*en bloc*) or may not (overlap) wait until all digits have been received before commencing connection set-up, but if it does, advantage could be taken of the *en bloc* address information in preventing the switched network being used unnecessarily on incomplete dialling. If it does not, and begins connection set-up when sufficient information has been received to determine the routing from that point, the network would be used unnecessarily on incomplete dialling, the register, originating or subsequent depending on the arrangements, being forced released on time-out. *En bloc* of the address information at originating registers is liable to result in postdialling delay penalty on multilink connections even though overlap operation may apply at transit registers, a penalty that occurs with either link-by-link or end-to-end interregister signalling. Depending on the number of links in the connection, the penalty could be particularly severe with link-by-link forward and backward signalling. The penalty arises, but not so severely, with forward signalling only link-by-link.

With nonuniform numbering schemes, it is the practice not to complicate registers for complete network intelligence in regard to numbering and an originating register would not always be aware of complete dialling. It could be made aware, however, in certain circumstances such as receipt of a maximum total number of digits, a trunk area code for a destination local area having a uniform number of digits in its local linked numbering scheme. This could be handled by the originating register:

(i) operating overlap, beginning connection set up when sufficient address information has been received to enable the routing to be determined from that point, or
(ii) operating *en bloc*, waiting a certain time (say 5 s) after receiving a certain minimum number of digits to make arbitrary decision of complete number.

The latter is not normally acceptable for postdialling delay reasons. In either case, with incomplete dialling, the connection would be partially set up necessitating an incoming end time-out to indicate appropriate action (equipment congestion tone to caller, a backward 'address incomplete' interregister signal to release the registers).

Both uniform and nonuniform numbering schemes permit adoption of any one of the signalling modes (*a*)–(*d*) above, but should *en bloc* apply, any form of compelled signalling (fully or semi) would result in postdialling delay penalty. As any form of compelled signalling implies backward signalling, the magnitude of the postdialling delay penalty would also be related to the link-by-link, end-to-end signalling choice.

A network with a uniform numbering scheme in combination with the adoption of complete *en bloc* of the address information would wish to avoid backward interregister signalling which would indicate a preference for the pulse mode (*a*) in the unacknowledged form, as in the Bell system. A nonuniform numbering scheme would normally imply backward interregister signalling to control the transfer of address information as complete *en bloc* would not normally apply, which would indicate adoption of some form of backward acknowledgment interregister signalling (*a*)–(*d*) above as appropriate.

(*ii*) *Forward and backward interregister signalling:* Backward, in addition to forward, interregister signalling is preferred for reasons additional to the control of the transfer of address information and is preferred regardless of whether the numbering scheme is uniform or nonuniform. The backward signalling could be utilised to acknowledge functional signalling which would indicate adoption of any one of the modes (*a*)–(*d*) above, but (*a*) in the acknowledged mode.

(*iii*) *Link-by-link or end-to-end signalling:* With backward interregister signalling, link-by-link signalling greatly increases the holding time of transit registers which is normally unacceptable (Section 10.3). Thus link-by-link signalling usually implies forward interregister signalling only and the avoidance of backward signalling. This would indicate avoidance of any form of compelled signalling and thus adoption of the pulse mode (*a*) in the unacknowledged form. End-to-end signalling implies the inclusion of backward signalling, which would indicate adoption of any one of the modes (*a*)–(*d*) above, but (*a*) with backward signalling.

(*iv*) *Facilities:* Backward interregister signalling enables subscriber or system facility information to be transmitted over the network as appropriate:

(i) Calling subscriber facilities (class of service, type of call, etc.) and various system facilities (calling line identity, etc.) may be programmed in, and the information obtained from, registers. Such information may be transferred forward over the network from an originating register in response to backward request signals.
(ii) Various items of information regarding the called subscriber (free, busy, etc.) may be sent back from a terminating to an originating register.
(iii) Backward signalling to indicate appropriate supervisory tone

condition by signal over the switched network, the actual tone being generated at the originating exchange.

Note. It is advised that the ring tone be transmitted over the switched network and not indicating this condition by electrical signal, although a 'called subscriber free' status signal may be transmitted back for reasons other than causing ring tone to be generated at the originating exchange for transmission to the caller. Reliance on a 'ring tone' backward signal would require a line split at the originating exchange when transmitting ring tone to the caller, and would introduce the danger of a quick verbal answer being clipped with the possibility of abandoned call should the verbal answer not be repeated.

Such a facility exploitation approach, considered to be a necessary requirement for modern networks, requires backward, in addition to forward, interregister signalling which would permit adoption of any one of the modes (*a*)–(*d*) above, but (*a*) with backward signalling.

(*v*) *Postdialling delay:* For minimum postdialling delay, the complete connection set-up signalling process and the release of registers to switch the speech path through, should be as fast as possible. In the unacknowledged pulse-address signalling mode (*a*), the time to send a digit is made up of a digit pulse duration (typically 80 ms) plus the interval (typically 80 ms) between pulses: total 160 ms. In the fully compelled mode (*b*), the time is made up of four propagation times plus four signal recognition times. Assuming a 10 ms propagation time and 40 ms recognition (10 ms receiver response, 30 ms signal recognition), then a compelled cycle to transfer one digit would be 200 ms, a time, however, that is propagation-time-dependent. Individual acknowledgment for each digit is required with compelled, but individual acknowledgment, group acknowledgment or no acknowledgment may apply with pulse.

The fastest interregister address information transfer is link-by-link forward signalling only, preferably with overlap operation at all the registers on the connection. This would imply adoption of pulse-signalling mode (*a*) in the unacknowledged form. As concluded elsewhere, however, the inclusion of backward signalling is preferred which normally implies end-to-end signalling. In this circumstance, and as a generalisation, the pulse (*a*) and fully compelled (*b*) modes have much the same speeds when the propagation time is reasonably short, but the compelled would be increasingly slower with increase in propagation time. The semicompelled modes are faster than the fully-compelled (*b*), and could be faster than pulse (*a*) over a certain range of short propagation times, but would be slower at the longer propagation times.

For general application over a wide range of propagation times it is considered that the pulse mode (*a*) would be faster than the fully- and semicompelled modes (*b*)–(*d*) and thus preferred in regard to postdialling delay.

(*vi*) *Reliability:* Interference (noise, interruptions, etc.) to inter-register signalling can arise from (i) switching, and (ii) transmission equipment. In regard to (i), switch and relay contacts and surges from switch and register equipments may give interference coincident with the m.f. signal to energise an m.f. receiver and not allowing time for receiver recovery to accept the signal. In regard to (ii), transmission noise (crosstalk, intermodulation products, impulse noise, etc.) and interruptions of the signalling path may also occur to interfere with signalling. Other transmission impairments such as delay distortion, attenuation distortion, frequency deviation, etc. are not regarded as being interference and are safeguarded against in the normal design of both pulse and continuous interregister signalling systems.

Continuous signals have theoretical merit in regard to interference and interruptions. Under interference conditions, the system could simply wait for the signal to settle down to a 2-and-2 frequencies-only persistence check for a valid signal. Interruptions of certain duration could be tolerated without loss of signal. Care, however, must be taken in the exploitation of these possibilities to avoid further delays to a basically slow compelled signalling technique. The permitted settling-down period should not be unduly long. Interruptions are more liable to coincide with a relatively long continuous signal, and safeguarding against relatively long interruptions would slow the normal function of detection for true cessation of signal in the compelled sequence.

In pulse signalling with reasonable pulse durations there is less possibility of a 2-and-2 only persistence check being successful in interference conditions. There is also less possibility of an interruption being coincident with a pulse, but if coincidence did occur the signal could well be lost.

Reliability is also related to the total number of signals per call connection set-up. In general, a compelled sequence, fully or semi, would require a greater number of signals than noncompelled pulse, tending to a lowering of reliability owing to the greater signalling requirement. This, however, may not be the case with unacknowledged pulse, link-by-link, signalling, as here the complete address information must be sent to transit registers which could result in a significant total signalling requirement.

Finally, unlike pulse, fully compelled does not measure signal durations and the simpler arrangements contribute to reliability.

On balanced assessment, it is considered that continuous fully-compelled is marginally more reliable than semicompelled and pulse in interference conditions, and, overall, is marginally more reliable.

For all signalling modes, end-to-end signalling is more onerous, and thus less reliable, than link-by-link, particularly in interference conditions. Typically, the minimum operate level of receivers has a significant bearing on reliability in interference conditions. As an example, the R1 link-by-link interregister signalling system specifies $-31\,dBm0$ per frequency for nonoperation and $-22\,dBm0$ for operation, whereas the R2 end-to-end system specifies $-42\,dBm0$ and $-35\,dBm0$, respectively, and it is clear that system R1 would be more reliable in this regard.

Choice of m.f. signalling mode

Factors other than those discussed (e.g. maintainability, cost, etc.) would influence the choice, but those discussed are sufficiently indicative. No one mode has clear and overriding advantages over others, except perhaps the preference for backward, in addition to forward, interregister signalling. Indeed, at present, it would perhaps be difficult to support the introduction in any network of interregister m.f. signalling which did not incorporate both forward and backward signalling, which would imply end-to-end signalling. Interregister m.f. signalling systems have tended to be designed to meet the characteristics and requirements of individual networks, taking account of the pros and cons of various factors such as pulse, continuous, reliability, etc. as discussed, the various weightings being individual administration assessment. This accounts for the different types. If m.f. signalling rationalisation is desired, it is thought that this would indicate the adoption of an end-to-end, forward and backward signalling, compelled, for wired-logic common control switching system networks (i.e. the R2 system). The fully-compelled mode would apply for reasonable propagation times (say up to 30 ms) and the pulse acknowledgment semicompelled for the longer times (e.g. satellite circuits). CEPT has recommended system R2 for use in European national networks and the system is rapidly gaining acceptance in many other countries. As R2 was formulated as a common system catering for the different requirements of many national networks, particularly in regard to numbering schemes, the situation could arise when its application in a particular network may not be the optimum arrangement for that network, e.g. if the transit registers always require a fixed and uniform number of address digits to determine the forward routing, group, as distinct from the R2 per-digit acknowledgment, could well be a preferred arrangement for that network. Also, there would be little point in adopting per-digit acknowledgment when complete *en bloc* of the address information applies. Nevertheless, the rationalised concept of the R2 system interregister signalling has significant merit, as many national networks have nonuniform numbering schemes, often differing from each other.

Many national networks are not of a size and complexity as to make the inflexibility of the R2 end-to-end signalling too serious a penalty to signalling evolution. Very large and complex networks with uniform numbering schemes present different problems. Here, signalling evolution flexibility would be an important factor, with a preference for link-by-link signalling to allow the flexibility. For this, and other reasons, the Bell system USA network, which is by far the largest national network in the world, is presently link-by-link forward interregister m.f. signalling only (system R1).

Nevertheless, while recognising the problems and preferences of large, complex networks, it is considered that backward, in addition to forward, interregister m.f. signalling is desirable and indeed necessary to allow a network to exploit its full potential. This raises the obvious problem of combining backward signalling with link-by-link signalling without excessive holding time of transit registers, and thus undue register provision. These are conflicting requirements in present analogue, wired-logic control, switched networks, which accounts for the adoption of end-to-end and forward and backward interregister m.f. signalling by most administrations, with acceptance of the signalling evolution constraint.

10.3 Link-by-link and end-to-end interregister signalling

Interregister m.f. signalling, being inband, may be applied link-by-link or end-to-end signalling on multilink connections.

Link-by-link
Here the outgoing and incoming registers on a link, interchange signals on that link only in the process of connection set-up (Fig. 10.3). The mode:

(*a*) Eases signal detection as the signal level range, receiver minimum-response threshold, delay distortion, attenuation distortion, noise, etc. are for one link only and thus no additional requirements arise on a multilink connection relative to a single link

(*b*) Increases the holding time of transit registers when backward interregister signalling applies as all registers on a multilink connection must be line-associated to pass the final backward signal. The increased holding time has reaction on expensive register provision

(*c*) Slows the transfer of signals on a multilink connection and, in particular, the slow transfer of the backward signalling may increase the postdialling delay.

If backward signalling does not apply (e.g. as in system R1), a register may release as soon as it has transferred the complete information forward. This avoids increased holding time of registers and increase in postdialling delay.

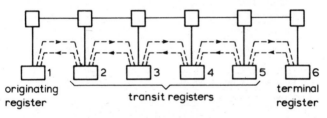

Fig. 10.3   *Link-by-link interregister signalling*
Note: Backward interregister signalling may or may not apply

End-to-end

Here, each transit register receives only the address digit(s) required for the transit selection to be performed, with the originating register always in control of the call set-up as it progresses towards the called party. As in normal practice it is preferred not to complicate registers for complete network intelligence; the originating register must be prepared to send partial address information on request by backward interregister signals, sometimes repeatedly, and backward interregister signalling is a requirement with end-to-end signalling. A transit register drops from the connection when the connection is extended forward from that transit point, progressive end-to-end signalling occurring as the connection is progressively built up. The originating and terminal registers only are line-associated when the connection set-up is complete and thus when the final backward interregister signal (the register dismissal signal) is required to be transferred (Fig. 10.4). The mode:

(i) conveniently gives originating (leading) register control of connection set up, which allows such facilities as reroute, repeat attempt etc., from the originating centre

(ii) reduces the holding time of transit registers, and thus register provision, relative to link-by-link with backward signalling

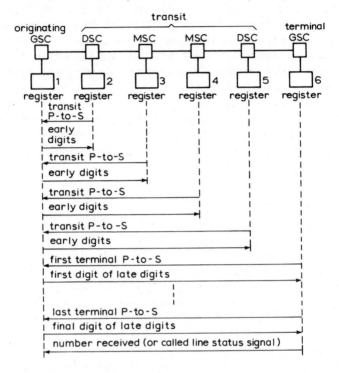

Fig. 10.4   *End-to-end interregister signalling*

(iii) simplifies transit registers when dedicated transit switching units apply as these registers are required to deal with part of the address information only, e.g. the trunk area code in s.t.d. Should it be preferred that switching units, and thus the registers, be general-purpose, dedicated functions would not arise, and in the transit mode of operation, such registers, while equipped to deal with the full address information, receive and function on part

(iv) by avoiding reception, storage and retransmission of through signals at transit points, minimises the probability of functional failure and can provide faster signalling

(v) relative to link-by-link with backward signalling, the faster signalling, particularly the backward, reduces the postdialling delay, the registers being released and the speech path switched through in faster time

(vi) complicates signal detection and m.f. receiver design as the dynamic range is wider relative to link-by-link to cater for the wider variations of transmissions impairments to signalling owing to the multilink condition.

It will be understood that various degrees of end-to-end signalling build up as the connection set-up progresses, the final end-to-end signalling state being between the originating and terminating registers in a given network. If end-to-end interregister signalling applies in both the international and national networks, it is logical that this should be in sections, one section in the international and another in the national, network. This is preferred as it is preferred that international switching units be true buffer interface switching points between the national and international networks.

Assessment

Theoretically, link-by-link signalling is always preferred in any signalling system in any network mainly to allow for flexibility for signalling evolution in a network, a new signalling system having the capability of being applied on individual link(s) without conflict. Existing end-to-end signalling provision in a network would be a constraint on this. This philosophy, however, must be weighed against any significant penalties link-by-link signalling may have in particular signalling systems. In line-signalling systems, link-by-link signalling does not result in a significant penalty and is invariably adopted; indeed, link-by-link signalling is implicit in certain line-signalling techniques. In interregister signalling, with both forward and backward signalling which is preferred, the penalty in particular of the increased holding time of transit registers is considered to be sufficiently severe as to make link-by-link signalling unacceptable in most national networks. There is no penalty in this regard when backward signalling does not apply and here the signalling is link-by-link. The inclusion of backward interregister signalling is preferred, however,

(*a*) to control the transfer of address information
(*b*) to invite the transmission of information from registers to achieve facility exploitation on a network basis
(*c*) to transfer appropriate supervisory tone indication over the switched network as electrical signals
(*d*) to facilitate retransmission on error detected, etc.

Thus interregister signalling systems are end-to-end signalling in the majority of cases, which is contrary to the desired philosophy of link-by-link for signalling evolution, but is accepted, particularly when a network is not too big nor too complex. It is sometimes reasoned that, as interregister signalling is common, and not per-speech circuit, provision, the penalty of lack of flexibility for signalling evolution is not so severe compared with line signalling, but this is simply a question of degree. Clearly, the size of a network would be a factor in the consideration and a very large complex network would be more reluctant to the adoption of end-to-end signalling than a smaller less complex network.

10.4 Control of address information transfer

Address information may be transferred between registers as complete *en bloc*, partial *en bloc*, or requested digit-by-digit. In interregister signalling systems now in use, the complete *en bloc* mode is associated with link-by-link signalling, and the partial *en bloc* and requested digit-by-digit modes associated with end-to-end signalling.

Complete en bloc link-by-link
The Bell system N. American network adopts complete *en bloc*, the interregister signalling being pulse, link-by-link, forward signalling only, i.e. system R1.

Complete *en bloc* implies that the register at the calling subscriber's exchange assembles the complete address information (local or toll) before commencing connection set-up. Transit registers could operate in a similar manner, but as there would be little point in this, and as there would be a significant increase in the postdialling delay, they operate in the overlap mode, which is the simultaneous reception of digits from a preceding register and the sending of digits onwards to the next register in the routing. With complete *en bloc*, the control of transfer is by means of a backward proceed-to-send signal, sent when the next register is line-associated and prepared to accept address information. The proceed-to-send signal is included in the various Bell system line-signalling systems, there being no backward interregister signalling. A toll transit register receives the complete national significant number in response to the proceed-to-send signal. A terminal toll register receives the called local number only in response to the proceed-to-send signal, the Number Plan Area (NPA) code digits not being sent to the destination NPA on the reasoning that the toll routing is complete.

As one type of proceed-to-send signal only applies, there is no discrimination in this signal with respect to transit or terminal (it would be difficult to include two different types of proceed-to-send signal in the line-signalling system) and the necessary discrimination is performed by the register immediately preceding the terminal toll register having knowledge that a transit or a terminal toll route is taken in the forward routing, path-of-entry discrimination to the next exchange thus applying. In basic philosophy of course the NPA code could be sent to the destination NPA terminal toll register, which would avoid the necessity for discrimination. The terminal toll register would then effectively perform the discrimination in that in receiving its 'own address', toll terminal conditions would be recognised. Receipt of own address could be regarded on the one hand as being a validity check in that the routing had reached a correct destination, but on the other hand as receipt of unnecessary and redundant information.

The called local linked number is transferred between the registers in the local area, also under the control of line proceed-to-send signals.

Partial en bloc and digit-by-digit, end-to-end
Partial *en bloc* refers to the sending of well defined digit groups under the control of backward signals from the incoming register. In digit-by-digit sending, each digit is invited-out separately from an outgoing register by request signals from an incoming register, per-digit acknowledgment signalling applying. Both modes are invariably associated with end-to-end signalling.

With end-to-end signalling of the form now used, a transit register requires the relevant digits only (typically the trunk area code) to enable the forward routing to be determined. A terminal register requires the address digits (typically the called local linked number only) to enable the called party to be connected. As the transit and terminal registers differ in their received address digits requirement, some form of transit/terminal discrimination is implicit with end-to-end signalling to enable the relevant digits to be invited-out from the originating register. This usually implies separate transit and terminal proceed-to-send signals (one type sent back from the transit register, and a different type from the terminal register) which are included in the backward interregister m.f. signalling implicit with end-to-end signalling. Figs. 10.4, 10.7 and 10.11 show some typical signal sequences.

With connection set-up complete, the registers are released by the final backward interregister signal (the register dismissal signal) to switch the speech path through. Should the end-to-end signalling extend down to the subscriber end exchange (which would then be the terminal exchange in the end-to-end signalling) the register dismissal signal is usually a called line status signal (called line-free, busy, etc.), which signal would also imply address complete. If, in the evolution of a network, the end-to-end interregister signalling is applied to part of the network, say, on the trunk network only (in which case the destination trunk exchange would be the terminal exchange in the end-to-end signalling), it may not be possible for this terminal register to determine the called line condition, nor address complete. In this event, the register dismissal signal, typically indicating number received only, would imply limited facilities applying, typically as in the UK MF2 system.

The detail of transit/terminal discrimination varies depending upon the particular interregister m.f. signalling system, typically:

(i) In system R2, transmission of a first digit from the originating register to a transit or terminal register occurs without a backward-request signal from that particular register (the 'giving' technique), subsequent digits being invited-out one-by-one as required by successive request signals (the 'asking' technique) from the receiving register. An immediately preceding register, originating or transit, determines the first digit to be sent to the next register, transit or terminal, from the originating register. Thus a transit register requests the first digit to be sent to the next register. Should the immediately preceding register be the originating, the first digit from this register is sent without a request signal. The immediately preceding register thus performs the transit/terminal discrimination, and the arrangement requires continuous compelled signalling, fully or semi.

(ii) In the UK MF2 system, a receiving transit register returns a transit proceed-to-send, and the terminal register returns a terminal proceed-to-send signal (the 'asking' technique). The Socotel system adopts much the same principle.

There are other possibilities.

As a matter of principle, it is preferred that, for simplicity, transit registers should not require network intelligence in the sense of deter-mining, by analysis, the number of digits to be requested to determine the forward routing. This is facilitated by transit registers in a given network always receiving the same number of digits. For some calls in nonuniform numbering scheme networks, a transit register may well receive more digit(s) than strictly necessary to determine the forward routing, but this is of little consequence. This concept has particular relevance when simple, rationalised, dedicated trunk transit registers are desired, as, typically, in the present UK trunk transit network, the MF2 signalling system applying.

Terminal registers would also be simplified by a uniform number of digits concept. Alternatively, the local linked number could be requested digit by digit.

10.5 General arrangement interregister m.f. signalling system

As stated, interregister 2/6 m.f. signalling may be link-by-link or end-to-end, forward signalling only, or both forward and backward signalling. The frequencies may be 120 or 200 Hz spaced, 200 Hz spacing being employed when forward signalling only, or when signalling in the two directions and the direction of transmission of certain signals changed as the call set-up proceeds and thus fewer signal frequencies used (French Socotel system, Section 10.9). The 200 Hz spacing eases the receiver frequency-detection process relative to 120 Hz. The same frequencies 200 Hz spaced could be employed in the two directions for wholly 4-wire working, both the switching and transmission being 4-wire. The speech circuit bandwidth is divided for forward and backward signalling and 120 Hz spacing adopted for 2-wire working (2-wire circuits or 4-wire circuits with 2-wire switching). For application flexibility, 120 Hz spacing is usually adopted for both 4-wire and 2-wire working, forward and backward signalling and directional highpass and lowpass filters equipped at the 2-wire situations (Fig. 10.5).

The 2/6 m.f. capability is usually required in the forward direction, but should a network not require all the 15 possibilities of 2/6 m.f. in the backward direction, 2/5 m.f. (or 2/4 m.f.) may apply for equipment economy, but the potential for 2/6 m.f. is not disturbed. As the lowest frequency (f_{12}) is more subject to transmission impairment relative to higher frequencies, this is the first choice for signal frequency omission f_{11} being the next. For this reason, the frequencies (f_1–f_6) in the higher band are transmitted forward and the lower band frequencies (f_7–f_{12}) backward.

A 2/6 m.f. signal is transmitted in compound form by simultaneous operation of two of the transmit gates. After frequency detection at the receive end, the signal appears as simultaneous energisation of two d.c. stages on which a 2-and-2 only persistence check is performed as signal recognition for valid signal.

The pushbutton telephone set is also based on m.f. signalling (Section 2.3), but the 2(1/4) m.f. coding system adopted is different from the 2/6 m.f. of interregister signalling. This loss of compatibility is of no consequence since the exchange must always function as a buffer between subscriber signalling systems and interexchange signalling systems.

10.6 Bell system interregister signalling (R1)

This system is applied in the N. American network, the associated line-signalling systems being Bell system d.c., outband, v.f., or p.c.m. as appropriate. The combination of the Bell interregister signalling and the Bell SF (1 v.f.) line signalling, analogue and digital (Section 8.7),[2] is specified as system R1 (Regional system 1) in the CCITT series. It will be understood of course that there is no technical reason whatever to limit line signalling to v.f.

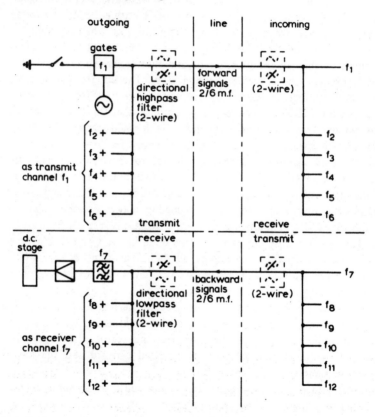

Fig. 10.5 *General arrangement m.f. signalling system*

Bell system interregister signalling[2–5] is 2/6 m.f., unacknowledged pulse, link-by-link, forward signalling only, frequencies 700 to 1700 Hz, 200 Hz spaced, and uses 12 signals only, the three spare signal codings being available for special purposes (Table 10.1).[3–6] The analogue m.f.

Table 10.1   *Bell interregister m.f. signal code*

Signal	Frequencies (compounded)
	Hz
KP (start of pulsing)	1100 + 1700
digit 1	700 + 900
2	700 + 1100
3	900 + 1100
4	700 + 1300
5	900 + 1300
6	1100 + 1300
7	700 + 1500
8	900 + 1500
9	1100 + 1500
0	1300 + 1500
ST (end of pulsing)	1500 + 1700
Spare	700 + 1700
Spare	900 + 1700
Spare	1300 + 1700

system is used in the digital application, the m.f. signals being bit-encoded as for speech.

The complete address information is assembled at the originating register and, in response to the proceed-to-send signal, is transmitted *en bloc* preceded by the KP signal and terminated by the ST signal, KP and ST being generated by the register on subscriber dialling. Overlap operation applies at transit registers. The incoming m.f. receiver is inhibited to address-digit signal response until it receives the KP signal, which thus serves to condition the receiver and to minimise interference effects prior to address-digit signal reception. This contributes to address-digit signalling reliability, which is further aided by the KP pulse being longer than address-digit pulses. The ST signal informs transit and terminal registers that no more digits are forthcoming, the originating and transit registers releasing when ST is transmitted forward. All registers release in succession as a multilink connection set-up proceeds forward.

A 2-and-2 only persistence check, minimum period 30 ms for the address digits and ST signals and 55 ms for KP, for signal applies at the d.c. output stage of reception, an m.f. pulse signal being recognised as valid if the two frequencies arrive within 6 ms of each other and then persists for the check period. If either frequency is received alone for longer than 6 ms, or should the 2-and-2 only check indicate a nonvalid signal for any other reason, reorder tone is returned to the caller.

Registers release in the following abnormal conditions:

Outgoing register

(*a*) nonreceipt of a proceed-to-send (start dialling) signal within 5 s of receipt of a delay-dialling signal

(*b*) nonreceipt of a delay-dialling signal within 5 s of outgoing circuit seizure

(*c*) an unexpected off-hook line signal subsequent to the recognition of an on-hook start-dialling signal, but prior to the completion of register outpulsing (this may occur in certain condition of double seizure)

(*d*) on the overall 4-minute register time out.

Incoming register

(i) nonreceipt of the KP signal within 10–20 s of register seizure

(ii) nonreceipt of address digit and ST signals 10–20 s after KP, or 10–20 s after receipt of various address digit signals

(iii) error detected by the 2-and-2 only persistence check.

Reorder tone is returned to the caller on register abnormal release.

Relevant data

Transmit:

Signal durations KP 100 ± 10 ms

All other signals 68 ± 7 ms

Intervals between all signals 68 ± 7 ms

Signal frequency tolerance ± 1·5% of nominal

Signal level − 7 ± 1 dBm0 per frequency

Receive:

Signal frequency ± 1·5% ± 10 Hz of nominal

Absolute power level N of each received frequency within the limits $(− 14 + n \leqslant 0 \leqslant + n)$ dBm, where n is the relative power level at receiver input. Assuming a zero-loss circuit, these limits give a margin of ± 7 dB on the nominal absolute level of each received signal. While this margin permits m.f. pulsing on circuits having switch-to-switch losses of 14 dB, the limit for operating sensitivity of the receiver is − 22 dBm per frequency.

Difference in level between the two frequencies comprising a received signal not to exceed 6 dB.

As there is no backward interregister signalling, which would normally include proceed-to-send signalling, the delay-dialling and proceed-to-send signals must be returned by the line-signalling systems. This complicates these systems somewhat, not only in the arrangements to give these signals, but also in the safeguards required in the conflict that the return off-hook delay-dialling and the subsequent on-hook proceed-to-send introduces in other requirements of the line signalling such as double seizure detection on bothway working (see typically Section 8.7).

Preferences have been advanced in Sections 10.2 and 10.3 for the inclusion of backward interregister signalling and the consequent adoption of end-to-end signalling, which features conflict with Bell system practice. The choice is clearly influenced by the understandable importance attached to signalling evolution in the very large and complex N. American network, combined with the postdialling delay problems of the complete *en bloc* of the address information. While the requirements are met in the Bell system interregister signalling for the particular conditions applying and in the background of the preferred approaches, the arrangements adopted in the solution do not permit achievement of other desired network requirements, particularly network-facility exploitation, whose requirements necessitate backward interregister signalling. In this regard it will be noted that the information transferred between registers is a minimum in the Bell system. Thus while Bell (R1) interregister signalling meets the present requirements and conditions of particular network(s), it is considered that it would have limitations for general application to networks, and would not be as suitable as a system with both forward and backward signalling.

10.7 CEPT interregister signalling (R2)

10.7.1 General

The combination of this CEPT standardised interregister m.f. signalling system and the associated outband line-signalling system, analogue and digital, (Section 9.5), is specified as system R2 (Regional system 2) in the CCITT series.[7] The analogue R2 interregister signalling is used in the digital application, the analogue m.f. signals being bit-encoded as for speech. It will be understood of course that there is no technical reason whatever to limit line signalling to outband. CCITT signalling-system specifications specify basic principles and requirements to be met, and do not embrace design detail. As system R2 is postulated for a wide field of application, the design realisation of the system may vary greatly, but the basic arrangements of Figs. 10.1 and 10.5 apply.

R2 interregister signalling is 2/6 m.f., continuous compelled, end-to-end, forward and backward signalling, overlap register operation (including originating), with frequencies 120 Hz spaced:

Forward: 1380, 1500, 1620, 1740, 1860, 1980 Hz
Backward: 1140, 1020,   900,   780,   660,   540 Hz

The CCITT is now proposing to specify a semicompelled version for application to long-propagation time circuits (e.g. satellite). The specification detail for this version is not yet complete, but will be based on minimum change to the basic fully compelled system.

System R2[7-12] is visualised for international, regional and national application. The signal repertoire is considerable, embracing all three fields of application, taking account of repertoire requirements for various national networks. A main aim of the system is flexibility to cater for the different conditions of different national networks, particularly in regard to different facility requirements and different numbering schemes. This accounts in large part for the features of the system.

In a particular national application, the signal repertoire adopted would be the administration's choice from the signals available in the specified system, with adequate spare capacity being available for national network signal(s) requirements not included in the specification, the basic principles of R2 being the vehicle for transferring the required signals. If limited backward signals are required nationally and 2/6 m.f. backward not found necessary, administrations are free to adopt 2/5 m.f. (or 2/4 m.f.) backward as appropriate. Equipment would not be provided for the signal channels not used, but the electrical potential of the system would still be 2/6 m.f. backward. The backward signal frequencies omitted always start at the lowest, i.e. in order 540 Hz, 660 Hz as required. 2/6 m.f. forward signalling would always apply. In national application, if desired, administrations could clearly adopt line signalling as dictated by the transmission media, providing the R2 line signals are given.

10.7.2 R2 interregister signal code

Both the forward and backward frequency combinations have primary meanings, which, by the use of certain backward signals, may be changed to secondary meanings. Each signal is acknowledged in the fully compelled sequence, the backward acknowledgment signals carrying additional information requesting transmission of the next forward signal or the condition of the called subscriber's line. On seizure of the outgoing circuit, an originating register automatically sends the first forward signal without requiring a backward request, and on recognition of this an incoming register returns an acknowledgment which gives instruction as to the next forward signal required (Fig. 10.6).

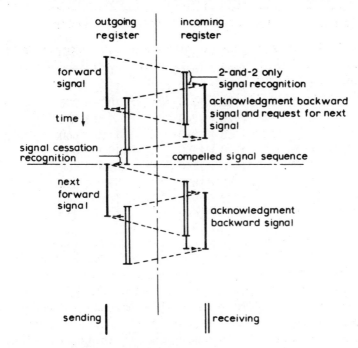

Fig. 10.6   *Compelled signalling procedure*

The last acknowledgment signal from a transit register also carries an instruction as to the first forward signal required by the next register, thus requesting transmission of this signal from the originating register. Tables 10.2–10.6 detail the R2 interregister signal code:

(*a*) Group I (Table 10.3) are the primary meanings of the forward signals.
(*b*) Group II (Table 10.4) are the secondary meanings, the change from primary to secondary being commanded by backward signals A-3 or A-5. Secondary meanings can change back to primary meanings only when the original change from primary to secondary was in response to signal A-5.
(*c*) Group A (Table 10.5) are the primary meanings of the backward signals.

Table 10.2   *R2 m.f. signal allocation*

Signal number	Forward signals of groups I and II frequency combination (compounded)	Backward signals of groups A and B frequency combination (compounded)
	Hz	Hz
1	1380 + 1500	1140 + 1020
2	1380 + 1620	1140 + 900
3	1500 + 1620	1020 + 900
4	1380 + 1740	1140 + 780
5	1500 + 1740	1020 + 780
6	1620 + 1740	900 + 780
7	1380 + 1860	1140 + 660
8	1500 + 1860	1020 + 660
9	1620 + 1860	900 + 660
10	1740 + 1860	780 + 660
11	1380 + 1980	1140 + 540
12	1500 + 1980	1020 + 540
13	1620 + 1980	900 + 540
14	1740 + 1980	780 + 540
15	1860 + 1980	660 + 540

(*d*) Group B (Table 10.6) are the secondary meanings, the change from primary to secondary being indicated by the backward signal A-3. There is no change back to primary once the change to secondary has been made.

In international application, the sending of the R2 signals is in a defined sequence (Section 12.6), the first forward signal giving routing information as follows:

(i) The country code indicator, followed by the country code itself, for transit. In addition to transit indication, the indicator gives information as to whether an echo-suppressor is required or not (col. *a* Table 10.3).

(ii) The language (semiautomatic working) or the discriminating (automatic working) digit, either digit indicating terminal. Neither the country code indicator nor the country code are sent to terminal registers.

In national application, administrations adopt a signal repertoire

Table 10.3   *R2 forward signals group I*

Signal designation	When first signal on an international circuit (a)		When other than the first signal on an international circuit (b)	See note
I-1	Language digit:	French	Digit 1	
I-2		English	2	
I-3		German	3	
I-4		Russian	4	
I-5		Spanish	5	1
I-6	Spare (language digit)		6	
I-7	Spare (language digit)		7	
I-8	Spare (language digit)		8	
I-9	Spare (discriminating digit)		9	
I-10	Discriminating digit		0	
I-11	Country code indicator, outgoing half-echo suppressor required		Operator Code 11	2
I-12	Country code indicator, no echo suppressor required.		Operator Code 12, or request not accepted.	2, 3, 4
I-13	Test call indicator (call by automatic test equipment)		Code 13 (call to automatic test equipment)	
I-14	Country code indicator, incoming half echo suppressor required		Incoming half echo suppressor required	2, 4, 5
I-15	This signal not used		End of pulsing (Code 15)	6

required for individual national networks. In principle, all frequency combinations, primary and secondary, have unique signal meanings whether used internationally or nationally. Fig. 10.7 shows a typical signalling sequence in a national application of R2 interregister signalling, a sequence that may be different for different networks.

Notes on Tables 10.3–10.6 (international application)

Table 10.3
(1) On terminal calls, col. (*a*) signals I-1–I-10 are the first signals transmitted, the country code indicator and the country code (both sent

Table 10.4   *R2 forward signals group II (calling party's category)*

Signal designation	Signal	See note
II-1		
II-2		
II-3	Signals assigned	1
II-4	for national use	
II-5		
II-6		
II-7	Subscriber (or operator without forward transfer facility)	
II-8	Data transmission call	
II-9	Subscriber with priority	
II-10	Operator with forward transfer facility	
II-11		
II-12		
II-13	Spare signals for national use	1
II-14		
II-15		

to transit exchanges) not being sent to the terminal international exchange.

(2) Col. (*a*) signals: It may be decided by bilateral agreement that signal I-11, when sent as the first signal, shall serve as a country code indicator instead of signal I-14 to indicate that the first international transit exchange must insert an outgoing half-echo suppressor. If the connection passes through two or more international transit exchanges, signal I-11 is not sent beyond the first transit exchange. Signal I-12 is used solely when no echo suppressor has to be inserted on the international connection. Signal I-14 sent as the first signal, serves as the country code indicator and shows that the connection requires echo suppressors and that the outgoing half-echo suppressor has already been inserted. In response to a signal A-14, the only meaning of signal I-14 is that an incoming half-echo suppressor is necessary.

(3) Col. (*b*) signals: An outgoing international R2 register which receives

Table 10.5 *R2 backward signals group A*

Signal designation	Signal	See note
A-1	Send next digit (n + 1)	1, 2
A-2	Send last but one digit (n − 1)	1, 2
A-3	Address complete, changeover to reception of B signals	3, 4, 9
A-4	Congestion in national network	4
A-5	Send calling party's category	5
A-6	Address complete, charge, set up speech conditions	4, 9
A-7	Send last but two digits (n − 2)	1, 2
A-8	Send last but three digits (n − 3)	1, 2
A-9	Spare for national use	6
A-10	Spare for national use	
A-11	Send country code indicator	
A-12	Send language or discriminating digit	2, 7
A-13	Send location of outgoing R2 international register	2
A-14	Request for information on use of echo suppressor (is an incoming half echo suppressor required?)	8
A-15	International exchange congestion	4

signal A-9 or A-10, the use of which is national, or which receives by signal A-13 a request for identification to which it is unable to reply, indicates that it cannot answer the request by transmitting I-12.

(4) Col. (*a*) signals: The sending of signals I-12 or I-14 may be repeated as often as necessary, on request by signal A-11.

(5) Col. (*b*) signals: I-14 is sent in response to signal A-14.

(6) Col. (*b*) signals: I-15 is also used to indicate (in response to A-13) that transmission of the code identifying the location of the outgoing international R2 register is terminated.

Table 10.4

(1) The outgoing international R2 register which receives one of these signals converts it into a signal II-7, II-8, II-9 or II-10.

Table 10.5

(1) Last received digit $= n$.

(2) Reply expected in the form of a Group I forward signal.

(3) Reply expected in the form of a Group II forward signal.

(4) This signal may be sent:

(i) either as an acknowledgment of any forward signal, or

(ii) automatically in pulse form when there is no forward signal.

(5) This signal, used to acknowledge a Group I forward signal, requests transmission of a Group II signal. It may be followed by any other A signal, but the latter will be linked to the sequence of Group I forward signals already received and will automatically cause the forward signals to revert to their Group I primary meanings.

(6) Reply by signal I-12 (request not accepted) (see Note 3 Table 10.3).

(7) The outgoing international R2 register is informed by the first signal A-12 that an international circuit connecting to a terminal international exchange has been made.

(8) This signal is used at an international exchange (incoming) where it is possible to insert an incoming half-echo suppressor. It is sent to acknowledge the discriminating digit or the language digit and the reply is received:

(*a*) signal I-14 when an incoming half-echo suppressor is required, or

(*b*) next digit of the address information when no incoming half-echo suppressor is required.

(9) When the terminal exchange is unable to send detailed information on the condition of the called subscriber's line, signal A-3 followed by a Group B signal do not apply, and signal A-6 is used.

Table 10.6

(1) Any Group B backward signal acknowledges a Group II forward signal and is always preceded by an A-3 signal which indicates that the incoming register has received all the Group I forward signals it requires from an international R2 register.

(2) Signal B-1 is always interpreted by the outgoing international R2 register as signal B-6.

(3) After recognising B-2 or B-8, the outgoing international R2 register clears forward and causes the transmission of a recorded announcement or an appropriate tone. If the destination national network cannot recognise transferred subscriber, or subscriber's line out of order, A-3 may be followed by B-5 instead of by B-2 or B-8 to ensure that an appropriate tone is sent to the caller.

(4) If the destination national network can only distinguish called line-free or busy, A-3 is followed by B-3 when line-busy and by B-6

Table 10.6   *R2 backward signals group B (condition of called subscriber's line)*

Signal designation	Signal	See note
B-1	Spare for national use	2
B-2	Subscriber transferred	3
B-3	Subscriber line busy	4
B-4	Congestion (encountered after change from group A to group B signals)	5
B-5	Vacant national number	6
B-6	Subscriber line free — charge	4, 7
B-7	Subscriber line free — no charge	7
B-8	Subscriber line out of order	3
B-9	Spare for national use	8
B-10	Spare for national use	
B-11	Spare for international use	
B-12	Spare for international use	
B-13	Spare for international use	
B-14	Spare for international use	
B-15	Spare for international use	

when free (or A-6 only shall be sent without being followed by a Group B signal so that the caller may receive tone or recorded annouce-ment sent by the incoming equipment).

(5) When the congestion condition is encountered following the change-over from Group A to Group B signals, B-4 is sent on the conditions for A-4.

(6) After recognising B-5, the outgoing international R2 register clears the forward connection and causes transmission of an appropriate tone to the caller.

(7) After recognising B-6 or B-7, the outgoing international R2 register sets up speech conditions to enable the caller to receive ring tone.

(8) Signals B-9 and B-10 are always interpreted by the outgoing inter-national R2 register as signal B-5.

10.7.3 *Basic philosophy of the R2 interregister signal code*

(*a*) The R2 interregister signal code is framed to take account of various conditions arising with R2 in the international network and R2 variants in national networks.

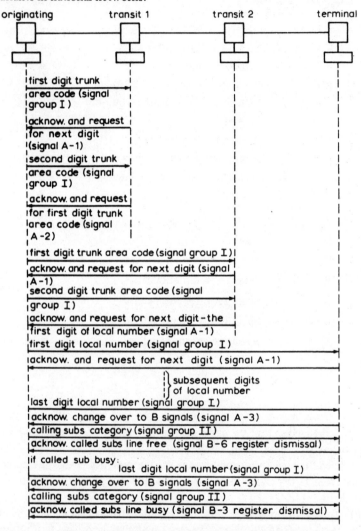

Fig. 10.7   *Typical national network R2 interregister signalling sequence. (assuming two-digit trunk-area code)*

(*b*) All the information necessary for setting up the connection is transmitted by Group I signals. An outgoing R2 register sends the first signal spontaneously immediately after outgoing circuit seizure, this being possible as the signal is continuous. All other signals are sent in reply to appropriate Group A backward signals.

(*c*) All incoming R2 registers (transit and terminal) receive the first signal without request from the register receiving this signal, which indicates transit or terminal. Transit is indicated by the country code indicator I-11, I-12 or I-14 and terminal by the language digit I-1 to I-5 or the discriminating digit I-10. This obviates the need for two different line-seizure signals, transit and terminal. The R2 register is required to determine transit or terminal in the forward routing, and the last acknowledgment signal from a transit register requests the transmission of the first signal to be received by the next register.

(*d*) The country code transit indicator serves to condition the receiving register to analyse for transit routing, normally on the country code, and also conveys relevant echo-suppressor information:

I-12, which may be requested as many times as required by A-11, is used when no echo suppressor has to be inserted.

I-14 is used when a half-echo suppressor is required as an outgoing half-echo suppressor has already been inserted. When sent in reply to A-14, I-14 has as its sole meaning 'incoming half-echo suppressor required'.

I-11 indicates to the first transit exchange that an outgoing half-echo suppressor should be inserted and that no half-echo suppressor has been inserted at the outgoing exchange. I-11 is not sent beyond the first transit exchange and the outgoing exchange, having sent I-11 once, sends I-14 if it is again asked by A-11 for the transit indication.

(*e*) Signal I-15, end-of-pulsing, is used to indicate the end of the identification procedure (see (*l*)), and in the international semi-automatic service that there are no more digit signals to follow.

(*f*) Information concerning the nature of the call can be transmitted on the network by the Group II forward signals, and sent in reply to A-3 or A-5. The signals cater for both international and national calls. II-1 to II-6 are assigned for national use as follow:

II-1 subscriber-initiated call (national)
II-2 priority call (national)
II-3 maintenance-equipment call (national)
II-4 spare (national)
II-5 operator-initiated call (national)
II-6 data-transmission call (national)

When an international outgoing R2 register receives a Group II signal with national meaning from the national network, the signal is converted into an internationally accepted signal as follows: II-1 to II-4 and the national spares II-11 to II-15 into II-7, II-5 into II-7 or II-10, and II-6 into II-8.

(g) The potential for error-detection and correction by retransmission exists. If the time allowed for a signal to settle down to 2-and-2 only is relatively long, it would not normally be necessary to request a repeat of a digit n signal as the signal would remain online until acknowledged. A short waiting time would strengthen the need for error-correction and if adopted is achieved by returning A-2 requesting digit $n - 1$ and acknowledging this by A-1 requesting digit $n + 1$, to receive digit n. This procedure is not applicable when the digit n is the first in the store of the outgoing register.

(h) The concept of requesting send last but so many digits by various backward A signals facilitates the successive transmissions of the country code indicator and the country code to successive transit registers in a flexible manner. The same flexibility applies in national application, the transit registers receiving the trunk area code.

(j) The condition of the incoming-exchange switching equipment or of the called line can be transmitted back on the network by Group B signals, the outgoing register then taking appropriate action. These signals are sent to acknowledge receipt of Group II forward signals after changeover to B signals has been requested by A-3, which signal must always precede the B signals.

(k) In some circumstances the signal A-3 changeover to B signals may be sent before the connection is made with the called line. Congestion in selection stages can therefore still occur after changeover from A to B signals. A similar situation may arise when interworking system R2 with other signalling systems. B signals may then be sent from a centre of a higher category in the network hierarchy than the exchange where the called line is connected, so that congestion in a group of circuits can be encountered after changeover to B signals. Since A-4 is no longer applicable in these cases, B-4 is sent.

(l) Certain identification procedures can be transmitted over the network. An international transit register, or a register in the destination country, can request, by A-13, the location of an international R2 register as soon as at least one forward signal has been received from that register. The international outgoing register replies with the first digit of its own country code. A further digit of the country code is sent in reply to each subsequent request by A-13. Further requests by A-13 may elicit successive digits of the trunk code of the exchange where the international R2 register is situated. When all the digits required to indicate the location have been sent, the next A-13 signal is acknowledged by the end of pulsing signal I-15.

The calling subscriber's number may be transmitted over the connection, for example by repeating A-5, or by the use of A-9 or A-10. This procedure is limited to national networks, and international registers prevent its operation on international circuits by replying with I-12 (request not accepted).

10.7.4 Operational features
Various features of system R2 are made flexible so as to cater for different conditions arising in different applications, typically:

Determination of number complete: Any one of the following criteria may be used to determine whether or not the address information received by an incoming register is complete, depending on the following conditions:

(i) Analysis of the number received. This is applicable when the connection set-up extends to the exchange at which the called line terminates.

If the incoming register is equipped to determine the condition of the called line, signal A-3 is returned on receipt of the last digit. A Group B signal is then sent indicating the condition of the called line. If the incoming register is not equipped to determine the condition of the called line, signal A-6 is returned and no Group B signal follows.

(ii) Criteria given by the switching equipment subsequent to the m.f. register.

To avoid delay in sending the answer signal, no B signal is sent when the called line is free. A-6 is sent to set up speech conditions.

(iii) When given, receipt of I-15, the end of pulsing signal.

I-15 is always sent in the international semiautomatic service, and is also sent in certain procedures such as identification (Section 10.7.3 note (*l*)). It is not normally sent in the transfer of the called party's address information in the automatic service. I-15 is acknowledged by A-1, A-3, A-4, A-6 or A-15.

(iv) The assumption, after a time delay, typically 5 ± 1 s, that no further address digits will be received.

A-6 is returned as a pulse signal when this condition applies.

Termination of m.f. signalling when a connection cannot be completed: A register ceases m.f. signalling immediately any condition preventing call set-up is recognised, signal A-4, A-15 or an appropriate B signal being returned.

Transmission of pulse signals: Particular conditions can arise when the fully-compelled sequence cannot be used, it then being necessary to have the option in the system of sending a backward pulse signal without prior receipt of a forward signal, e.g.

(*a*) when an incoming register, after acknowledging a forward signal, is unable to complete the call (e.g. congestion) and the next forward signal does not appear

(*b*) when the address complete signal A-6 is sent, the last forward signal having already been acknowledged.

The problem concerns signals A-3, A-4, A-6 and A-15, it being required that these be pulse (150 ± 40 ms) in those circumstances not allowing the fully-compelled sequence. No forward signal is sent by the outgoing register on receipt of A-4, A-6 or A-15, the pulse signals releasing the registers. On receipt of A-3 pulse, the outgoing register sends a Group II signal forward, the incoming register acknowledging this in the normal compelled manner by sending a B signal.

Setting up of speech conditions: The last backward signal releases the outgoing and incoming registers to set up speech conditions by switching the speech path through. This register-dismissal signal will vary depending upon the conditions of a particular network and is normally A-6 or a B signal.

Release of transit registers: For international R2 transit exchanges, it is specified that the last forward signal received by a transit register be acknowledged by a backward signal inviting a specific signal which is the first forward signal to the succeeding register. The following backward signals are used:

(i) A-11 if the next exchange is international transit, which causes I-12 or I-14 (country code indicators) to be sent by the outgoing register, this being the first signal to the next transit register.

(ii) A-12 if the next exchange is international terminal, which causes the language digit (semiautomatic) or the discriminating digit (automatic) to be sent by the outgoing register, this being the first signal to the international terminal register.

The basic R2 concept permits the termination of m.f. signalling and the setting up of speech conditions at transit exchanges in any desired way to suit the requirements of particular networks, but the adoption of the international arrangements for national networks would give uniformity and ease of interworking with international R2.

Abnormal release of registers: National application-register time-outs would be a matter for individual administrations, but the recommended international time-outs are specified as follows:

An outgoing international R2 register times-out:

(a) 15 ± 3 s during the sending of forward m.f. signals, which time-out delay is a function of the time required in the extreme for the switching procedures in a transit exchange.

(b) Not less than 24 s during intervals when no forward signalling is sent, which time delay is a function of the maximum interval between the dialling of two successive digits and the time-out delay of incoming registers.

On time-out, appropriate tone is returned to the caller and the register released.

An incoming international R2 register, transit or terminal, times out in 8–24 s, this time delay being a function of:

(i)   the maximum intervals between the dialling of two successive digits
(ii)   the maximum time required to set up a connection
(iii) the incoming register being required to be released before the expiry of the outgoing register time-out
(iv)   the interval between register seizure and the receipt of the first forward signal
(v)   the interval between two successive signals in the forward direction.

On time-out:

(a) The congestion signal is sent in pulse form, which prompts release of the international connection.
(b) The incoming register and other equipment in the incoming exchange is released.
(c) The incoming circuit is blocked until the clear-forward (release) signal is received.

Interworking between international system R2 and national systems derived from it: National systems derived from system R2 may, or may not, have the full 2/6 m.f. capacity in either direction, but more usually would in the forward direction. International R2, having 2/6 m.f. in each direction, readily interworks with national R2 systems having the same capacity. The system can also be adapted for interworking with national R2 with less than 2/6 m.f. in the backward (and in the forward if this ever arose) direction.

On a routing outgoing national network – international network –

incoming national network, it is logical for signal transmission and other reasons that the overall multilink route route be divided into sections, each being end-to-end signalling in its own right. In this event:

(i) The outgoing register in a given end-to-end section must be able to recognise at least all the backward signals in that section. Every incoming register in that section must be able to recognised at least all the forward signals used on that section and directed to that register.
(ii) When the number of signals provided is not the same on all parts of the route, the division is logically made at a connection point between links having different numbers of signals, and thus at the interface international gateway exchanges in the originating and destination countries.

The R2 system has a considerable end-to-end signalling capability, allowing the theoretical possibility of extending the R2 end-to-end signalling into the destination national network to an extent depending upon the national network transmission characteristic. Assuming international circuits of nominal loss 0·5 dB and deviation 1 dB, and a four-link international routing, the nominal transmission loss between the incoming international gateway exchange and any R2 register in the destination country should not exceed 11·4 dB for a country using three 4-wire switched links at most, and 11 dB for four. For the general case, however, it is preferred that the end-to-end signalling be divided at the incoming, as well as at the outgoing, gateway exchange.

10.7.5 R2 interregister signalling on satellite circuits

With the fully-compelled mode, the speed of address-information transfer is slow on long-propagation circuits such as satellite, which increases the postdialling delay and the CCITT is at present considering the adoption of a semicompelled mode as an option to be applied on such circuits. In the approach, the forward signals are continuous and the backward acknowledgment pulse (75 ms or 150 ms, yet to be decided) instead of continuous (Section 10.2*c* and Fig. 10.2*c*), approximately two propagation and two signal recognition times being involved per signal sequence. This approach does not involve major changes to the basic fully-compelled system and is at present being considered for this reason. Any weakness the backward pulse signals may have compared with continuous is judged to be acceptable in the interest of improving the speed of information transfer. The detail is not yet finalised, but it is clearly desirable that register-holding on the satellite links should be such that networks, including cases where the R2 fully-compelled sequence is used, after a satellite link be buffered against the satellite link connection set-up delays.

Long propagation times will also have reaction on the R2 line signalling system (Section 9.5.2). The T1 and T2 times of the clear forward/release guard sequence, and the double seizure detection time on bothway working, which times are propagation-time-dependent, would need to take account of the longer propagation time of satellite circuits. The increase in these times is yet to be decided.

Relevant data

Transmit:
Absolute power level of each signal frequency $-11 \cdot 5$ dBm ± 1 dB

Signal frequency tolerance ± 4 Hz

Time interval between start of sending of each of the two frequencies not to exceed 1 ms

Interval between cessation of each frequency not to exceed 1 ms

Receive:
Receiver response range -5 dBm to -35 dBm

Difference in level between the two frequencies of a signal not greater than 5 dB for adjacent frequencies and 7 dB for nonadjacent. In application this allows a 4 dB attenuation distortion of the end-to-end chain for two adjacent frequencies and a 6 dB distortion for two nonadjacent, provided the level of the weaker signal frequency is not lower than -35 dBm at the receiver input.

Receiver not to recognise a signal of two frequencies of level -5 dBm and a duration less than 7 ms

Receiver not to recognise a signal of two frequencies having a difference in level of 20 dB or more

Received signal frequency variation ± 10 Hz

The sum of the operate and release times to a two frequency signal not to exceed 80 ms (the operate and release times are not specified separately)

Receiver not to release to interruptions to signal 7 ms or less

System malfunction to signal interruptions greater than 7 ms (typically 20 ms at administration's choice) is prevented by further logic elements

10.7.6 *General comment on R2 interregister signalling*

The R2 interregister signalling system is postulated as a flexible general-purpose system for a wide field of application, catering for the different conditions likely to arise in different networks. Individual administrations would adopt features to meet the requirements of their own networks, but within the basic principles of the R2 system. This could well mean that application of R2 in a particular network may not give optimum interregister signalling for the conditions of that network, e.g. the numbering scheme may be such that group acknowledgment as distinct from per-signal acknowledgment, would meet the requirements. This, however, must be balanced against the potential of system R2 for rationalised interregister signalling, which is clearly very desirable, and in this regard system R2 has significant merit.

10.8 UK MF2 interregister signalling system

10.8.1 General

This system is used on the UK hierarchical trunk-transit network, with worst routing of five links. Register-translator controlled 2-wire step-by-step switching (2-motion type switches) may apply at some existing early type first-level trunk exchanges (Group Switching Centre (GSC)). The higher level exchanges (District Switching Centre (DSC), Main Switching Centre (MSC)) are 4-wire switched common control dedicated trunk-transit switching units. As the m.f. signalling path at the originating GSC may sometimes include 2-motion-type switch flexible wiper-to-bank contacting, problems arise as a result of intermittent disconnections of the contacts, and transients which interfere with m.f. signalling. Further, in m.f. signalling, transients may arise on the signalling path due to the switching equipment, particularly when the registers are initially line-associated. The severe interference conditions in this particular application, combined with the dedicated trunk-transit registers always receiving a fixed number of digits, thus admitting the possibility of a preference for group as distinct from per-signal acknowledgment to improve the postdialling delay, had considerable influence on the choice of m.f. signalling method. The system was produced prior to the availability of the R2 system, although the fully compelled mode was considered as a possibility, but not adopted for reasons of safeguard, and speed of information transfer, in the severe interference conditions. In the result the MF2 system was designed for the particular conditions applying and pulse signalling adopted in combination with measures to prevent simulation of signals and to minimise errors arising from mutilation of genuine signals. These measures are based on a guard tone, a particular 2/6 m.f. combination in each direction, and functioning as a prefix.[13,14]

The UK MF2 system is guarded pulse, end-to-end signalling, 2/6 m.f.

forward and backward signalling, overlap operation at all registers, with signal frequencies (same as system R2) 120 Hz spaced:

Forward: 1380, 1500, 1620, 1740, 1860, 1980 Hz
Backward: 1140, 1020, 900, 780, 660, 540 Hz

The UK national numbering scheme is nonuniform, and the first one, two or three digits of the national significant number may determine the trunk call routing, depending upon the call destination. End-to-end signalling applies (Fig. 10.4) and it is arranged that the trunk transit registers always receive the first three digits (A, B, C) of the national significant number in response to a backward transit proceed-to-send (send early digits) signal, which simplifies these dedicated registers. The local area numbering schemes are variable, the terminal registers receiving a varying number of digits to connect the call, starting with the second (B) or the third (C) digit of the national significant number depending upon the destination local area, these being requested one by one by the terminal to the originating register by means of repeated terminal proceed-to-send (send late digits) signals. All the initial three digits A, B and C could of course be sent to the terminal trunk register should this ever be a future requirement.

A form of group acknowledgment applies in the pulse signalling adopted, as a proceed-to-send signal from a transit or a terminal register is an effective acknowledgment for the receipt of the group of digits A, B and C by the immediately preceding trunk transit register. Certain exchanges, which may be either GSCs or local exchanges dependent on a GSC are limited in the facilities provided, and can be involved on outgoing and incoming calls requiring the use of MF2 at the GSC. Such exchanges may not be able to determine the condition of the called line, in which case the signalling sequence is terminated by sending the backward number received (register dismissal) signal which also indicates limited facilities to the originating register, and would not necessarily imply 'address complete'. When the condition of the called line can be determined, the called line status signal (and not the number received) is returned, which signal performs register dismissal and implies address complete. Also, limited facility exchanges may not be able to make use of class of service information, in which case this is not requested in the backward signalling.

10.8.2 Guarded pulse m.f. signalling

Various safeguarding features to protect against interference are known, typically the KP signal in the Bell system to condition the receiver for m.f. reception, continuous compelled signalling to allow time for a signal to settle down to the required 2-and-2 only persistence. It was considered, however, that such features were not sufficiently powerful in the interference conditions applying and a guarded pulse technique was developed to ensure reliability in the particular conditions of interference from contacts in the m.f. signalling path (Fig. 10.8). A particular 2/6 m.f. combination is used as a guard tone, one in each direction, the pulse information signal being transmitted without gap of signal energy (Fig. 10.2e).

In theory the guard tone would be continuous in each direction for as long as the registers are line-associated and interrupted to pass the pulse signal. In this concept, false interrupt disconnections would have no signal meaning, which would contribute to signalling reliability and simplicity. This implementation, however, would overload line transmission systems at the signal level desired (Section 7.8) and the practical realisation varies the theoretical concept, the guard being a mixture of continuous and pulse prefixes, with the liability of false interruptions having signal meaning and thus appropriate safeguards necessary.

In the design, when any one frequency is detected by the m.f. signal receiver, all other signal channels are biased so that with a level difference between frequencies in excess of 12 dB, the lower level frequency is not detected. This satisfies the requirement that a signal frequency should be rejected if its level is 15 dB or more below that of another simultaneously received frequency, and provides improved immunity against error conditions arising from transients occurring during the transmission of signals.

All signal channels are open when no energy is transferred, and when a guard tone is received and recognised (20 ms 2-and-2 only) all other signal channels are inhibited for as long as the guard tone is received. If during the recognition time, any other (interfering) channel has an output of sufficient magnitude meeting the above threshold condition, the recognition timing is cancelled, to start again on reappearance of the 2-and-2 only guard. Thus interferring frequencies do not result in error conditions during the guard tone period. An incoming register returns continuous guard tone when line associated, and after recognition, the outgoing register applies continuous guard tone as acknowledgment. A 20 ms 2-and-2 only recognition of this at the incoming end indicates that association of registers, and other relevant equipment, has been completed, and that the transmission path in both directions is satisfactory for the transmission of signal pulses. Thus switching equipment transients, etc., would be terminated before effective signalling begins. This continuous guard prefix is compelled. Recognition of the backward continuous prefix prompts the sending

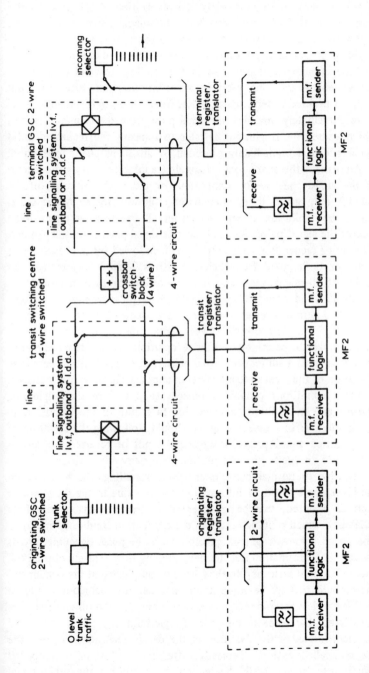

Fig. 10.8   *UK MF2 trunk-transit network signalling and switching arrangements*

of the forward continuous prefix and recognition of this prompts the cessation of the backward prefix and the immediate transmission of the backward pulse signal (proceed-to-send in the above case). Recognition of this prompts the cessation of the required forward continuous prefix and the immediate transmission of the required forward signal pulse. Subsequent guard tone prefixes are continuous or, after the first terminal proceed-to-send signal, pulse. When pulse, an 80 ms guard tone pulse immediately precedes the signal pulse from the terminal register, which causes the originating register to transmit the requested pulse signal which is immediately preceded by an 80 ms guard tone pulse.

All, except the guard tone, channels are inhibited during the receipt of the guard tone prefix, continuous or pulse. On cessation of the 2-and-2 only guard tone, immediately followed by an 80 ms signal pulse, the inhibition is removed from all the receiver channels to receive the signal pulse. Recognition of a signal pulse is thus dependent upon prior recognition of a prefix guard, followed by the cessation of either or both guard frequencies. A 20 ms 2-and-2 only recognition applies to the signal pulse and error conditions brought in should this not be achieved during the pulse. After signal pulse reception, the logic is reset to permit receiver response to a subsequent guard prefix only. The use of the guard prefix between signal pulses permits the receiver output stages to be inhibited after recognition of a signal pulse and to remain so until a subsequent signal pulse is received. This prevents false operation of the receiver to interference appearing between pulse signals.

Error conditions prompt a repeat attempt from the originating register and in addition to error due to failure of the 2-and-2 only recognition, error conditions apply should, following the cessation of a recognised guard prefix, a signal pulse not be received and stored within 135 ms due to line interruption, or for any other reason.

Relative to normal pulse interregister signalling, the MF2 system sacrifices some speed of information transfer due to the safeguarding features adopted, this being accepted in the interests of signalling reliability, which is of high order in the application conditions applying. The group acknowledgment, possible with the pulse signalling, avoids the speed of information transfer being unduly slow. The guard prefixes, and consequent logic, add to the cost and complexity of the system and there could perhaps be some doubt whether the technique would be justified in less interference prone conditions. On the other hand, it is considered that some of the type of logic built into the system could be incorporated with advantage in the design realisation of, say, the R2 compelled system. Typically, after recognition of a compelled signal, interference could be ignored by inhibiting the other signal channels for as long as the two frequencies of the recognised signal are still being received. Otherwise, interference after signal recognition could result in false signal cessation and false signal meaning.

10.8.3 MF2 interregister signal code
This is detailed in Tables 10.7–10.9.

Table 10.7 *MF2 m.f. allocation*

Signal number	Forward signals frequency combination compounded	Backward signals frequency combination compounded
	Hz	Hz
1	1380 + 1500	1140 + 1020
2	1380 + 1620	1140 + 900
3	1500 + 1620	1020 + 900
4	1380 + 1740	1140 + 780
5	1500 + 1740	1020 + 780
6	1620 + 1740	900 + 780
7	1380 + 1860	1140 + 660
8	1500 + 1860	1020 + 660
9	1620 + 1860	900 + 660
10	1740 + 1860	780 + 660
11	1380 + 1980	1140 + 540
12	1500 + 1980	1020 + 540
13	1620 + 1980	900 + 540
14	1740 + 1980	780 + 540
15	1860 + 1980	660 + 540

Notes

Table 10.8
(1) The limited class of service applies when some existing exchange equipments are not capable of dealing with the full facility repertoire.
(2) A send class of service signal in the backward direction causes one class of service signal to be sent forward. Some forward signal numbers have both primary and secondary meanings and to allow for this, the send class of service signal also functions as a shift signal.

Table 10.9
(1) Initially 2/5 m.f. backward signalling applies (with 2/6 m.f. forward). The system design allows equipment for the sixth backward frequency 540 Hz to be added at a future time.

Table 10.8   *MF2 forward signals*

Signal number	Signal
14	Guard prefix
1	Digit 1
2	2
3	3
4	4
5	5
6	6
7	7
8	8
9	9
10	0
	Class of service
2	General ⎫ when facilities
8	Coin box ⎬ limited by existing
3	Operator ⎭ equipment
1	General ⎫
4	Coin box ⎪
6	Operator ⎬ full
5	Trunk offering ⎪ facilities
7	Reverted call ⎪
9	Barred trunk ⎪
10	Barred international ⎭
11	Spare
12	Spare
13	Spare
15	Spare

(2) Backward signals in the call signalling sequence transmitted prior to and including the initial terminal proceed-to-send signal 4 are each preceded by a continuous guard prefix and a continuous guard acknowledgment received. Backward signals transmitted after the initial terminal proceed-to-send are each preceded by a pulse prefix guard. This procedure is necessary since, prior to the initial terminal proceed-to-send signal, the originating register is unable to determine the source of the signal and must therefore be prepared to treat all received guard prefix signals as continuous and requiring a continuous guard acknowledgment. On receipt of the initial terminal proceed-to-send signal, the

Table 10.9 *MF2 backward signals*

Signal number	Signal
9	Guard prefix
3	Transit proceed-to-send (send early digits)
1	Send class of service
4	Initial terminal proceed-to-send (send late digits)
4	Subsequent terminal proceed-to-send (send late digits)
5	Congestion
7	Called line free (ordinary)
10	Called line free (coin box)
8	Called line busy
2	Spare code (called line unobtainable)
6	Number received (limited facilities)
6	Early repeat attempt
3	Late repeat attempt
11	Spare
12	Spare
13	Spare
14	Spare
15	Spare

originating register is programmed to receive and send pulse guard prefixes thereafter.

(3) A pulse guarded backward signal requiring a response causes the originating register to transmit a pulse guard prefixed signal. When no response is required to the backward signal (e.g. called line free) the originating register does not return a guard prefix.

(4) Signal numbers 3 and 6 have both primary and secondary meanings, the shift being dependent upon the signal sequence, e.g. signal 3, the transit proceed-to-send being transmitted from a transit register and the late repeat attempt from the terminal register.

(5) When sent, the send class of service signal is sent back from a terminal register. Should the terminal register not be able to make use of class of service, the backward send class of service signal is not sent and the signalling sequence from the terminal register commences with the initial terminal proceed-to-send signal.

(6) When the condition of the called line can be determined, the registers are released and the speech path switched through by the called subscriber's status signal free or busy (signals 7, 10 or 8) or the spare code (signal 2) from the terminal register, the number

received (signal 6) not being sent. Should the terminal register not be able to identify called subscriber status or spare code, these signals are not sent and the registers released by the number received signal, which also indicates limited facilities to the originating register.

10.8.4 Outline of MF2 operation

Signalling commences with the sending of a continuous guard backward prefix from the incoming transit register when this is associated due to the line signalling seizure signal from the originating exchange. The originating register replies by sending a continuous guard forward prefix as acknowledgment. Recognition of this at the incoming end indicates that the signalling path is proved and the backward guard prefix is replaced without gap by the transit proceed-to-send (send early digits) signal pulse. On detecting the cessation of the backward guard prefix, the m.f. receiver is conditioned to respond to the information pulse signal and on recognition of this, the originating register responds by terminating the forward guard prefix and transmitting a sequence of signals consisting of 80 ms pulses of digit signals (A, B and C digits) interspaced with 80 ms pulses of forward guard prefix (Fig. 10.9). These early digits determine the switching path to be followed to provide access to the subsequent link. On congestion at the transit exchange, an 80 ms pulse congestion signal is returned to the originating register after an exchange of continuous guard prefixes. Clear down of relevant equipment at the transit and originating exchanges follows on congestion and the originating register initiates a second attempt to complete the call. Error conditions at the transit exchange prompt the return of the early repeat attempt signal after an exchange of continuous guard prefixes.

The same sequence occurs at subsequent transit exchanges, a transit proceed-to-send signal being an effective acknowledgment that the early digits requested by the previous transit proceed-to-send signal were received.

On association of the terminal register, the initial signals (those transmitted prior to the second terminal proceed-to-send signal) cause an interchange of continuous guard prefixes over the connection, the initial information signals being typically 'send class of service' followed by the first 'terminal proceed-to-send'. The originating register responds to the send class of service signal by sending the appropriate class of service signal forward. Should the terminal register be unable to deal with class of service of the caller, the request signal is not sent and the first backward signal will be the first terminal proceed-to-send signal.

On recognition of the first terminal proceed-to-send signal, the originating register terminates the forward continuous guard prefix and transmits the first late digit 80 ms pulse signal. The originating register transmits further late pulse digit signals one by one, each preceded by a forward guard pulse prefix (80 ms) in response to the second and subsequent terminal proceed-to-send signals. A terminal proceed-to-send signal is an effective acknowledgment that the digit signal requested by the immediately previous terminal proceed-to-send signal was received and recognised.

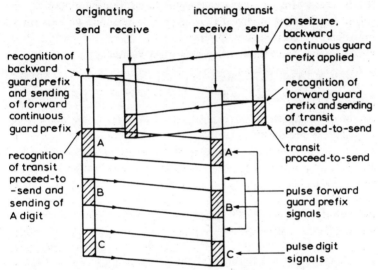

Fig. 10.9   *UK MF2 sequence in response to transit proceed-to-send*

Final stage signals (called line free, called line busy, number received, congestion, spare code), the pattern varying according to the type of terminal exchange, are pulse guard prefixed and do not provoke a response signal from the originating register. The final backward signal will be one of the signals, 2, 3, 5, 6, 7, 8 or 10 to release the registers. The number received backward signal 6 is sent to release the registers when, owing to limited facilities at the terminal exchange, no other signal is available for the purpose. The congestion signal 5 prompts a second attempt from the originating register, this being the limit.

Error conditions
An error is detected at the originating exchange, and busy tone returned, under the following conditions:

(*a*) Failure to receive a backward information signal within 4·2 ± 0·2 s of the association of the m.f. equipment with the outgoing line.

(*b*) If, prior to the recognition of the first terminal proceed-to-send signal, a backward information signal is not recognised within $4 \cdot 2 \pm 0 \cdot 2$ s of the recognition of the previous information signal.

(*c*) If, subsequent to recognition of the first terminal proceed-to-send signal a backward information signal is not recognised within $2 \cdot 1 \pm 0 \cdot 1$ s of the recognition of the previous information signal. When a backward signal requests digit information, timing of the above period is delayed until the digit information is available for transmission.

(*d*) If the 2-and-2 only persistence signal recognition is not met.

(*e*) If no information signal is recognised within 135 ± 25 ms of the cessation of a received guard prefix

(*f*) If a signal is not allocated an identity is received.

An error is detected at an incoming exchange, transit or terminal, and a repeat attempt signal returned, under the following conditions:

(i)   if the 2-and-2 only persistence check for information signal recognition is not met

(ii) if no information (including digit) signal is recognised within 135 ± 25 ms of the cessation of a received guard prefix

(iii) if a signal not allocated an identity is received.

Errors detected at a transit exchange prompt the return of the early repeat attempt signal. Errors detected at the terminal exchange prompt the return of the early repeat attempt (continuous guard prefix) or the late repeat attempt (pulse guard prefix) depending on whether the error arose before or after the first terminal proceed-to-send signal.

Receiver arrangements

Design detail of m.f. receivers varies depending upon administration approaches and the conditions to be met. It is of interest, however, to describe the basic arrangements of the UK MF2 receiver, it being understood that this would be typical and for indication only. Further logic elements are required for the m.f. sending, but these are not described in the interest of presenting principles rather than the design detail of a complete system.

Fig. 10.10 shows the arrangements for receiving and detecting the m.f. signals in the 1380–1980 Hz band, similar arrangements applying for the lower band. The directional filter prevents backward signals in the 540–1140 Hz band, returned via the 4-wire/2-wire line termination at the 2-wire switched originating exchange, from entering the signal channels. Each signal channel filter is tuned to the appropriate signal frequency with pass band ± 20 Hz. The output of each filter drives a

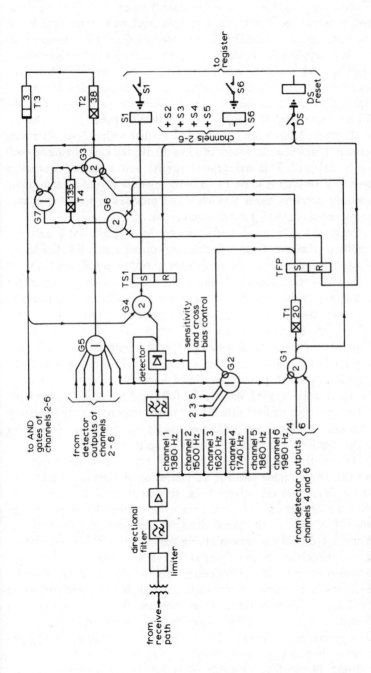

Fig. 10.10   *UK MF2 system receiver and logic*

detector stage, which gives a d.c. output when the a.c. input signal level exceeds a predetermined threshold level value controlled, in the first instance by a sensitivity control connected to each channel detector. As the applied signal builds up, the diode network crossbias control, coupling each channel sensitivity circuit, becomes operative. The sensitivity of each signal detection stage is reduced and becomes a function of the crossbias voltage developed in the channel with the highest power level at its input. The crossbias is applied by this channel to all other channels, and is arranged so that with a level difference between frequencies in excess of some 12 dB, the lower frequency will not be detected. This arrangement ensures that a signal frequency is rejected if its level is some 15 dB or more below that of another simultaneously received signal frequency and provides improved immunity against transients during signal transmission.

On receipt of a guard prefix, the corresponding 2-and-2 only d.c. condition (channels 4 and 6 in this case) operates gate G1, the output of which primes the prefix persistence check timing element T1 and inhibits gate G3. Should, during the 2-and-2 only persistence check period, any other channel gives an output of sufficient magnitude to overcome the threshold of the channel detector, this other channel will give an output from gate G2 to inhibit gate G1. The prefix condition is then removed from timer T1, which restores. After a 2-and-2 only prefix recognition of 20 ms, output from T1 sets toggle TFP, the operation of which primes gate G3 in preparation for the response to the signal pulse applied via gate G5 following cessation of the prefix. TFP output also inhibits gate G2 to prevent response to any frequency other than the two guard prefix frequencies and in this way, interfering frequencies have no affect after prefix recognition.

On cessation of the prefix, the inhibition on gate G3 is removed and the 38 ms timing element T2 primed on appearance of a channel output from gate G5. After 38 ms, an output from T2 operates the pulse timing element T3, producing a 3 ms pulse. T3 output primes gate G4 of each of the signal channels, permitting any output from channels 1–6 which is present during the operation of T3 to be stored on the associated channel toggles TS1–TS6, respectively. The continuous outputs of TS1–TS6 operate corresponding reed relays S1–S6, the contacts of which connect earth d.c. signals to the register, where the 2-and-2 only persistence check, nominal 20 ms, and the detection of error conditions, for the signal information pulses, is performed. After processing the signal pulses, the register resets the receiver toggles to permit response to a subsequent incoming guard prefix.

Timer T4 provides for a 135 ms period for the acceptance of any signal condition following a cessation of the prefix signal, and protects against interruptions due to line breaks or for any other reason. In the event of a signal pulse not being stored within the 135 ms, T4 inhibits gate G3 of each channel detector stage, which inhibition remains operative until reset by the register.

Relevant data

Transmit:

signal level − 8 dBm0 ± 1 dB per frequency

duration pulse information signal 80 ± 15 ms

duration pulse guard prefix 80 ± 15 ms

signal frequency tolerance ± 0·3% of nominal

Receive:

response range of originating and terminal receivers − 8 dBm to − 22 dBm per frequency

response range of transit receiver − 4 dBm to − 22 dBm per frequency

maximum level difference 3 dB between adjacent channel frequencies and 8 dB between nonadjacent

maximum time displacement between the two frequencies of a signal 5 ms

bandwidth channel filters ± 20 Hz

10.9 French Socotel interregister m.f. signalling system

This is a continuous fully compelled system having some resemblance to system R2, but the difference is such that it cannot reasonably be regarded as a variant of R2, it is a different type system.[15] Both the forward and backward information signalling use the code combinations in the basic 2/6 m.f. mode, the same frequencies being used in the two directions, the acknowledgment in the two directions being a single frequency carrying no signalling information. The basic objective of the system is to reduce the total number of signal frequencies relative to R2, the reduction permitting a wider frequency spacing of 200 Hz relative to the R2 120 Hz. The system is suitable for either 4-wire or 2-wire operation without sacrifice of information transfer capability, the facility potential being the same as that of system R2. Changeover operation controlled by the registers enables the same signal frequencies to be used in the two directions.

Economic merit is claimed relative to R2 in that less equipment is

Table 10.10   *Socotel system French application signal code*

Signal	Code a Access code		Code b Address information	
Hz				
Forward				
700 + 900	a1	Regional call	b1	Digit 1
700 + 1100	a2	Spare	b2	2
900 + 1100	a3	National call	b3	3
700 + 1300	a4	Spare	b4	4
900 + 1300	a5	Two digit call (special services)	b5	5
1100 + 1300	a6	Spare	b6	6
700 + 1500	a7	Spare	b7	7
900 + 1500	a8	Spare	b8	8
1100 + 1500	a9	International call (semi-automatic)	b9	9
1300 + 1500	a0	International call (automatic)	b0	0
Backward	Code A		Code B	
700 + 900	A1	Request access code and first four digits	B1	Called sub free charge
700 + 1100	A2	Requests last digits	B2	Called sub free no charge
900 + 1100	A3	Change over to B signals	B3	Called sub busy
700 + 1300	A4	Spare	B4	End of selection
900 + 1300	A5	Spare	B5	Spare
1100 + 1300	A6	Normal transit	B6	Spare
700 + 1500	A7	Spare	B7	Spare
900 + 1500	A8	Spare	B8	Spare
1100 + 1500	A9	Spare	B9	Spare
1300 + 1500	A0	Congestion	B0	Absent sub

required due to the fewer frequencies, full potential 6 + 1 relative to the 6 + 6 of R2, and the wider frequency spacing easing the signal channel filtration. The system is end-to-end signalling with overlap operation at the registers, and has the potential for three groups (A, B, C), each 2/6 m.f., in both the forward and backward directions, a shift technique applying. The full potential frequencies are 700, 900, 1100, 1300, 1500, 1700 Hz for the 2/6 m.f. signal information plus the acknowledgment frequency 1900 Hz. Like system R2, the Socotel system concept is a flexible system for general interregister signalling application, particular networks adopting arrangements, in particular the number of signal information group(s) and size of group, as required.

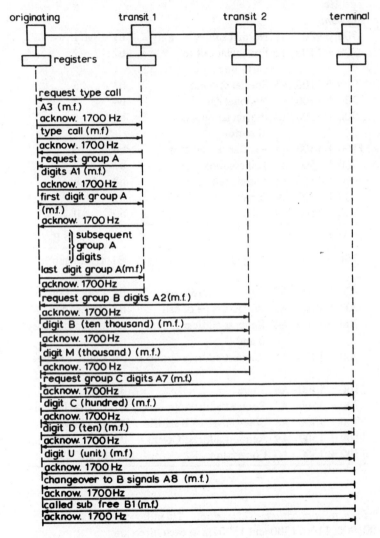

Fig. 10.11 *Typical signal sequence Socotel system (Spain)*

As far as is known, the Socotel system application is at present limited to the French and Spanish national networks.

Table 10.11   *Socotel system Spanish application signal code*

Signal		Code a Type of call		Code b Address information
Hz				
Forward				
700 + 900	a1	Normal sub	b1	Digit 1
700 + 1100	a2	Provincial call to party line	b2	2
900 + 1100	a3	Special services	b3	3
700 + 1300	a4	National call	b4	4
900 + 1300	a5	Provincial outside of sector	b5	5
1100 + 1300	a6	Assistance operator	b6	6
700 + 1500	a7	Toll coinbox	b7	7
900 + 1500	a8	Absent sub	b8	8
1100 + 1500	a9	Spare	b9	9
1300 + 1500	a0	Spare	b0	0

Backward		Code A		Code B
700 + 900	A1	Request address group A	B1	Called sub free charge
700 + 1100	A2	Request address group B	B2	Congestion
900 + 1100	A3	Request type of call	B3	Absent sub
700 + 1300	A4	Request address groups B and C	B4	Called sub busy
900 + 1300	A5	Request address Group D	B5	Called sub free no charge
1100 + 1300	A6	Spare	B6	Spare
700 + 1500	A7	Request address group C	B7	Changed number
900 + 1500	A8	Changeover to B signals	B8	Dead level
1100 + 1500	A9	Request address Group E	B9	End of selection
1300 + 1500	A0	Congestion	B0	Spare

French application: This adopts 2/5 m.f. with signal frequencies 700, 900, 1100, 1300 and 1500 Hz in each direction.

Acknowledgment 1900 Hz in each direction
Transmitted signal level − 8·7 ± 1 dBm0 per frequency
Receiving range − 4·3 dBm to − 34 dBm
Table 10.10 shows the signal code.

Spanish application: Signal frequencies (2/5 m.f.) 700, 900, 1100, 1300, 1500 Hz in each direction.

Acknowledgment frequency in each direction 1700 Hz (note that the French acknowledgment is 1900 Hz).

The Spanish national significant number is of the type XY AB MCDU and address digits are grouped as follows for the transfer over the interregister signalling system:

Group A   XYABM
     B   BM
     C   CDU
    BC   BMCDU
     D   PXYAB (P is a digit added by the switching equipment on national calls and having a type of call meaning)
     E   MCDU

Groups D and E are only used between secondary toll exchanges. For local (or provincial) calls, Group A is YABM (7 digit areas) or ABM (6 digit areas).

Table 10.11 shows the signal code and Fig. 10.11 a typical signal sequence in the Spanish network, reflecting the digit group principle.

While the Socotel system has the merit of fewer frequencies and wider frequency spacing, the registers are complicated by the necessary changeover arrangements. Further, as the acknowledgments carry no other signalling information, for a given amount of information transferred for a given numbering scheme, the Socotel system requires additional signals in the total process of address information transfer relative to a system (e.g. R2) in which the acknowledgment signals carry other signalling information of the request nature. This increases the holding time of the registers and the post dialling delay. The increase is modest, but would need to be taken into account.

Should a flexible, general application interregister signalling system be desired, it is considered that this would indicate system R2 rather than the Socotel system, the preference being based in the main on the request signalling capability of the acknowledgments.

References

1 WELCH, S.: 'The signalling problems associated with register controlled automatic exchanges', *Proc. IEE*, 1962, **109**B, pp. 465–475

2 CCITT: Green Book, **6**, Pt. 3, Recommendations Q310-Q331, 'Specification of signalling system R1', ITU, Geneva, 1973, pp. 553–557
3 DAHLBOM, C. A., HORTON, A. W., and MOODY, D. L.: 'Multifrequency pulsing in switching', *Electr. Engg.* (USA), 1949, **109**, p. 505
4 DAHLBOM, C. A., HORTON, A. W., and MOODY, D. L.: 'Application of multifrequency pulsing in switching', *Trans. AIEE*, 1949, **68**, Pt. 1, p. 392
5 'Notes on distance dialling'. Blue Book, Section 5 'Signalling', (AT & T Co., 1975), pp. 33–37
6 BREEN, C., and DAHLBOM, C. A.: 'Signalling systems for control of telephone switching', *Bell Syst. Tech. J.*, 1960, **39**, pp. 1381–1444 and Bell System Monograph 3736, 1960
7 CCITT: Green Book, **6**, Pt. 3, Recommendations Q350-Q368 'Specification of signalling system R2', ITU, Geneva, 1973, pp. 589–636
8 BAGER, R., and CARLSTRÖEM, P.: 'L. M. Ericsson's multifrequency code signalling (MFC) system', *Ericsson Rev.*, 1961, **38**, pp. 101–105
9 VRIES, W. C., et al: 'The multifrequency code (MFC) system', *Ingenieur*, 1961, **73**, Nos. 15, 24 and 30
10 HERTOG, M, den.: 'Interregister multifrequency code signalling for telephone switching in Europe', *Elec. Commun.*, 1963, **38**, 1, pp. 130–164
11 HERTOG, M, den.: 'Mehrfrequenzcode – Wahlverfahren', *Elktr. Nachrichtenwesen*, 1963, **38**, 1
12 GASSER, L., and RAHMIG, G. A. W.: 'Ergebnisse von betriebsversuchen mit mehrfrequenz-codewahlverfahren (MFC – wahl)', *ibid.*, 1964, **39**, 4, pp. 550–565
13 MILLER, C. B., and MURRAY, W. J.: 'Trunk transit network signalling systems – multifrequency interregister signalling', *Post Off. Electr. Eng. J.*, 1970, **63**, Pt. 1, pp. 43–48
14 HORSFIELD, B. R.: 'Fast signalling in the U.K. telephone network', *ibid.*, 1971, **63**, Pt. 4, pp. 242–252
15 GAZANION, H., and LEGARE, R.: 'Systems de signalisation Socotel', *Commutat & Electron.*, 1963, **4**, p. 32

Chapter 11

Common channel signalling

11.1 Introduction

Conventional signalling on speech path systems (d.c., v.f., outband, m.f.) have a number of limitations:

(*a*) Relatively slow signalling. This is usually tolerable for telephony purposes in most networks in view of the acceptable postdialling delays, but the application of computer-like techniques (stored program control) to the control of switching introduces new factors.

(*b*) Limited information capacity.

(*c*) Limited capability to convey signalling information which is not call related and the inability of some systems to signal during the speech period.

(*d*) The signalling systems tend to be designed for specific application conditions, which results in a relatively large number of different systems in the one network with consequential economic and administrative problems.

(*e*) Expensive due to the per-speech circuit provision of most of the systems.

Conventional line-signalling systems incorporating decadic address signalling are slow and inflexible, but meet the needs of most direct-acting noncommon control switching systems. Wired-logic common control switching systems demand more flexibility, more facilities and the application of interregister signalling systems as discussed, which systems are common provision, line-signalling systems being necessary to deal with the supervisory signals. Time-sharing of conventional analogue line-signalling systems is not usually practicable and they are invariably per-speech circuit provided, and expensive in total network signalling cost.

Stored program (processor) control (s.p.c.) of switching and networks has prompted a reappraisal of the signalling technique. Processor control makes possible the concentration of signalling logic for a large number of information circuits, e.g. speech, with consequent cost-reduction. Unless care is taken, however, an exchange processor may be given an excessive load in merely scanning speech circuits to detect the signalling condition, with a consequent loss of capacity for other functions, such as switching control, to be performed by the processor. It would also be necessary for the processor, or other suitable interface, to translate the various analogue conditions of speech-path signalling to digital, and vice versa. Thus with s.p.c., it is inefficient for the processor which works in the digital mode, to deal with signalling on the speech path. A much more efficient way of transferring information between s.p.c. exchanges is to provide a bidirectional high-speed data link between the two processors over which they transfer signals in digital form by means of coded-bit fields. A group of circuits (many hundreds) thus shares a common channel signalling (c.c.s.) link in the time-shared mode. In preferred c.c.s., all the signals between two exchanges are passed over a signalling link which is separate from speech, c.c.s. thus replacing speech-path signalling systems such as d.c., v.f., outband and m.f. While c.c.s. could be applied to non-s.p.c. exchanges, the need to provide means of directing the information contained in the data link signals to or from the registers, and, in the case of tandem connections, to transfer such information across the exchange, would tend to make c.c.s. uneconomic; c.c.s. is thus mainly suited for s.p.c. exchanges.

Many administrations are at present programming c.c.s. application, but owing to the well known inertia of networks to change, it will require a number of years for c.c.s. to make significant application impact in a network, and conventional signalling systems will exist, and continue to be applied, for a long time to come.

11.2 Basic common channel signalling

General

With c.c.s., independent signalling is performed in the respective signalling directions on the respective signalling channels, the two signalling channels comprising the c.c.s. link (Fig. 11.1). Modems are used to transmit and receive serial binary data over analogue transmission channels, the multiplex performing this function on digital transmission channels. C.C.S. has the following merits:

(i) signalling is completely separate from switching and speech transmission and thus may evolve without the constraints normally associated with such factors

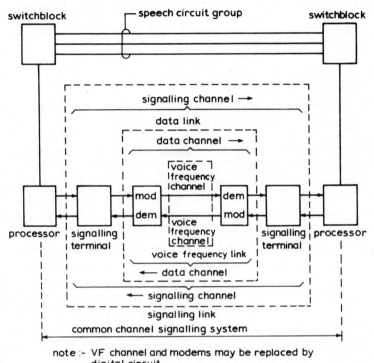

note :- VF channel and modems may be replaced by digital circuit

Fig. 11.1 *Basic schematic common channel signalling*

(ii) significantly faster signalling

(iii) potential for a large number of signals

(iv) freedom to handle signals during the speech period

(v) flexibility to change or add signals

(vi) potential for network centralised-service signalling (e.g. network management, network maintenance, centralised call accounting, etc.)

(vii) is economic for large speech circuit groups

(viii) can be economic for the smaller speech circuit groups owing to the quasiassociated and dissociated signalling capabilities

(ix) unlike v.f. signalling, signalling line splits are not necessary; this eliminates the problem arising on v.f. signalling systems where a fast nonrepeated verbal answer response may be clipped, or lost, owing to the speech path being split on electrical-answer signal transmission.

It was the recognition by the CCITT of the signficance of this problem in the international service which produced the initial thoughts for a c.c.s. system of the type under discussion.

(x) allows possibility for signalling rationalisation in networks.

C.C.S., however, gives rise to requirements which do not arise with signalling on speech-path systems such as:

(*a*) high order error rate performance
(*b*) signalling link security backup
(*c*) assurance of speech-path continuity as, unlike speech-path signalling, c.c.s. does not establish speech path integrity.

A c.c.s. system serving many speech circuits must have a much greater signal dependability than signalling on speech-path systems, as random errors on the signalling link would disturb an appreciable number of signals, and therefore speech circuits. For this reason, provision must be made in c.c.s. to control errors, and as a generalisation, an undetected error rate of the order of 1 in 10^8 to 1 in 10^{10} is desired. Automatic diverting of signalling traffic to an alternative signalling facility occurs on excessive error rate, or on complete failure of a c.c.s. link.

Time-shared signalling requires each signal to have identification of the speech circuit to which it belongs, the identification being on a time basis when a speech circuit has exclusive use of a signalling facility, as in built-in p.c.m. signalling (Section 6). This time-assigned signalling cannot apply when, as in c.c.s., a speech circuit does not have exclusive use of a signalling facility in the time-shared signalling, and a signal is labelled to give the speech circuit identity. With circuit labelling, the coding of a bit-field in the signal itself gives the address of the speech circuit, the number of bits depending on the number of speech circuits to be identified. This depends on the maximum number of speech circuits per c.c.s. link-loading adopted, which is influenced by the capacity of the c.c.s. link to transfer signals. This, in turn, depends on the signalling bit rate and the size of signal unit.

With c.c.s., the speech-path connection may be set up in parallel with the signalling connection set-up, or retrospectively. The former is adopted for simplicity and assurance of speech-path availability. Speech-path availability could be assured with the retrospective set-up by marking, but not switching, relevant speech circuits in the connection build up, but there appears to be little point in this.

General features of c.c.s.
The following general features apply:

(*a*) Information may be transferred as signal messages of varying bit length, or by defined signal units. Signal units will be assumed for the present.

(*b*) Each signalling channel operates in the synchronous mode with its continuous bit stream divided into contiguous signal units which all contain the same number of bits for the same application or service. The two signalling channels of a link need not necessarily be synchronised to each other, and, if not, drift may arise between the signal units in the respective directions, which, in certain implementations of c.c.s. (e.g. CCITT 6 system) may require compensation arrangements.

(*c*) Synchronisation (idle) units are transmitted when message signals are not being transmitted, to maintain signal unit synchronism on the signalling channel.

(*d*) The signal unit is divided into a number of constituent bit fields each having its own function in the system, typically, heading, signal information, circuit label, parity check, etc.

(*e*) As the c.c.s. link is time-shared, each message signal requires identification of the speech circuit (and thus the call) to which it belongs. This is by means of a circuit label bit-field of size depending on the number of speech circuits to be identified.

(*f*) Unlike time-assigned, channel-associated, time-shared p.c.m. signalling (Sections 6.2 and 6.3), a speech circuit does not have the exclusive use of a signalling facility in time-shared circuit-label addressed c.c.s. and queueing delays arise; message signals being placed in a queueing store and offered in turn for transfer over the signalling channel. Thus the heavier the signal loading the greater the queue, and the queueing delay, and the slower the speed of signal transfer.

(*g*) As c.c.s. signals do not prove the continuity of the speech path selected, other arrangements (e.g. per-call continuity check, routine testing of idle paths) must be made.

(*h*) Errors are liable to occur in the signal-transfer process and some form of error control is required as the uncorrected error-rate of transmission plant is usually unacceptable for c.c.s. Error detection is based on redundant coding, the parity check bits being, typically, part of each signal unit. Error correction can be, and usually is in telephony c.c.s., by retransmission. Despite the incorporation of error control, undetected errors may arise. Even with a high degree of error correction a signalling link could be unusable for varying periods, which requires signalling security backup.

(*i*) For signalling security, signals may be directed, by automatic procedures, from a regular signalling link to an alternative signalling facility when an excessive error rate, or complete failure, of the regular link is detected.

(*j*) On multilink connections, signalling information is transmitted on a link-by-link basis, the signals being transferred from one link to the next only after processing. This is an inherent feature of c.c.s.

(*k*) Since the circuit label may require a substantial proportion of the bits available in a signal unit, comparatively few bits may remain for coding the information a message has to convey. It follows that when the information is even moderately extensive, and particularly when it has a data content which can vary from one message to another, it cannot be transmitted efficiently by a succession of single signal unit messages. Instead, the initial circuit label carrying signal unit may be augmented by one or more subsequent signal units in which all the available bits can be used for carrying the information, thus producing an efficient multiunit message. Thus lone signal units (LSU) and multi-unit messages (MUM) may apply with c.c.s., and in the latter case the initial signal unit (ISU) carries the circuit label bit-field for the whole MUM. Each SU in a MUM carries its own check bit-field.

11.3 Association between c.c.s. and speech (or equivalent) networks

The signal messages relating to a given group of speech circuits between two switching centres using a c.c.s. system can be transferred in the following ways:

(*a*) *Associated mode of operation:* The signal messages are transferred between the two switching centres over a c.c.s. link which terminates at the same switching centres as the group of speech circuits to which the signalling link is assigned (Fig. 11.2*a*). The term 'associated signalling' applied to c.c.s. should not be confused with speech path, or channel, associated signalling used to describe signalling systems in which the signals are passed over the speech paths comprising the connections to which they relate, or over paths which are permanently and individually related to the speech paths.

(*b*) *Nonassociated mode of operation:* The signal messages are transferred between the two switching centres over two or more c.c.s. links in tandem, and thus on a different routing from the relevant speech-circuit group, the signal messages being processed and forwarded through one or more intermediate signal-transfer points. It follows from this definition that there may be a range of nonassociated modes of operation which vary in the degree of rigidity imposed on the choice of the path used by the signal messages. The extremes of this range can be described as the fully dissociated mode and the quasiassociated mode.

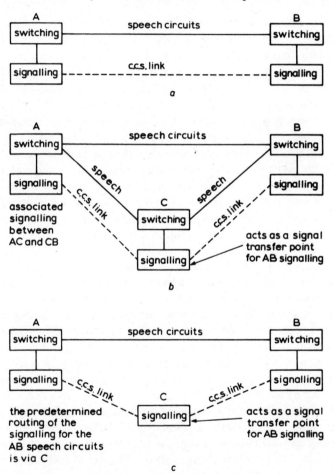

Fig. 11.2 *Examples of associated and quasiassociated signalling*
a Associated signalling between A and B
b Quasiassociated signalling between A and B
c Quasiassociated signalling between A and B

(c) *Fully dissociated mode of operation:* The signal messages are transferred between the two switching centres via any available path in the switching network according to the routing principles and rules of that network. The great flexibility in the routing of signal messages demands a more comprehensive message-addressing scheme than is needed for associated signalling. This is the ultimate of the principle of separate paths for signalling, and a completely separate c.c.s. network has been suggested.[1]

(*d*) *Quasiassociated mode of operation:* The signal messages are transferred between the two switching centres over two or more c.c.s. links in tandem, but only over predetermined paths and through predetermined signal transfer points. The predetermined routing permits the same method of addressing (circuit labelling) the signal messages as in the associated mode of operation.

The constituent signalling routes of a quasiassociated signalling relationship may be associated-signalling in their own right (Fig. 11.2*b*), in which case the constituent routes carry the quasiassociated signalling traffic in addition to the associated-signalling traffic. Alternatively, the constituent signalling routes need not be associated-signalling in their own right (Fig. 11.2*c*). When the constituent signalling routes (AC and CB) in the quasiassociated signalling relationship are associated-signalling in their own right (Fig. 11.2*b*), the circuit labels for the speech-circuit group AB must be different from those of speech-circuit groups AC and CB. Thus for the necessary signalling discrimination, the circuit labelling of signalling routes AC and CB must be increased to accommodate for this. When the constituent signalling routes AC and CB in the quasiassociated signalling relationship are not associated-signalling in their own right (Fig. 11.2*c*), the circuit labelling concerns the speech circuit group AB only in the typical case shown.

Quasiassociated signalling may be adopted when a speech-circuit group is too small to justify economic application of associated-signalling The mode may also be used for signalling security backup for associated-signalling and for backup for another quasiassociated signalling relationship, but the circuit labelling tends to be complex in the latter case.

With quasiassociated signalling, the number of signal transfer points in the signalling path for a group of speech circuits between two switching centres should be as few as practicable to minimise the signalling time of those circuits, and to minimise the total signal processing load of the network. Normally one or two signal transfer points should suffice in a quasiassociated signalling relationship.

Signal transfer point
In a nonassociated mode of operation, a signal transfer point is a signalling centre which forwards signal messages received on one c.c.s.

link for onward transmission over another c.c.s. link. It follows from this that there is no need for a signal transfer point to have any connection with a switching centre (Fig. 11.2*c*). Alternatively, it may have (Fig. 11.2*b*).

The main functions of a signal transfer point are:

(i) to analyse the circuit label and the priority indication of every received signal message to determine the routing and, hence, to offer the message to the proper signalling link for onward transmission of the message taking account of the appropriate priority
(ii) in doing so, it may be necessary to change (translate) the circuit label of the message according to some preset rules
(iii) if for some reason, a signal transfer point is unable to transfer signal messages, a procedure is desired to notify the preceding signalling centre(s) so that the signal messages may be sent via an alternative route, if available.

While c.c.s. has the potential for the various signalling modes discussed, it is thought that a combination of associated and quasi-associated signalling would meet the needs of most networks. The predetermined signal routings possible with these modes would be attractive in facilitating an orderly and efficient traffic-dimensioned c.c.s. network.

11.4 Network centralised service signalling

Compared with wired-logic, processor (s.p.c.) control of switching has the potential for a far more sophisticated and greater degree of network-centralised services, and it is thought that this will be the undoubted trend. The combination of s.p.c. and c.c.s. will enable networks to be controlled and exploited to a greater extent than previously attainable. Many administrations are at present actively studying the extent and desired features of centralised services, typical services being:

> Speech-network management. Typically to initiate temporary changes of traffic routing patterns for reasons of congestion, catastrophic failure, etc.
> Signalling network management
> Network maintenance
> Centralised call accounting etc.

Fig. 11.3 shows a possible basic arrangement in simplified form.

Various implementations are possible, e.g. the various services may or may not be combined in the one centre, and the centres could be hierarchical (local, regional, national). The processors (CC) at relevant exchanges could be connected to relevant centralised service centres by data-signalling links, the signalling not being concerned with call handling. Should a hierarchical service-centre structure apply, further data links would interconnect the hierarchical centres as appropriate.

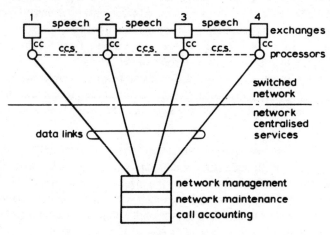

Fig. 11.3   *General arrangement network centralised services*

The present interest is in the relevant signalling arrangements in the c.c.s. environment and it is thought that the switched-network c.c.s. links should be capable of carrying centralised service in addition to the normal call-handling signalling to allow for the following possibilities:

(*a*) switched-network c.c.s. links used as security backup against failure of the centralised-service data links, e.g. on failure of the data link from exchange 1, the centralised service signalling passed on the c.c.s. link between exchanges 1 and 2 and then on the data link from exchange 2 (Fig. 11.3)
(*b*) switched-network c.c.s. links utilised as data links to carry the centralised-service signalling *en route* to the service centres
(*c*) when an exchange does not have direct data-link access to the service centres, access is obtained by means of a switched-network c.c.s. link via an exchange having direct data-link access.

It is clearly desirable that the switched-network c.c.s. links should have adequate spare capacity to carry the centralised service in addition

to the call-handling signalling, the centralised-service signalling con-
forming to the switched-network c.c.s. signalling features (error control,
signal unit size, etc.) in this situation. It would not be essential for
the centralised service data links to be of same type as the switched
network call handling c.c.s. links, but in certain circumstances there
could be merit in uniformity if they were.

11.5 CCITT No. 6 signalling system

11.5.1 Basic concepts.
The CCITT 6 signalling system was the first c.c.s. system to be specified, designed and tested.[2,3] While specified primarily for international
use, it is equally suitable for national application and some administrations are presently programming its use in their national networks.[4,5] Initially, system 6 was specified for 2·4 kbit/s analogue
application. Subsequently, digital versions 4 kbit/s and 56 kbit/s were
included in the specification but without change to the initial basic
signalling arrangements formulated for analogue. Other signalling bit
rates may be adopted as desired (e.g. 4·8 kbit/s analogue).

It is convenient to present the pioneer system 6 solutions to the
many problems arising in c.c.s. as a general basis before discussing
improved solutions which may be adopted in some areas, improvements
that arise from the greater knowledge now available in the c.c.s. art and
from the natural evolution of the technique.

Each signalling channel is operated synchronously but the two
channels are not synchronised with each other, and a continuous stream
of data flows in both directions. The data stream is divided into signal
units (SUs) of 28 bits each, of which the last 8 are check bits, all SUs
being of the same length for ease of synchronisation in particular. The
SUs in turn are grouped into blocks of 12, the 12th and last SU of
each block being an acknowledgment SU coded to indicate the number
of the block being transmitted and carrying the acknowledgment i.e.,
the number of the block in the other direction being acknowledged,
and whether or not each of the 11 SUs of the block being acknowledged was received without detected errors. Blocks are sequenced
numbered but individual message SUs in the block are not, being
identified by their position in the block. The first 11 SUs of a block
consist of message or synchronisation SUs; the latter, transmitted only
in the absence of other signalling traffic, facilitate achieving or maintaining SU and block sychronisation, and are coded to indicate the

number of the position they occupy within the block to facilitate locating the acknowledgment SU.

A signalling bit-rate of 2·4 kbit/s adopted for the analogue version is the maximum rate for transmission with acceptable error rate over a phase-equalised speech-band channel. At this bit-rate, one c.c.s. link can carry all the signals required by some 1500–2000 speech circuits in the telephony service without excessive delay to individual signals. Compared with the 2·4 kbit/s analogue version, the 4 kbit/s and 56 kbit/s signalling bit-rate digital versions allow, in theory, more speech circuits to be served per c.c.s. link due to the reduced emission time of the signals. In practice, the same size circuit label bit-field applies for both the analogue and digital versions in the international specification. National network variants of the specification may, of course, extend or reduce the size of the circuit label bit field as required. Optional 4·8 kbit/s analogue is presently being considered.

The signals, which may be LSUs or MUMs, are formatted in the processor function and delivered to an output buffer, which delivers the highest-priority signal awaiting transmission to a coder in serial form in the next available time slot. Each SU is encoded by the coder by the addition of check bits in accordance with the check-bit polynomial. The signal is then modulated and transmitted over the link in serial form. At the receive end, serial data is delivered to a decoder where each SU is checked for error on the basis of the associated check bits, SUs received with detected errors being discarded. Message SUs which are error-free are transferred to an input buffer after deletion of the check bits, the input buffer delivering the SUs to the processor function for appropriate action.

A transmission path error rate of 1 in 10^6 to 1 in 10^7 was assumed and as the requirement of system 6 was to achieve an undetected error rate of 1 in 10^{10} for significant signals and 1 in 10^8 for others, error control was adopted with error detection by redundant coding and error correction by retransmission. Message SUs are retransmitted on error detected, synchronisation SUs are not. A data channel failure detector complements the decoder for the longer error bursts. On excessive error rate, or on complete failure, automatic procedures direct the signal traffic from a 'failed' link to an alternative signalling facility.

11.5.2 Signal codings

Basic philosophy

The binary codings of bit fields in c.c.s. allows potential for a considerable signal repertoire depending upon the bit field, and SU, size.

Information is transferred in system 6 by means of one (LSU) or more (MUM) SUs, a SU being the smallest defined group of bits on the signal channel, and, in system 6, contains 28 bits (20 information, 8 check). A LSU may transmit either a single telephone signal, a system control signal (e.g. ACU) or a centralised service signal. Basically, a MUM, which conveys a number of related signals (e.g. address digit signals), may consist of any desired number of SUs, but in system 6 consists of 2, 3, 4, 5 or 6 SUs, the constituent SUs of a MUM being identified as ISU (the first) and SSU (the second and any following). In the c.c.s. art (but not 6) the FSU is the final SU of a MUM, the FSU being a SSU and is so called in system 6. The ISU, SSU (and FSU) facilitate MUM length indication in the signalling system. If the MUM is speech-circuit related, the circuit label is included in the ISU only.

Basically, a defined c.c.s. system can be regarded as being merely a means of transferring bit-coded messages, individual administrations being free to adopt their own codings as appropriate. It is logical, however, to adopt some uniformity in the codings and constituent bit-field sizes, for ease of interworking between national, regional and international applications of system 6, the approach that has been adopted. The CCITT specification details the signal codings for the international application, with adequate spare capacity allocated to cater for particular national network requirements not included in the international specification. As there is a high degree of commonality in international and national signalling requirements, considerable uniformity will apply which is the objective, but it must be accepted that national network variants may well arise.

(a) Telephone signals

LSU (lone signal unit): A number of bits within each SU must be allocated to distinguish the specific message being transmitted and in the telephony LSU nine bits perform this function, divided into two fields; heading field bits 1–5 and signal information bits 6–9 (Fig. 11.4*a*). The LSU Fig. 11.4*a* is also the basic format of the ISU. The two bit-fields permit ease of administration of the signals, a heading being allocated to a class of signals, e.g. address digit signals, individual signals within each class being distinguished by the relevant coding of the signal information field. These two bit-fields more than cover existing requirements (Table 11.1) and leave capacity for new signals and classes of signals not yet defined.

Where appropriate, the coded bits 1–5 also give indication of the signalling direction (forward, backward) of the signal identified by the signal information bits 6–9 (e.g. heading code 11011 indicates that all the signals under this heading are backward signals – Table 11.1). While the heading generally consists of bits 1–5, there are two exceptions:

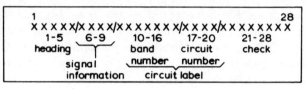

a

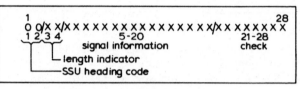

b

```
        1                                                          28
ISU   1 0 0 0 0/0 0 0 0/X X X X X X X X X X X/X X X X X X X X
        heading   signal      circuit label          check
        code      information
        ISU       code IAM
1st.
SSU   O 0/0 1/X X X 0 0 0 0 0 X X X 0 0 0 0/X X X X X X X X
        *  **             other routing                 check
                          information

2nd
SSU   O 0/0 1/X X X X/X X X X/X X X X/X X X X/X X X X X X X
        *  **    1st      2nd     3rd     4 th      check
                         address digits
              *   SSU heading code
              * *  length indicator
```

c

Fig. 11.4   *CCITT 6 system: typical message signal formats*
    a Basic format of LSU and ISU of MUM
    b Format of SSU of MUM
    c Example of 3-unit MUM

(i) all SSUs of a MUM are identified by the same 2-bit heading code 00 (bits 1–2), Fig. 11.4*b*.
(ii) the acknowledgment signal unit ACU is identified by a 3-bit heading code 011 (bits 1–3), Fig. 11.5*a*.

The heading codes (bits 1–5) are allocated:

00	SSU
01000 01001 01010 01011 }	spare
011	ACU
10000	ISU of IAM or of MUM
10001 10010 10011 10100 10101 10110 10111 }	SAM (one-unit message or multiunit message)
11000 11001 11010 11011 }	international telephone signals
11100	spare
11101	signalling system control signals (except ACU) and management signals
11110 11111 }	spare

The signal information codes (bits 6–9) are allocated as shown in Table 11.1.

MUM (multiunit message): A MUM, which may be address messages, centralised service messages, or messages conveying other types of information, consists of an ISU and a number of SSUs. The basic format of the ISU is the same as that of the LSU (Fig. 11.4a) the heading code identifying the ISU. SSUs are identified by the heading code bits 1–2 and have a 2-bit field (bits 3–4) coded to indicate the number of SSUs in the MUM (Fig. 11.4b), thus serving as MUM length indication. In system 6, an IAM MUM may be 3–6 SUs and all other MUMs 2–5 SUs, and each SSU of a MUM carries the same relevant length indicator coded as follows:

Number of SSUs in MUM	Length indication in SSU (bits 3–4)	
	IAM MUM	Other MUMs
1	–	00
2	01	01
3	10	10
4	11	11
5	00	–

This leaves 16 bits (bits 5–20) in each SSU to carry information (Fig. 11.4*b*).

The term MUM should not be confused with the term block in the block transmission of system 6. The block is simply a group of 12 consecutive SUs, which may be LSUs, a mixture of LSUs and SYUs, a mixture of MUM (or part MUM) and LSUs, two MUMs, etc.

IAM (initial address message): The IAM is the first message of a call connection set-up. Unlike conventional speech path signalling systems, a discrete seizure signal is not required in c.c.s., the IAM performing this function. As the IAM conveys information (e.g. routing) in addition to address digits, more than one SU is always necessary and the IAM on any call is always a MUM, consisting of an ISU (Fig. 11.4*a*) and SSUs (Fig. 11.4*b*). As in SSUs of MUMs in general, bits 1–4 of SSUs of IAMs give SSU identification and MUM length indication. The information bit-field (bits 5–20) of SSUs 2–5 of IAMs in the international system 6 is subdivided into four 4-bit parts so that four address-digit signals can be conveyed in each SSU. The information bit-field (bits 5–20) of SSU 1 of IAMs is used for routing information required in the setting up of connections in the international system 6. This field could be used for other purposes in regional and/or national application of system 6 as required. Also as in SSUs of MUMs in general, SSUs of IAMs do not require the 5-bit heading or the circuit label bit-fields as these items of information are already contained in the ISU.

In the detail, the codes used in the IAM are (Table 11.1):

(i) *ISU:* The 5-bit heading code 10000 (bits 1–5) in combination with the signal information code 0000 (bits 6–9) identify that the ISU is the ISU of an IAM.

(ii) *SSU 1:* The heading code 00 (bits 1–2) identifies the SSU and bits 3–4 are coded to give the appropriate length indication. In the international application, the special routing information required in the setting up of connections is as follows:

Bit 5 Country code included or not in the IAM. This feature can be utilised to inform an incoming international exchange to function as terminal or as transit.

Bit 6 Nature of circuit indicator. In particular, this bit is used in satellite routing restriction and records if a satellite link has been used.

Bit 7 Echo-suppressor control. Whether or not an echo suppressor is required.

Bit 8 Spare (reserved for international)

Bits 9–12 Spare (reserved for regional and/or national)

Bits 13–16 Calling party's category, such as operator, ordinary sub-scriber, data service, etc.

Bits 17–20 Spare (reserved for regional and/or national)

(iii) *SSUs 2–5 – telephone call:* As in SSU 1, the heading code 00 (bits1–2) identifies the SSU and bits 3–4 give the appropriate length indication. The four 4-bit parts of the signal information field bits 5–20 contain address-digit signals in sequence, bits 5–8, 9–12, 13–16, and 17–20, respectively, being coded as follows (Table 11.1):

0000	filler (no information)
0001	digit 1
0010	digit 2
0011	digit 3
0100	digit 4
0101	digit 5
0110	digit 6
0111	digit 7
1000	digit 8
1001	digit 9
1010	digit 0
1011	code 11 operator
1100	code 12 operator
1101	spare
1110	spare
1111	ST (end of pulsing)

The filler code 0000 is used where needed to complete the signalling information of the last SSU of the IAM if this SSU is partially used for carrying address digits.

Table 11.1 *CCITT system 6 allocation of heading and signal information codes*

bits 6–9 \ bits 1–5	0000X	0001X	0010X	0011X	01000	01001	01010	01011	011XX	10000	10001	10010	10011	10100	10101	10110	10111	11000	11001	11010	11011	11100	11101	11110	11111	bits 1–5 / bits 6–9
0000	SSU				ISU of MUM	ISU of MUM	ISU of MUM	ISU of MUM		ISU of IAM	ISU of SAM1	ISU of SAM2	ISU of SAM3	ISU of SAM4	ISU of SAM5	ISU of SAM6	ISU of SAM7	ISU of MUM	ISU of MUM	ISU of MUM	ISU of MUM	ISU of MUM	ISU of MUM	ISU of MUM	ISU of MUM	0000
not 0000					LSU	LSU	LSU	LSU		ISU of MUM	lone SAM1	lone SAM2	lone SAM3	lone SAM4	lone SAM5	lone SAM6	lone SAM7	LSU	LSU	LSU	LSU	LSU	LSU	LSU	LSU	not 0000
0000	One SSU or five SSUs (1AM only)	Two SSUs	Three SSUs	Four SSUs	Reserved for regional and/or national use	Reserved for regional and/or national use	Reserved for regional and/or national use	Reserved for regional and/or national use	ACU													Reserved for regional and/or national use		Reserved for regional and/or national use	Reserved for regional and/or national use	0000
0001											digit 1	digit 1	digit 1	digit 1	digit 1	digit 1	digit 1	RLG		COT	AFC		NMM			0001
0010										Reserved for regional and/or national use	2	2	2	2	2	2	2	ANC		CLF	AFN					0010
0011											3	3	3	3	3	3	3	ANN	SEC	FOT	AFX					0011
0100											4	4	4	4	4	4	4	CBI	CGC		SSB					0100
0101											5	5	5	5	5	5	5	RAI	NNC		VNN		SNM			0101
0110											6	6	6	6	6	6	6	CB2			LOS		regional and/or national			0110
0111											7	7	7	7	7	7	7	RA2			SST					0111
1000											8	8	8	8	8	8	8	CB3	CFL							1000
1001											9	9	9	9	9	9	9	RA3								1001
1010											0	0	0	0	0	0	0				ADC					1010
1011																				BLO	ADN		MBS			1011
1100																				UBL	ADX		SCU			1100
1101																				BLA	ADI		SYU			1101
1110																			COF	UBA			regional and/or national			1110
1111											ST	ST	ST	ST	ST	ST	ST			MRF						1111

Signal abbreviations, system 6

ACU	acknowledgment signal unit-error control
ADC	address complete signal, charge
ADI	address incomplete signal
ADN	address complete signal, no charge
ADX	address complete signal, coin box
AFC	address complete signal, subscriber free, charge
AFN	address complete signal, subscriber free, no charge
AFX	address complete signal, subscriber free, coin box
ANC	answer signal, charge
ANN	answer signal, no charge
BLA	blocking acknowledgment signal
BLO	blocking signal
CB1–3	clear back signal No. 1–No. 3
CFL	call failure signal
CGC	circuit group congestion signal
CLF	clear forward signal
COF	confusion signal
COT	continuity signal
CSSN	circuit state sequence number
FOT	forward transfer signal
IAM	initial address message
ISU	initial signal unit
LOS	line-out-of-service signal
LSU	lone signal unit
MBS	multiblock synchronisation signal
MRF	message-refusal signal
MUM	multiunit message
NMM	network management and maintenance signal
NNC	national network congestion signal
RA1–3	reanswer signal No. 1–No. 3
RLG	release guard signal
SAM1–7	subsequent address message No. 1–No. 7
SCU	system control signal unit
SEC	switching equipment congestion signal
SNM	signalling network management signal
SSB	subscriber busy signal (electrical)
SST	subscriber transferred signal
SSU	subsequent signal unit
SU	signal unit
SYU	synchronisation signal unit
UBA	unblocking acknowledgment signal
UBL	unblocking signal
VNN	vacant national number signal

Fig. 11.4*c* shows a typical three-unit IAM.

In bothway operation of c.c.s., a double seizure is detected by an exchange when it receives an IAM for a speech circuit for which it has sent an IAM.

At the analogue signalling bit-rate 2·4 kbit/s, each SU has an emission time of approximately 11·7 ms and with a three-unit IAM, four address digits can be transmitted in some 35 ms. Typically, a ten-digit national number, if transmitted *en bloc*, would require a five-unit IAM and transmitted in some 58·5 ms at a rate of 170 digits per second. This is much faster than any speech-path telephony-signalling system currently in use. Digital c.c.s. transfers address information at still faster speeds.

SAM (subsequent address message): A SAM, which may be a LSU or a MUM, is used to transmit additional address information not available for transmission when the IAM is formed, or (but not in international system 6) when the transmission of address information after the IAM is controlled by backward request signals. The format of an LSU SAM and of the ISU of a MUM SAM, is the same as that of the normal LSU (Fig. 11.4*a*). The LSU SAM carries one address digit in the signal-information field-bits 6–9. The ISU of a SAM does not carry an address digit, bits 6–9 being coded 0000 filler (no information). The format of an SSU of a MUM SAM is the same as that of the SSU of any MUM (Fig. 11.4*b*) except that, unlike SSU 1 of an IAM (which does not carry address information in international system 6), SSU 1 of a SAM may carry address information in the information field bits 5–20. Thus all SSUs of SAMs may carry four address digits each, bits 5–20 being subdivided into four 4-bit parts. The 4-bit address digit fields are coded as for SSUs 2–5 of an IAM to transmit address information. The Code 11 and Code 12 operator codes are not used in SAMs, as this information will have been given by the IAM. The filler code 0000 is used, where needed, to complete the signal information field of the last SSU of a SAM.

Heading codes (bits 1–5) in the range 10001–10111 are used in LSU, or the ISU of, SAMs, depending upon the sequence number of the SAM concerned (Table 11.1), i.e.

> 10001 first SAM
> 10010 second SAM
> 10011 third SAM, etc.

This sequence numbering of SAMs is necessary in system 6 as messages can get out of sequence at the processor function in the working of the c.c.s. system, and the sequence numbering facilitates the restoration of

SAMs to correct the sequence at the processor. It is preferred to limit the number of SAMs on a call connection set-up, but if more than seven are sent, the sequence is recycled so that the eighth SAM uses code 10001.

(b) Management signals
In international system 6, management signalling may include network management, network maintenance and signalling network management, the signals being transferred as LSUs or MUMs. The management signalling detail is not yet resolved for international system 6 and specific formats are not yet specified. The basic format of the normal LSU (Fig. 11.4*a*) will apply for management LSUs and ISUs of management MUMs, and the heading code 11101 (bits 1–5) is assigned. Appropriate coding of the signal information field (bits 6–9) distinguish the various categories of management signalling, i.e. network management, maintenance, etc. (Table 11.1).

Most management signals will require a band number (bits 10–16 Fig. 11.4*a*) to route the signal information on the switched network via appropriate signal transfer points. As management signals will not require a circuit number in the band, bits 17–20 (Fig. 11.4*a*) are used for management information.

(c) Signalling system control signals
Control signals in system 6, always LSUs, are necessary for the proper functioning of c.c.s. systems. They are not related to telephone signalling information and are thus not speech-circuit related. In system 6, the control signalling comprises:

  (i) acknowledgment ACU signals in the error control
 (ii) synchronisation (idle) SYU signal units
(iii) multiblock monitoring and acknowledgment signals MBS for multiblock synchronisation
(iv) system control signal units SCU, a group of signals concerned with signalling link failure and the various consequential procedures.

The heading field (bits 1–5 Fig. 11.4*a*) is coded 11101 for signalling system control signals, except the ACU which is coded 011 in heading bits 1–3. Signal information bits 6–9 are coded to distinguish the category:

        1100 signalling link failure procedure signals (SCU)
        1101 SYU
        1011 MBS

Bits 10–20 give further signalling information in each category, bits that would otherwise be used for the circuit label in speech-circuit-related LSUs.

SYU (synchronisation signal units): This consists of a 16-bit fixed pattern for bit (analogue c.c.s.) and SU synchronisation, and a 4-bit field for block synchronisation, the latter identifying the location of the SYU within the block (i.e. 1st, 2nd . . . or 11th unit). The SYU (Fig. 11.5*b*) completes with the eight check bits. The particular 16-bit fixed pattern of the SYU presents an easily recognisable pattern with a variation at the end to aid in its detection. It also provides six-bit transitions to aid the attainment of bit-synchronisation by the modems in analogue c.c.s. (Section 11.5.5(*a*)), bit-synchronisation being maintained by the transition between dibits.

MBS (Multiblock synchronisation signal unit): The original system 6 specification, limited to 2·4 kbit/s analogue c.c.s., allowed for a maximum of eight blocks in the error-control loop. This was adequate for a maximum delay of 740 ms (single-hop satellite) and accounted for the 3-bit (8 possibilities) block-completed and block-acknowledged fields in the ACU. The revision of the specification to include 4 kbit/s and 56 kbit/s digital versions was coincident with a requirement to meet a maximum error-control loop delay of 1200 ms (double hop satellite), requiring more than eight blocks in the error-control loop at all the signalling bit rates specified i.e. 2·4, 4 and 56 kbit/s. The revised specification treats eight consecutive blocks as a multiblock and allows for up to 32 multiblocks. Thus the maximum number of blocks in the error-control loop is 256 (3072 SUs), which is adequate for a 1200 ms loop-delay signalling link at 56 kbit/s in the telephony service. The full 256-block capability need not be handled in all applications. Block memory may be limited to that required for the expected range of loop delays and signalling bit-rates at which the system is applied in particular situations.

The concept of the multiblock gives rise to the requirement for two signals of a control nature (Fig. 11.5*c*) i.e.

(*a*) multiblock monitoring signal, required on links where the number of blocks in the error-control loop exceeds eight, and sent to check multiblock synchronism

(*b*) multiblock acknowledgment signal, sent in response to the multiblock monitoring signal and used by the terminal receiving it to verify multiblock synchronism.

Of the **MBS** as shown in Fig. 11.5*c*:

Bits 10–12 are coded 000 for the monitoring signal and 100 for the acknowledgment.

Bits 13–17 are coded to indicate the sequence number of the multiblock in which the multiblock monitoring signal is sent.

Bits 18–20 are coded to indicate the sequence number of the block in which the multiblock monitoring signal is sent (or placed into the output buffer).

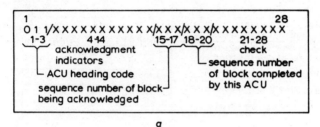

a

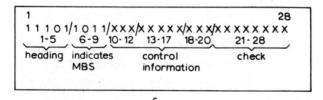

b

c

Fig. 11.5   *CCITT 6 system: typical control signal formats*
   a Format of ACU
   b Format of SYU
   c Format of MBS

The above covers the main signals of CCITT system 6; the international specification details the format and codings of other signals included in the specification.[2] Fig. 11.6 shows a typical signalling sequence with system 6.

11.5.3 Error control

In system 6, provision is made in every SU to detect errors due to noise, line faults or transmission faults causing a mutilation of bits.

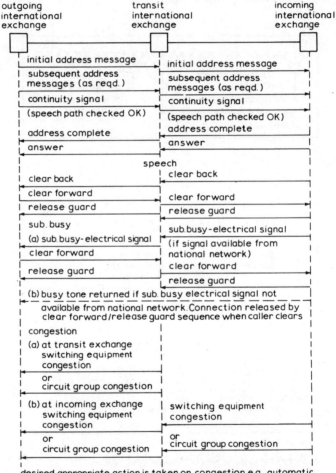

Fig. 11.6   *Typical signal sequence CCITT 6 system*

Error detection is by redundant coding, a number of check bits being generated in the form of a cyclic code,[6] each SU having an 8-bit check field. The generator polynomial selected for system 6 to generate the check-bit pattern provides maximum protection for short noise bursts affecting up to four consecutive bits.[7] To detect for longer noise bursts, a data channel-failure detector supplements the error detection by the check bits. This detects, and takes effect after a delay of about 5 ms, the following:

(*a*) failure of the data carrier or an excessive noise level relative to the carrier level in analogue c.c.s.

(*b*) loss of frame alignment in digital c.c.s.

Not all errors will be detected by the error control, undetected errors will arise.

Error detection: Error detection is performed by check bits (bits 21–28) in each SU, with coders and decoders equipped at the transmit and receive terminals, respectively, for the purpose. The principle of operation is that the digital sequence constituting the message is treated as a polynomial and is divided at the transmitter and at the receiver by a preagreed number. The remainder is transmitted from the transmitter to the receiver as check bits. The process is implemented using a modulo-2-division by a shift register with a number of stages equal to the number of check bits. For a given number of check bits C, there are 2^{C-1} different generating polynomials which can be chosen for the divisor, these polynomials having different characteristics which are indicated by their modulo-2 factors.[6]

Error burst patterns can also be considered as polynomials. Any error burst whose polynomial is the same as the generating polynomial is undetectable as the division process will change the quotient by one bit, but not alter the remainder. If the length of the error polynomial is less than the generating polynomial, the remainder will be changed and the error detected. When the error-burst polynomial has more terms than the generating polynomial, the errors will only be undetected if the generating polynomial forms a factor in the error-burst polynomial. In system 6, the coder generates eight check bits based on the polynomial[7]

$$P(X) = (X + 1)(X^7 + X^6 + X^5 + X^4 + X^3 + X^2 + 1)$$
$$= X^8 + X^2 + X + 1$$

A characteristic of cyclic codes[6] is their inferior protection in the case of a slip in frame synchronisation in p.c.m. Security is maintained by arranging for the check bits C to be inverted before transmission and reinverted before checking at the receiver. If there has been a slip, the reinversion is applied on an incorrect sequence of bits.

When the decoder at the receive terminal has received all 28 bits of the SU after the check bits have been reinverted, it will indicate whether or not the signal has been checked as correct. This information is stored for inclusion in the acknowledgment indicator field of an ACU to be emitted in the return direction.

The cyclic code for error detection is supplemented by a data-channel failure detector (Section 11.8). Indication of data-channel failure owing to unsatisfactory transmission conditions causes rejection of SUs in the process of reception, and during the alarm condition all SUs are rejected (i.e. acknowledged as corrupted) regardless of check.

The system 6 specification requires that an undetected error rate of not more than one erroneous SU out of each 10^8 transmitted should result in a false operation and not more than 1 in 10^{10} should cause malfunction such as false metering or false release. The specification is worded in this way as it was anticipated that most signals with undetected errors would result from the disturbance of idle units which predominate even during the busy hour. It was also evident that many false signals would cause no trouble as they would be addressed to nonassigned circuit labels, or would occur in conditions in which they were meaningless.

Error correction: Error correction is by retransmission in system 6, which necessitates a retransmission store at the transmit end, correctly received SUs being cleared from the store. On error-detected, the SU must be discarded at the receive end and the information retransmitted. This requires a method of acknowledging signals and a means of identifying SUs. The latter is achieved in system 6 by transmitting SUs in blocks and allocating identifying numbers to the blocks, individual SUs can then be identified by position within a block. Blocks are of a 12-unit fixed length, comprising 11 SUs (which may be message or synchronising) and an acknowledgment unit (ACU) in the 12th position (Fig. 11.7). Positive/negative block acknowledgment is employed. A positive acknowledgment indicates that the relevant SU has been received error-free, a negative acknowledgment indicates error detected and is thus a request for a retransmission.

Each ACU (Fig. 11.5*a*) contains:

(*a*) a 3-bit fixed pattern (011) to identify the unit as an ACU
(*b*) a 3-bit code to identify the block of 12 units completed by this ACU (providing for a count of 8)

(*c*) a 3-bit code to identify the block being acknowledged by this ACU (providing for a count of 8)

(*d*) the signal acknowledgment mechanism: eleven bits are allocated to identify the SUs of blocks received in the other direction on a one-to-one basis, zero (0) or (1) being written in the ACU depending on whether the corresponding SU was received error-free or in error

(*e*) 8 check bits.

As a MUM may spread over two blocks, break-in to a MUM may take place due to an ACU, whose position in the block is fixed.

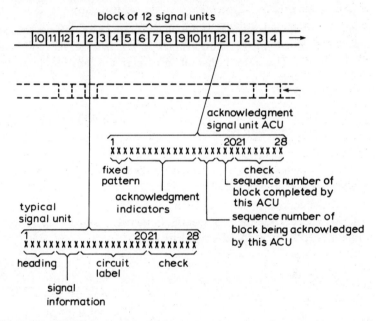

Fig. 11.7 *CCITT 6 system: block signalling*

With the block structure used, the block sequence numbers must repeat every 1120 ms at 2·4 kbit/s. In the original specification this was sufficiently long at 2·4 kbit/s to ensure that under normal conditions, there could be no ambiguity as to what is being acknowledged, even when a single-hop satellite is included in the signalling link in the telephony service (but see Section 11.5.3 'Influence of digital versions system 6 on the error control').

Retransmission procedures: The following procedures apply in error conditions:

(i) When a SU is corrupted, the appropriate acknowledgment indicator in an ACU is set negative. This procedure is also used if for any
reason the receive terminal cannot handle a SU.

(ii) When the corrupted SU is a LSU, that LSU is retransmitted (in
another block).

(iii) When a SU (or SUs) of a MUM is corrupted, the whole MUM is
retransmitted. This avoids associating SSUs with a corrupted ISU.

(iv) Corrupted SYUs (idle units) are not retransmitted.

(v) It will be noted that a received ACU is not itself acknowledged.
If an ACU, always known as such by its 12th position in the block,
is corrupted, it is assumed that all its acknowledgment indicators are
negative and all message SUs in the block waiting to be acknowledged
are retransmitted. In the limit, a block of 11 message SUs could be
retransmitted. This means that on occasions the same SU may be
received error-free more than once.

Processor reasonableness checks: The following irregularities on the
same call may arise with system 6:

(*a*) Unreasonable messages:
   (i) in incorrect sequence, as may arise on retransmission due to
   error detected
   (ii) an incorrect signal direction.
(*b*) Duplicated (superfluous) messages, as may arise owing to corruption of an ACU, or to drift compensation.

To resolve such irregularities, special processor procedures are
defined and are included in reasonableness check tables which cover all
possible stages in the signalling sequences. The reasonableness check
tables for system 6 are based on the logical signal processes and
sequences which should apply on a call.[8] Typically, if two IAMs are
received owing to corruption of an ACU or to drift compensation, the
two are compared. If they are identical, one is discarded, and if they are
different, a confusion signal can be sent.

The action taken by the processor on unreasonableness detected
depends on the nature of the irregularity, the various actions being:

(*i*) *Rejecting:* Messages or SUs recognised to be unreasonable or
superfluous are discarded.

(*ii*) *Waiting:* Unreasonable messages or SUs which may become
meaningful at a later stage of the signal sequence are provisionally
held. The waiting time should be longer than the retransmission delay
of the delayed message. The provisionally-held SUs are processed if

the arrival of retransmitted signals within the waiting period makes them meaningful. Otherwise, if they are still meaningless at the end of the waiting period, they are rejected with the exception of the cases where the held signal is a clear forward. In this case, the release-guard signal must be sent.

(*iii*) *Clearing:* If owing to an abnormal signal sequence an ambiguous situation arises which would result in a circuit being held unduly for a prolonged period, the circuit is cleared by the release sequence.

(*iv*) *Confusion signal:* If none of the above actions is suitable for resolving the situation created by the irregularity, the confusion signal is returned. On receipt of this, the distant exchange sends the clear-forward signal to release the connection and on certain irregularities an automatic repeat attempt is made to set up the call. The confusion signal will not be sent subsequent to sending the address complete signal or other signal causing the clearing from the store of address and routing information at the preceding exchange.

The irregularities arise in the message-transfer process of the signalling system. Some reasonableness checking is always necessary at the processor regardless of the type of error control involved in the message-transfer control. It is a question of degree. Error-control methods which do not allow incorrect sequence and duplicated messages to be passed from the signalling terminal to the processor would permit relatively simple reasonableness-checking covering undetected errors in the main. The system 6 error-control method allows out-of-sequence and duplicated messages to be passed to the processor, resulting in significant irregularity in addition to that due to other causes, such as undetected errors. As a result, the reasonableness checks associated with system 6 are complex and load the processor.

Drift compensation: As the two signalling channels of a signalling link are not synchronised with each other, the usual condition for telephony c.c.s., the two signalling terminals may get out of step. The following applies in system 6:

(*a*) The faster terminal may be ready to send an ACU before it has a block to acknowledge. In this case the ACU of the previous block is repeated.

(*b*) The slower terminal may find it has two blocks to acknowledge when it is ready to send an ACU. In this case it will omit to acknowledge one block altogether, the 12th unit being sent to complete the block but will not contain acknowledgment information. The far end then retransmits all message SUs of the unacknowledged block.

Influence of digital versions system 6 on the error control

The revision of the orignal analogue 2·4 kbit/s signalling specification
to include 4 kbit/s and 56 kbit/s digital versions was coincident with
a requirement to meet a maximum error-control loop delay of some
1200 ms (double-hop satellite). This requires more than eight blocks
in the error-control loop at all the signalling bit-rates specified (2·4,
4 and 56 kbit/s). With the retention of the 3-bit block-sequence number
field in the ACU, an implicit approach was adopted, the block-
completed and block-acknowledged indications being based on a so
called 'dynamic indexing' plan.

The plan takes advantage of the fact that after a signalling link is
in synchronism, much of the information transmitted in the block-
sequence number fields of the ACU is redundant. Once block counts
have been established in the initial synchronisation, and as long as
synchronism is maintained, local block counts kept at each end are
entirely adequate. Block acknowledgment only, without block-
sequence numbering, would suffice. As, however, drift may arise due
to the two signalling channels not being synchronised with each other,
some redundancy in block-number information is necessary in each
block to permit compensation for drift and for detection of loss of
block synchronisation. The 3-bit block-sequence number fields in the
ACU performs this function.

Each signalling terminal has two block counters, each of up to
8-bit capacity (256 possibilities):

(*i*) *Block completed counter:* This indicates the sequence number
of the last block transmitted by the terminal. The last 3 bits of this
number are also sent in the ACU of the block in the 3-bit block completed field of the ACU.

(*ii*) *Block acknowledgment counter:* This is updated by the 3-bit
block-acknowledgment sequence number in the incoming ACU and
thus indicates the sequence number of the block being acknowledged
by the last received ACU. To keep it updated even when the ACUs are
corrupted, the block acknowledgment counter is also incremented
whenever the 12th unit of a block is received corrupted.

Thus the signalling terminal and/or processor keeps track of the
block numbers and identifies blocks in the error-control loop. The
subsequent return in the block-acknowledged field of a received ACU
of the last 3 bits of the maximum 8-bit block-completed number of
the block-completed counter in the ACU in the opposite direction gives
the block-acknowledgment indication to a transmit terminal. Should

the 3-bit block acknowledged number in the received ACU be different from the last 3 bits of the block transmitted, drift or loss of block sychronism will have occurred.

11.5.4 Analysis of the system 6 error-control method
The following main areas are analysed, many points of less significance being omitted:

(*a*) *Correct sequence of messages:* Messages can get out of correct sequence on retransmission and would be passed in incorrect sequence from the signalling terminal to the processor.

(*b*) *Duplicated messages:* Messages may be received correctly more than once for a number of reasons e.g.

(i) corruption of ACU, resulting in all message SUs in the block being retransmitted, could result in the retransmission of messages already received correctly
(ii) drift compensation resulting in a block not being acknowledged
(iii) loss of an ACU during failure conditions.

Such duplicated messages would be passed from the signalling terminal to the processor.

(*c*) *Reasonableness checks:* The incorrect sequence and duplicated messages passed to the processor require complex reasonableness checks at the processor to deal with the situation. This complicates the processor and increases the processor load.

(*d*) *Unnecessary retransmissions:* These can occur owing to

(i) complete MUMs being retransmitted on corruption of constituent message SU(s) of MUMs.
(ii) Corruption of message SU(s) in unrequested retransmissions.

SYUs are known to be such at the transmit end by virtue of their positions in the block and thus by their positions in the block store. Corrupted SYUs, while returning negative acknowledgments in the relevant indicator bits in the ACU, do not cause retransmission of SYUs, or any other signals. Thus unnecessary retransmissions do not occur on corrupted SYUs.

(*e*) *Unrequested retransmissions:* These are said to arise when acknowledgments are corrupted. All message SUs in a block are retransmitted on corrupted ACU. Approximately 50% of all retransmissions will be unrequested due to this cause.

(*f*) *Drift compensation:* While drift is always liable to occur when the two signalling channels of a link are not synchronised with each other, complex compensation procedures are necessary when the ACU is in a fixed position in a block.

(*g*) *Unique ACU signal unit:* The use of a complete signal unit (ACU) for acknowledgment reduces the traffic handling capacity of a link by some 8·5%.

(*h*) *Fixed position of ACU in block:* While the unique ACU has significant disadvantage in regard to unrequested retransmissions on ACU corruption, it has advantage when it is in a fixed position in a block. The SU will always be known as being an ACU and thus corruption and undetected errors can never be interpreted as being a normal SU. On the other hand, the fixed position of the ACU introduces complication when it splits up the SUs of a MUM. If a retransmission is requested for any of the message SUs comprising a MUM, then the whole MUM must be retransmitted. Thus, even though the correct receipt of the first part of a MUM in a particular block may be acknowledged by the appropriate ACU, the possibility of a retransmission cannot be ruled out until the receipt of the subsequent ACU in the next block which acknowledges the SUs forming the remainder of the MUM.

(*i*) *Delays:* These are relatively long and are

(i) up to almost two blocks delay when requesting retransmission
(ii) up to one or two blocks delay in retransmitting.

It is concluded that the system 6 error-control method is complex, and that far too much requirement is placed on the processor to enable the error control to function. It will be understood of course that the system 6 error control was formulated on the basis that the processor would be involved.

11.5.5 Synchronisation

The transmit and receive terminals on a c.c.s. signalling channel must be in synchronism to enable transmitted information to be correctly received. It is not essential that the two signalling channels of a link be synchronised to each other and are usually not in telephony c.c.s.

Four levels of synchronism are necessary in system 6: bit, unit, block and multiblock. The number of blocks (and multiblocks) in the error control loop at any one time will depend upon the signalling bit rate, the traffic load and the error control loop delay of the signalling link.

(a) *Initial synchronisation*

In analogue 6, bit synchronisation is provided by the data modem (modulator and demodulator) which uses a 4-phase, synchronous, differentially-coherent modulation method. The four phases each represent two bits of information (00, 01, 10, 11) called dibits.[9] The modem then presents a continuous serial bit stream to the signalling equipment which then must identify units, blocks and multiblocks. In digital 6, modems not being required, bit synchronism is assured by the synchronised digital transmission system.

The block completed and block acknowledgment counters are set to zero during the initial synchronisation procedure and are checked periodically using the multiblock monitoring procedure.

On start-up of a link, each signalling terminal transmits blocks consisting of 11 SYUs and an ACU. The SYU (Fig. 11.5b) includes a 16-bit fixed pattern (for bit − analogue − and unit synchronisation) and a 4-bit field (for block synchronisation). The coding of the 4-bit field identifies the location of the SYU within the block (i.e. 1st, 2nd . . . . or 11th unit). The particular 16-bit pattern of the SYU presents an easily recognisable pattern with a variation at the end to aid in its detection, and because the pattern provides six dibit transitions, to aid the attainment of bit synchronisation by the modems in analogue c.c.s. Bit synchronisation is maintained by the transition between dibits.

The ACUs are transmitted initially with the 11 acknowledgment indicators set to 1, and the block-completed and block-acknowledged 3-bit sequence number fields set to 000.

After bit-synchronisation has been established (in the demodulator of the modem for analogue c.c.s. and by the transmission system in digital c.c.s.) the incoming bit stream is monitored to find an SYU pattern. When found, and thus an SYU verified, its position within the block is known and the ACU position located. In due course, the ACU on the incoming channel should be correctly received with its block number. At this time, the acknowledgment indicators in the next ACU sent back are set to reflect any detected errors in the SYUs of the associated receive block. Both block-sequence number fields in the ACU sent back remain at 000.

The reception of at least two ACUs at the transmit terminal which check correctly and acknowledge one (or more) SYU as having been received correct at the other end indicates that both terminals are in bit, unit and block synchronism.

Block-sequence numbering is initiated by the block-completed counter, and the block-completed sequence number in the next outgoing ACU from the transmit terminal being set to 1. Thereafter the block-completed counter and the block-completed sequence number are incremented by 1 each time an ACU is transmitted.

When the terminal receives an ACU having a block-acknowledged bit-field other than 000, the block-acknowledged counter at that terminal is set to this number received. Thereafter the counter is updated by the appropriate block-acknowledged number each time an ACU is received.

When the block-acknowledged counter is advanced for the first time, the number of blocks in the error-control loop may be determined by subtracting the contents of the block-acknowledged counter from the contents of the block-completed counter. This gives the maximum number of blocks which may apply in the error-control loop of the signalling link concerned at the signalling bit-rate used. If the initial synchronisation procedure has indicated more than eight blocks in the error-control loop, the multiblock monitoring procedure is used once every cycle of the block-completed counter. In this case, the multiblock procedure is used for block synchronisation. On receipt of a multiblock acknowledgment signal, the multiblock and block numbers are compared with the contents of the block-acknowledgment counter. Multiblock synchronism is assumed to exist if the received number is within -4 to $+3$ of the contents of the block-acknowledgment counter.

(b) Resynchronisation

Unit resynchronisation: Loss of unit synchronism results in continuous failure of SUs to check correctly, and when this occurs a signalling terminal takes unilateral action to resynchronise to the incoming bit stream. In any ACUs transmitted during this procedure, all the 11 acknowledgment-indicator bits are set to 1 and the block-acknowledged and block-completed numbers incremented as in normal operation. When synchronisation is re-established on the incoming channel, which is detected by SUs being checked correctly, the acknowledgment indicators are set according to the incoming SUs, i.e. normal operation is resumed. The SU error rate monitor continues to count SUs in error during this procedure, and changeover results if synchronisation is not re-established before the changeover criteria becomes operative.

Block resynchronisation: Loss of block synchronism is recognised when

(i) a valid SU, which is not an ACU, is received in the 12th position in a block
(ii) an ACU is received in other than the 12th position in a block
(iii) the block-completed number is not the one expected.

On recognition of any one of the above, the terminal will stop sending message signals and send only SYUs and repeated ACUs, the latter meaning that the acknowledgment indicators and the block-acknowledged number from the previous blocks are repeated.

When the terminal has identified the unit position in a block, either by recognising the SYU number or by identifying an ACU in the 12th position, and has also received an ACU which checks correctly, resynchronisation has been achieved. After successful block resynchronisation, the block being transmitted is completed with SYUs and an ACU. At least one complete block of 11 SYUs is sent before normal operation is resumed.

The first ACU sent after resynchronisation has been achieved has:

(i) the 11 acknowledgment indicators all set to 1
(ii) the block completed number set to the next in sequence
(iii) the block acknowledged number corresponding to the latest ACU received.

After the completion of block resynchronisation, multiblock synchronism should be checked if the multiblock condition is applicable, i.e. more than eight blocks in the error-control loop.

Multiblock resynchronisation: If the multiblock numbers in a multiblock acknowledgment SU are not within -4 to $+3$ of the contents of the block-acknowledged counter, a new multiblock monitoring signal is sent. If the result of the second measurement is not within the above limit, multiblock synchronism has been lost, and can be regained by updating the contents of the block-acknowledged counter to the obtained result. When the second multiblock monitoring signal is sent, the terminal will send only SYUs and ACUs for three blocks. Normal operation is then resumed and all messages transmitted in the interval between the two multiblock monitoring signals are retransmitted. If multiblock sychronism cannot be regained, changeover is initiated.

11.5.6 System 6 analogue and digital versions

Signalling in the digital environment, either point-to-point p.c.m. or integrated digital-switching and transmission, can be achieved by the built-in, channel-associated, p.c.m. signalling technique (Sections 6.2 and 6.3). Such signalling is underutilised in the telephony service. Improved utilisation can be achieved by treating the signalling facility on speech-circuit group rather than on a per-p.c.m. system basis; and c.c.s. is preferred when s.p.c. applies.

The c.c.s. technique has potential for rationalised signalling for analogue and digital application and this concept is adopted. The basic features of the c.c.s. system apply on a common basis, the analogue and digital differences being limited to the signalling bit-rates. Ideally, the common signalling features should be optimum for both analogue and digital, but as this is difficult to achieve in all the feature areas, some penalty may well arise in either application.

The pioneering CCITT system 6 was produced primarily for analogue application and its digital versions are far from optimum. The evolution of networks to integrated digital has shown the need for a new rationalised c.c.s. system optimised for digital rather than analogue, and this new system is currently being studied by the CCITT (Section 11.11).

Derivation of signalling data-bit streams

2·4 kbit/s (analogue version) and 4 kbit/s and 56 kbit/s (digital versions) signalling bit-rates are presently specified for system 6. Optional 4·8 kbit/s analogue is presently being considered. The data-bit streams are derived from the modem (analogue) and from the multiplex (CCITT specified p.c.m. systems) for digital, as follows:

(a) Analogue 2·4 kbit/s[9]

A 2·4 kbit/s modem (modulator/demodulator) is used. The modulation technique uses phase-shift-keying to transmit serial binary data over the analogue signalling channel. The binary-data signal is encoded by first grouping it into bit pairs (dibits), each dibit being represented by one of four possible carrier phase shifts. Thus, the output from the phase modulator consists of phase-shifted carrier pulses at half the data bit-rate. The phase shift between two consecutive modulation elements contains the information to be transmitted.

The principal requirements of the modem used for system 6 are:

(i) use of differential 4-phase modulation
(ii) use of differential coherent 4-phase demodulation
(iii) full duplex operation over a 4-wire signalling link
(iv) a modulation rate of 1200 bauds
(v) a bit rate of 2·4 kbit/s.

(b) 1·544 Mbit/s digital transmission 4 kbit/s signalling bit rate
As now specified, 24-channel p.c.m. systems (1·544 Mbit/s) do not permit a 64 kbit/s channel bearer for signalling purposes. Here, for signalling purposes, a channel is derived over which a stream of pulses is transmitted at 4 kbit/s. The binary data from the signalling terminal is transferred serially at a data transmission rate of 4 kbit/s to the primary multiplex. Here, each bit of the data stream is successively inserted into the *S*-position (Section 6.2.3).

In the receive direction, the primary multiplex extracts the bits from the *S*-position and transfers them serially to the signalling terminal.

(c) 2·048 Mbit/s digital transmission
A 64 kbit/s channel bearer is available for signalling in 2·048 Mbit/s 30-channel p.c.m. systems (Section 6.3) and signalling bit-rates of 4 kbit/s or 56 kbit/s may apply. A digital interface-adapter is provided between the multiplex and the signalling terminal.

4 kbit/s signalling bit rate: The binary data stream from the signalling terminal is transferred serially to the interface-adapter. Here, the 4 kbit/s data stream is modulated on a 64 kbit/s bearer channel such that 16 bits of bearer channel correspond to one bit of the 4 kbit/s channel. The 64 kbit/s data stream is transferred serially to the 2·048 Mbit/s primary multiplex in alignment with an 8 kHz clock (octet timing). At the primary multiplex, the 16 bits corresponding to one signalling information bit are inserted into the designated channel time slot of two successive frames.

In the receive direction, the primary multiplex extracts the bits from the designated time slot and transfers them serially at 64 kbit/s to the interface-adapter. Here, the 16 bits corresponding to one signalling information bit are extracted after detection and the binary data is transferred serially to the signalling terminal at 4 kbit/s.

56 kbit/s signalling bit rate: The system 6 signal unit is 28 bits. As this is not a multiple of the 8-bit octet it has weaknesses in digital transmission. Four bits are added to the signal unit to increase its size from 28 to 32 bits, for the reasons of (*a*) *octet timing:* The signal unit and padding bits together occupy exactly 4 octets. This simplifies the synchronisation and resynchronisation procedures.

(*b*) *slip detection:* With appropriate coding, the four padding bits may be used to detect octet slips.

No signalling information is carried by the four padding bits. Thus with the system 6 28-bit signal unit, the signalling bit-rate is 56 kbit/s on the 64 kbit/s bearer. The bit pattern for the padding bits is 1010 . . . . . 1010, which allows for detection of octet slips. As the pattern and location of the padding bits interfere with check-encoding, the padding bits are added after encoding and removed prior to decoding.

The binary data from the signalling terminal is transferred serially to the interface-adapter. Here, the 28 bits of the signal unit are placed in bit positions 1–7 of four 8-bit bytes. These four bytes are transferred serially at the data transmission rate of 64 kbit/s to the 2·048 Mbit/s primary multiplex, in alignment with an 8 kHz clock (byte timing). At the primary multiplex, the four bytes are inserted into the designated time slot of four successive frames.

In the receive direction, the primary multiplex extracts the bits from the designated channel-time slot and transfers them serially at the data transmission rate of 64 kbit/s to the interface–adapter, in alignment with an 8 kHz clock. In the interface-adapter, the bits 1–7 of each 8-bit byte are transferred serially to the signalling terminal at the 56 kbit/s rate.

The padding bits do not apply for the 4 kbit/s digital version. They would not be necessary in signalling systems having a signal unit size a multiple of the 8-bit octet.

11.6 Common channel signalling loading

With c.c.s., the signalling is time shared and a speech circuit does not have exclusive use of a signalling facility, thus a queue is built up from which signals are transmitted in order of their time of arrival and of their priority. Thus the loading in terms of number of speech circuits served per c.c.s. link must reflect acceptable queueing delays. This loading will be influenced by a number of factors, in particular:

(*a*) the number of busy hour calls per speech circuit
(*b*) the number of signals involved per call
(*c*) the mix of LSUs and MUMs; the greater the number of MUMs

for a given total signal information transfer per call, the less the signalling load

(*d*) the signalling bit rate: the faster the bit rate, the less the signal emission time and the greater the permissible number of speech circuits served.

(*e*) the error-control method: typically, a noncompelled error control method permits more speech circuits served per c.c.s. link than a compelled method, other factors being equal.

It is thus not practicable to specify a general maximum limit of speech circuits served per c.c.s. link. An adopted limit must take account of the various conditions applying so that the total signalling load per link is held to a level which will maintain an acceptable queueing delay. It is generally accepted that the signalling load per link should not exceed of the order 0·4 Erlang in normal operation. This allows temporary signalling overload in normal operation and for increased load in abnormal conditions such as security backup for a failed regular signalling link, when the load of the failed link is transferred to the backup link, the queueing delay increasing in the abnormal condition.

It is also generally accepted that the maximum number of circuits served per c.c.s. link adopted in practice for a particular network should reflect the probable size of the speech circuit groups applying, and the signalling network security philosophy adopted, for that network. Typically, one administration may adopt a maximum loading of, say, 1500 circuits per c.c.s. link in normal operation, whereas another may wish to adopt a smaller number, say 500 circuits, which reflects the different network conditions.

Indications of the maximum number of speech circuits served per CCITT system 6 c.c.s. link in the telephony service with acceptable queueing delays, assuming 15 busy hour calls per speech circuit, 0·4 Erlang signalling link loading, and representative conditions, are:

2·4 kbit/s signalling bit rate	1500 circuits served
4 kbit/s	2500
56 kbit/s	20 000

It is stressed that these values must be regarded with some reserve in view of the various assumptions adopted, nevertheless the broad indication is reasonable and is of interest. The values for the new CCITT optimised digital c.c.s. system with signalling bit rates 2·4, 4 and 64 kbit/s (Section 11.11) will be different as much will depend upon the finally resolved method of transferring the signal information – by messages of variable bit length or by fixed bit length SUs. In either event however, there are indications that the number of signalling bits transferred per telephone call will be somewhat greater than in system 6, and with noncompelled error control the maximum number of speech circuits served per c.c.s. link may well be somewhat less than for system 6 at the same signalling bit rate and for the same assumed conditions. Nevertheless, at 64 kbit/s, it is thought that some 20 000 speech circuits would apply. Compelled error control, should this apply, will significantly reduce the values relative to non-compelled.

While in some circumstances some 2500 or 20 000 speech circuits could be served per c.c.s. link in the telephony service in normal operation, it is extremely unlikely, particularly for signalling network security reasons, that such high values would be adopted in practice. It is thought that a more realistic maximum adopted would be of the order 1500 circuits in normal operation. This implies that a c.c.s. link would be very lightly loaded in the telephony service in normal operation at the higher signalling bit rates 56 kbit/s and 64 kbit/s. The maximum number of circuits served per c.c.s. link for the switched circuit data service would very likely be significantly less than for the telephony service in view of the shorter call durations.

Circuit label capacity
The size of the circuit label bit field should allow for greater possibilities than the requirement for the maximum number of circuits served per c.c.s. link in normal operation. This is to allow for the increased load in the abnormal condition when a link functions as security backup for a failed link. The CCITT specification for system 6 provides for an 11-bit (7-bit band, 4-bit circuit number) circuit label bit field (potential to identify 2048 speech circuits). This may be varied in particular national applications of the system, typically, the Bell system national version of system 6 proposes a 13-bit (9-bit band, 4-bit circuit number) circuit label field (potential to identify 8192 circuits), which reflects the large size speech circuit groups applying in the N. American network.

It is considered advisable that the circuit label bit field be of size giving flexibility to cater for possible future requirements not yet defined. Thus the potential for a larger field than that provided by the CCITT specified system 6 is preferred, which approach is proposed for the new optimised digital c.c.s. system (Section 11.11). A national network not requiring the large number would have possibility to use spare circuit label bits for other suitable national purposes.

The circuit label bit field is divided into band number bits and circuit number bits. The band number identifies a group of circuits (typically a p.c.m. system), the circuit number identifies a particular circuit in a group. The banding concept minimises the memory required at a signal transfer point (s.t.p.) in the quasi- and dissociated-c.c.s. signalling modes. The s.t.p. processor translates on only the band number bit field and not on the complete circuit label field, as signalling routing only is required at s.t.ps. A disadvantage of banding is the loss of some label capacity since all speech circuit group sizes are not an integer multiple of band size. Usually the circuit number field is 4-bits, which identifies a circuit within a group (band) of up to 16 circuits. Should spare bits from a large circuit label field be used for other national purposes, these should be spared from the band, and not from the circuit number, bit-field.

11.7 Signalling link security and load sharing

Various security and load-sharing arrangements are possible within a particular c.c.s. system specification and in national application would be matters of individual administration's choice. Bilateral agreement on particular arrangements may apply in international application.

11.7.1 Security

The basic concept of securing a signalling route is to supplement its regular signalling facility with a readily accessible alternative brought in by automatic procedures. This requires the provision of at least two signalling facilities, termed 'regular' and 'backup' respectively for convenience.[10]

The backup may be:

(*a*) A standby nonsynchronised link switched into service when required. (e.g. a speech circuit nominated as the signalling backup).
(*b*) A synchronised signalling link in the same signalling relationship (e.g. regular and backup signalling links on an associated signalling route).
(*c*) A separate signalling relationship (e.g. quasi-associated signalling as a backup for associated signalling).

There are other possibilities.

Further signalling security can be achieved by fabricating a backup under the control of network management. As this would not be an automatic procedure, it would not be a function of the signalling system.

It is clearly preferable that the backup link(s) should have immediate signalling capability on failure of the regular link. Thus while (a) is a valid approach, the procedures required to prove the signal carrying capability of the backup takes time and adds to the complexity. Approaches (b) and (c) are preferred.

Inbuilt associated signalling security: This implies that an associated-signalling route has inbuilt features to achieve, by automatic procedures, the degree of signalling security required. The concept requires the provision of at least two signalling links in that associated signalling relationship, preferably diversely routed. The provision of more than one signalling link to achieve the security can be justified on the relative cost insensitivity of providing additional link(s) due to:

(i) the inherent 'common equipment' feature of c.c.s.
(ii) the application of such signalling to concentrated traffic
(iii) in the digital environment, the low cost of signalling time slots in digital transmission systems due to the cost sharing in multiplex transmission.

Associated signalling module security: There is an interplay between the maximum number of speech circuits to be served per c.c.s. link in normal operation and the size of the circuit label bit field to determine the number of signalling links which may apply in an associated signalling relationship. The term 'signalling module' is adopted in this regard. Typically, with a 6000 circuit label capacity, the maximum size signalling module would be four links for a maximum loading of 1500 speech circuits per c.c.s. link, and six links for 1000 circuits per link. In practice, not all modules in a network would be of maximum size, they would be of size dependent upon the sizes of the speech circuit groups served. Many associated signalling routes in many networks would consist of two signalling links only (1 + 1), which could be smaller than the maximum size signal module. The larger speech circuit groups would require more than two signalling links and the very large groups would require more than one signalling module.

Various security approaches may apply with the signalling module concept, but the following are thought logical:

(a) Any one signalling link in the module should be capable of being a backup for any other links in the module. This precludes a particular link being the single nominated backup link.

(*b*) When the module is bigger than $1 + 1$, the potential to allow for more than one failed link in the module should be admitted in the automatic security procedures. In theory, failure could be allowed to the limit of one signalling link carrying the full load of the module, which approach could result in significant increase in queueing delay. In practice, an administration may wish to limit the extent of permitted failure within a module to be covered by the automatic security procedures. In a two-link module of course, one link must carry the total signalling load on failure of one link.

(*c*) When a number of signalling modules are provided on a signalling route, each module should be autonomous, being self-contained in the sense that no traffic assigned to a module would ever be handled by another module. Each module would be inbuilt secured and would not rely on another module for signalling security.

The circuit label bit field of the international system 6 is of modest size (11 bits, 2048 circuits). This is mainly suitable for a $1 + 1$ signalling concept and would restrict the module size in a module concept. If necessary of course, national variants of the international system 6 could arrange for additional circuit label bit(s) within the 6 SU size, as instanced by the 13-bit circuit label field of the Bell system version of system 6 for application in N. America.

Quasiassociated signalling security: A number of possibilities arise, typically

(*a*) Assuming the quasiassociated signalling route to have associated signalling on the relevant constituent routes (Fig. 11.2*b*), and the associated signalling modules to have inbuilt security, the quasiassociated signalling route is then secured in the inbuilt concept without the necessity for further arrangements.

(*b*) Quasiassociated signalling without associated signalling on the constituent routes (Fig. 11.2*c*). Assuming quasiassociated signalling to be the principal signalling mode in a network, situations arise where it would not be of the type having associated signalling on the constituent routes of the quasisignalling relationship. A signalling module concept could apply for quasiassociated signalling of this type. The module size would be determined by the same considerations as apply for an associated signalling module, the module would have inbuilt security and all the considerations postulated for the associated signalling module would apply. The smallest quasisignalling module would be two links $1 + 1$.

Unlike the associated signalling module case, s.t.p. security considerations arise in quasiassociated signalling of the type being considered and the s.t.ps. may or may not be duplicated depending upon an administration's policy. If duplicated, the operation of the quasiassociated signalling module would need to take account of the duplicated s.t.ps., but the basic philosophy, and the basic working arrangements of the quasiassociated signalling module would not be disturbed with duplicated s.t.ps. An example of this type module is the proposed Bell system quasiassociated signalling in the N. American network (Fig. 11.8). Here it is proposed that the complete network comprises a number of signalling regions, quasiassociated signalling being applied between the regions and thus in an organised manner on a complete network basis, as distinct from quasiassociated signalling applied on particular signalling route situations and thus not as the principal mode for the network. Quasiassociated signalling is proposed as the main c.c.s. mode in the N. American network. In some situations, associated signalling may apply on particular speech circuit groups and when applied, the basic capacity of an associated signalling link is 1500 speech circuits maximum in normal operation at the 2·4 kbit/s signalling. Under fault conditions, an associated signalling link may carry twice this load, but with an increase in signalling delay.

This same capacity, 1500 speech circuits maximum, applies to the quasiassociated A-links (Fig. 11.8). A toll office of 3000 speech circuits or less would connect with the quasiassociated signalling network by providing a single pair of A-links, one link to each regional s.t.p. An additional A-link pair would be provided for each increment of 3000 speech circuits terminated in the toll office. Each A-link pair functions as an autonomous pair, having no operational interaction (such as failure backup) with other A-link pairs.

The interregional quasisignalling quad (four-signalling link module fully inbuilt secured) forms a functional unit serving up to 6000 speech circuits between a pair of signalling regions. Additional autonomous quads would be provided for each increment of 6000 interregional speech circuits. As mentioned, this arrangement requires a 13-bit circuit label field (9-bit band, 4-bit circuit number).

Exploitation of c.c.s. modes for signalling security: Signalling security can be achieved by employing a backup signalling mode of a different type, or another mode of the same type, from the regular signalling mode. Both inbuilt security and regular modes secured by backup modes can be applied in the one network. The following are possibilities:

(*a*) When it is not possible, or not economic, to inbuilt secure an associated signalling route (i.e. when a single associated signalling link applies), such a route could rely on a quasiassociated signalling backup for its security.

(*b*) A quasiassociated signalling backup for a regular quasiassociated signalling route, the associated signalling constituent routes of the regular route not being inbuilt secured.

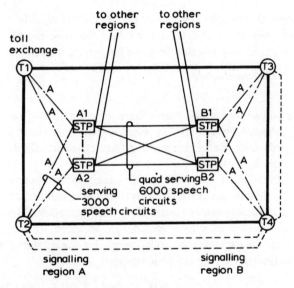

Fig. 11.8   *Bell system c.c.s. network configuration*
————— intraregional c.c.s. link
————— interregional c.c.s. link
--------- associated c.c.s. (as required)
========= speech circuit group

(*a*) is liable to arise in practice. (*b*) introduces circuit labelling problems, and routing complexities could arise should the quasi-associated backup be a regular quasiassociated signalling route in another relationship, with possible further quasiassociated backup. While not preferred, (*b*) could be applied under strictly controlled conditions within the circuit labelling capacity applying.

It is thought that the associated and quasiassociated signalling modes would be suitable for most national networks. Secured quasi-associated signalling could be applied in particular situations where associated signalling is not justified, or as in (*a*) above.

11.7.2 Load sharing

The technique of distributing the total module signalling traffic over the individual links is termed 'load sharing', it being understood that all the links, regular and backup, carry signalling traffic in normal operation.

There are two main categories of load-sharing mode:

(i) *Random mode*

This may be either

(*a*) messages distributed randomly on the links of the module
(*b*) all messages for a call transferred over the same signalling link, which link is randomly selected from the links in the module.

(ii) *Predetermined mode*

This may be either

(*c*) predetermined on a call basis. There are a number of variants of this basic approach, but all aim in principle that all messages for a call be transferred over the same signalling link. There is little difference between this and (*b*) above.

(*d*) predetermined on a speech circuit basis. All the speech circuits served by a module are distributed over all the signalling links in the module on a predetermined basis in normal operation. The circuit labels of a number of speech circuits are allocated to a particular link, and all messages for all calls carried by the allocated speech circuits are transferred over the same signalling link.

With (*a*), SUs or messages may get out of correct sequence on message transfer. With (*b*)–(*d*), all messages for a particular call are transferred over the same signalling link, which has potential to avoid out of sequence due to the load-sharing technique. SUs or messages may get out of correct sequence with certain error-control methods (e.g. system 6). With such methods, there is no compelling reason to adopt load-sharing techniques which aim to avoid out of sequence. Thus any one of (*a*)–(*d*) could be applied. With error control methods purposely designed to avoid out of sequence on message transfer, which is preferred, it would be illogical to adopt load-sharing technique (*a*), and here it is considered that the choice rests between (*b*) and (*d*).

Technique (*b*) has merits:

(i) The working links in a module would automatically take the load on signalling link(s) failure. Similarly when failed link(s) are restored to service.

(ii) Would automatically achieve the limit of one signalling link taking the whole load of a module.

(iii) Avoids a predetermined nominated link for security.

(iv) Meets the preferred principles of module signalling link security as discussed without requiring procedures beyond busying out failed link(s) and automatically diverting traffic from these link(s).

(v) Avoids a set program to allocate speech circuits to signalling links on a predetermined basis, which also avoids updating the allocation when speech circuit groups are extended.

(vi) Has the possibility of equalised distribution of signalling load on the working links in a module on signalling link(s) failure.

On the other hand (*b*):

(i) Necessitates an arrangement to ensure that all messages for a call are transferred over a randomly selected signalling link. This could be somewhat complex with certain approaches to processor system architecture.

(ii) On signalling link failure, requires a link to be nominated for the retrieval procedure if messages are to maintain correct sequence, with consequent complexity. For practical arrangements, it is often preferred to avoid nominated retrieval links, and retrieval on the link changed over to is considered preferable.

The features and merits of (*d*) are:

(i) Changeover to another link in a module on signalling link failure. The circuit labels of a failed link are transferred in total to another link (or alternatively, distributed over the remaining working links).

(ii) On restoration of a failed link back to service, all the circuit labels allocated to the failed link prior to the failure are transferred back.

(iii) Would automatically achieve, in potential, the limit of one signalling link carrying the whole module load on multilink failure.

(iv) Avoids requiring a nominated link for retrieval.

(v) Meets the principles of module signalling link security as discussed under the control of the changeover and changeback procedures.

(vi) Avoids any further arrangement in the processor to ensure that all messages for a call are transferred over the same signalling link. This feature is assured by the predetermined allocation of circuits labels to signalling links, which arrangement may be more capable of accommodating various approaches to processor function system architecture.

On the other hand (*d*):

(i) Necessitates an administrative procedure to allocate speech circuits to particular signalling links and to update the allocation on speech circuit group extension.

(ii) Compared with random selection, could result in a signalling link in a module carrying a significantly heavier load than others on signalling link(s) failure when all the circuit labels of a failed link are transferred to one working link. Depending upon the signalling load in normal operation, this could result in queueing delay problems in the abnormal condition of signalling link(s) failure, particularly at slow signalling bit rates.

On balanced assessment, there is probably little between load sharing techniques (*b*) and (*d*) and either could be applied when it is required to avoid out of sequence, which is preferred. Should simple arrangements be the main consideration, it is thought that this would be achieved by (*d*) in combination with the transfer of all the circuit labels of a failed link to one backup link. This could result in a significant increase in queueing delay in certain circumstances with certain size signalling modules depending upon the extent of multilink failure in a module permitted by the automatic security procedures.

11.8 Changeover, retrieval and changeback

The changeover, retrieval and changeback procedures consequent on signalling link failure are detailed and the principle only will be discussed. At this time, the procedures for international system 6 are the only ones available.[10] These could form the basis for procedures required for any future c.c.s. system(s). Administrations are free to adopt the CCITT international procedures for national application of international c.c.s. system(s), or, if desired, could adopt procedures more suited to the conditions of individual national networks, which approach would not disturb the basic specification of international c.c.s system(s).

Changeover
Signalling traffic is diverted from a regular link to an alternative signalling backup when the regular link is recognised as having failed. Failure may be caused by (Section 11.5.3):

(*a*) loss of the analogue data carrier or loss of the digital frame alignment
(*b*) continuous failure of SUs (or messages) to check correctly

(c) unacceptable intermittent failure of SUs, or messages, to check correctly

(d) loss of relevant synchronism for a certain duration.

Each signalling terminal in c.c.s. is equipped with appropriate monitoring equipment to recognise failure due to any one of the above causes. In system 6, the error rate monitor initiates changeover when:

(i) X consecutive SUs are received in error for 350 ms,

or

(ii) 2% of SUs are received in error out of Y SUs received.

X and Y are assessed as follows for the telephony service:

Signalling bit rate (kbit/s)	Number of SUs	
	X	Y
2·4	31 ± 1	2500
4	50	4200
56	700	60 000
64	800	68 500

On detection of failure, each terminal starts sending 'faulty link information' of appropriate form on the link just failed. Typically, in system 6, this could consist of alternate blocks of 11 changeover and 11 SYU signals plus ACU. If the terminal has lost synchronism, the normal synchronising procedure is started. Where applicable, the synchronising procedure on the backup signalling link is initiated.

Retrieval

Messages and acknowledgments may be lost in the transfer process on signalling link failure. Thus a terminal would not be aware whether or not all or some of the messages in the retransmission store had been received correct at the receive terminal. Retrieval is the process of recovering from this situation to avoid loss of messages on changeover. For simplicity, and also to avoid out of sequence messages, retrieval should be performed over one link, and for further simplicity, preferably over the backup link.

There are a number of retrieval possibilities, typically:

(a) Retransmission of all the contents of the retransmission store. This could result in duplicated messages, as arises in system 6.

(b) Retransmission only of messages known to have been received corrupted. This requires the receive terminal to inform the transmit terminal, by backward signalling over the retrieval link, as to the last message received correct over the failed link should cyclic retransmission be used (not in system 6).

(c) With SU (or message) sequence numbering, the retrieval retransmission carrying its original sequence numbering, (a) or (b) could apply and retrieval start and stop indicators would be necessary if retrieval is on a working link due to the differences in number sequences on the failed and backup links.

(d) With sequence numbering, the retrieval retransmission carrying the sequence numbering of the working backup link over which retrieval is performed. Assuming SU (or message) sequence numbering, (b) would apply and retrieval start and stop indicators would not be essential if retrieval is on a message, LSU, or MUM basis. A start retrieval indicator could be of value if retrieval is on an SU basis to cover the situation of a partial MUM having been received correct over the failed link.

There are other possibilities. The retrieval method adopted would depend upon preferences and circumstances.

Changeback

Once the backup facility has been taken into service, the regular signalling link should not be brought back into service for signalling traffic until it has been checked to give satisfactory performance for a period of about one minute. The proving period begins when either terminal has regained synchronism on the failed link.

End-of-failure monitoring equipment is provided at each terminal. In system 6, the failed link is not restored to service until a SU error rate of 0·2% or less has been achieved in the proving period. In the telephony service, the end-of-failure is assessed to have been achieved when not more than:

> 10 SUs at 2·4 kbit/s
> 16 SUs at 4   kbit/s
> 240 SUs at 56   kbit/s
> 266 SUs at 64   kbit/s

are received in error in the one minute proving period.

On end-of-failure recognised, a terminal will cease sending faulty link information. In system 6:

(a) The faulty link information is replaced by continuous blocks of SYUs (plus ACUs).

(b) To return to the regular link, an exchange, say A, initiating the changeback sends two load transfer signals on the regular link.

Exchange B responds with a load transfer acknowledgment signal on the regular link and transfers signalling traffic from the backup to the regular link. Receipt of one load transfer acknowledgment causes A to transfer signalling traffic from the backup to the regular link.

11.9 Continuity check of the speech path

Unlike speech path signalling, c.c.s. does not monitor the speech path and other arrangements must be made to assure speech path continuity (line plus switching), otherwise there could be possibility of a connection being set up on the signalling path but without the speech transmission facility. Speech path continuity assurance may be by:

(a) a per-call continuity check, or alternatively
(b) a statistical method of routine testing of idle transmission circuits and idle switching paths (and no per-call continuity check).

The per-call continuity check is made prior to the conversation. It may be loop link-by-link or end-to-end and consists of the transmission and detection of receipt of an audio frequency on the speech path. Typically, in system 6, the per-call check is line loop link-by-link with a 2000 Hz check tone frequency. A separate cross-office check is made.[11]

To admit the possibility of 2-wire circuits, which may arise in national networks, the per-call check could consist of different forward and backward check frequencies.

On 'check not OK' a repeat attempt is made to set up the call on an alternative speech path. To cater for intermittent faults, the original speech path is rechecked and passed to maintenance attention if still found to be faulty.

11.10 Signal priority

Transmission (queueing) delays can arise with c.c.s. As c.c.s. may have a considerable signal transfer requirement, embracing a number of different services and signals within the different service categories, delay to time sensitive signals or to time sensitive service categories could give rise to system problems and complexity. For this reason, some signals and some categories may be given priority over others for transfer purposes and the processor function is so programmed. The priority program will depend upon the characteristics of the c.c.s.

system, typically, with representative loadings a high speed 64 kbit/s c.c.s. system would transfer information at a faster rate and the queueing delay would tend to be less, relative to a slower speed system, and thus the extent of the priority requirement would tend to be less. As the arrangements for signal priority complicate c.c.s., it is logical that all efforts be made to minimise the priority requirement.

The following priority rules apply for international system 6:

(*a*) ACUs (12th position signal unit of each block) have absolute priority for emission at their fixed predetermined position.

(*b*) Faulty link information has priority over all signals except ACUs.

(*c*) The answer signal has priority over other waiting telephone signals and signalling system control signals except those in (a) and (b).

(*d*) All other telephone signals (LSUs and MUMs) and all other system control signals, except synchronisation signal units, have priority over management or other signals concerned with the bulk handling of information.

(*e*) Any signal which is to be retransmitted will take precedence over other waiting signals in the same priority category.

(*f*) Management signals have priority over synchronisation signal units.

(*g*) Synchronisation signal units have no priority.

It is the intention to minimise the priority requirement in the new optimised digital 64 kbit/s c.c.s. system. Typically, it could be reasoned with some logic that due to the high speed of information transfer, there would be no need to give priority to the answer signal. In slower systems, answer signal priority has been considered desirable to avoid speech clipping due to speech transmission path line splits on inter-working v.f. signalling systems.

Signal priority may be associated with the 'break-in' feature, which allows a priority message to break into the emission of a lower priority MUM to avoid delay in the emission of the higher priority message. The following break-in rules apply for international system 6:

(*a*) Potential for a priority LSU to break into a MUM is provided for in the specification of the system, but initially this will not apply except for ACUs.

(*b*) All telephone signals to have potential to break into a network management MUM or other MUMs concerned with the handling of bulk information.

The break-in feature gives rise to significant complication in c.c.s. systems and it is not proposed to adopt it for the new optimised

64 kbit/s digital c.c.s. system. To facilitate this, the maximum bit length of messages in the new system may well require to be limited.

11.11   CCITT No. 7 optimised digital common channel signalling system

11.11.1 Basic concepts

The pioneer CCITT 6 c.c.s. system was produced primarily for analogue application, the digital versions included in the specification being produced by subsequent study, but retaining the basic features of the analogue system. In the result, system 6 has limitations in the digital environment. In the background of experience gained in the production of system 6, and with recognition of the evolution of many networks from analogue to digital (integrated switching and transmission), the CCITT is presently engaged on the specification of a new c.c.s. system. This system will be optimised for the 64 kbit/s digital environment, will include other digital signalling bit rate(s), and analogue bit rate(s). Particular regard is being paid to national network requirements, and as it is aimed to produce a simpler and more flexible system than 6, a main desire for national application, there is little doubt that many administrations, programming c.c.s. application, will adopt this new system for their national networks in the s.p.c. digital (or analogue) environment. This CCITT study is not yet concluded, but it is clear from the approaches emerging that many features of the new system will be different from those of system 6, which, in part, reflects further knowledge in the c.c.s. art since system 6 was specified.

The c.c.s. technique has significant potential for signalling rationalisation and an objective of the new system is to use the same basic system for international, regional and national for a variety of services, digital and analogue. This would allow the potential, from the signalling aspect, for dedicated digital service networks (telephony, data, etc.) to evolve to a single multipurpose network, should this ever be a future requirement. While the new system will be optimised for digital, there is little doubt that it will be a superior system to 6 for analogue, as well as for digital. UK pioneered much of the new system (CCITT No. 7) and has adopted it for its System X digital network.

Rationalisation implies a flexible signalling system and it is proposed that the new system be based on a number of concepts to achieve this (Fig. 11.9).[12] These concepts cater for various services and applications without mutual interaction, and give potential to cater for future requirements not yet defined.

Commonality concept: Options incorporated in the same signalling system would permit reasonably optimum arrangements for different services and applications (international, regional and national) of the same service. A single c.c.s. system for various services within a particular network could also be realised by appropriate choice of options incorporated. Desired simplicity of national network c.c.s. is an example of the merit of the option approach, as here advantage could be taken of national network conditions, not present in, say, international, which could contribute to signalling simplicity. Typical options concern the signalling bit rate, the error control method, and message bit length, formats and codings for different services.

Functional concept: A functional concept embracing processor function, user function, and message transfer function, allows change to any one function without significant impact on the others. It involves, for example, flexible arrangements in the message transfer function to enable an error control option to be exercised without involving the processor. This implies that the error control should be associated with the signalling terminals, the processor never being involved.

User subsystem concept: This is part of the functional concept. User subsystems (separated parts of the processor) are functional parts which control the message requirements for particular services, typically, telephony call processing, data call processing, network management, network maintenance, centralised call accounting, etc. Such subsystems (Fig. 11.9) may be regarded as being the service dependent parts of the signalling system. They define message, signalling and call control procedures, and the function and coding of the signalling information for different services. Each user subsystem module deals with a particular service, and the concept allows the system to evolve in that additional services may be catered for by the addition of appropriate subsystem modules.

The message transfer function (the c.c.s. system) allows user subsystem modules held on different processors to communicate with each other by means of addressed messages. This processor subsystem addressing (type of service) directs the message to the appropriate subsystem module, the subsystem addressing identifying the service for which the message is intended. Should corresponding user subsystems communicate with each other for the interchange of information not related to call handling, such messages will be conveyed by the message transfer function.

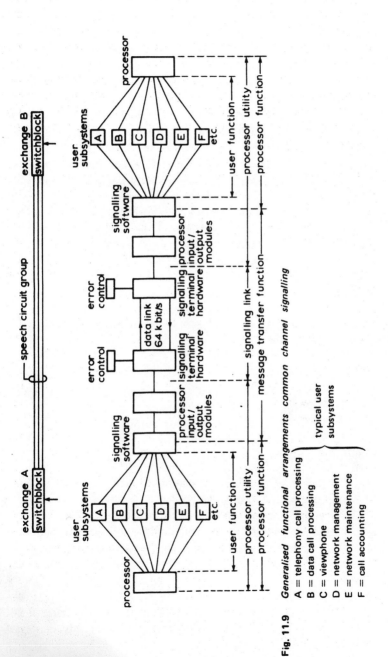

Fig. 11.9 *Generalised functional arrangements common channel signalling*

A = telephony call processing

B = data call processing

C = viewphone

D = network management   } typical user subsystems

E = network maintenance

F = call accounting

11.11.2 *Signalling bit-rate*

A 64 kbit/s bearer is conveniently available for c.c.s. in 30-channel digital transmission, and with a message bit length a multiple of the 8-bit octet, the signalling bit rate can be this bearer bit rate. There is no reason to depart from this rate for telephony c.c.s., indeed, at this rate the message signalling activity will be very light with the speech circuits per c.c.s. link loading likely to apply. Commonality c.c.s. for data services introduces further considerations as such services wish a faster connection set-up time than that usually acceptable for telephony. Relative to telephony, the data service may require a greater signalling load in terms of calls per second handled per c.c.s. link, which could slow data call connection set-up. Study concluded however that signalling bit rates greater than 64 kbit/s would not significantly reduce queueing delay, nor connection set-up time in the data service, and the new c.c.s. system will be optimised for 64 kbit/s. Optional signalling rates 4 kbit/s (for 24-channel digital) and 2·4 kbit/s (for analogue) will be used. Other rates could apply as appropriate (typically optional 4·8 kbit/s analogue).

11.11.3 *Error control*

The combination of the error control method and the various consequential procedures constitute perhaps the most complex part of c.c.s., and as a simple system was desired, the error-control problem was reassessed.

Forward error correction: Here, the receive end performs both error detection and correction, and error correcting codes ('Hamming' codes)[13] used, sufficient check bits being required to enable the original message to be reconstituted on error detected. Retransmission of messages does not arise, and as the method has some potential for simple arrangements, the following analysis is made in regard to c.c.s.:

(*a*) It is independent of the propagation time of signalling links.
(*b*) The procedure is simple; acknowledgment signalling and sequence numbering of messages not being required.
(*c*) Messages cannot get out of correct sequence, nor be duplicated.
(*d*) Error correction is reasonably rapid relative to correction by retransmission, which has merit on long propagation time signalling links.
(*e*) No retransmission store required.
(*f*) The automatic changeover procedure is simple and retrieval not required. This, however, is valid only when the possibility of loss of messages is acceptable when diverting signal traffic to an alternative signalling facility.

On the other hand, forward error correction:

(*g*) Requires far more redundancy in check bits relative to that required for error detection only (or for a given number of check bits it is possible to detect more errors than it is possible to correct). The coding and decoding systems for the check bits are relatively complex.
(*h*) Cannot request retransmission of a message which has been error detected but not corrected.
(*j*) Complicates the receiving terminal.

The following points are made in the consideration of the above:

(i) As c.c.s. is error-free for most of the time, penalties (*g*) and (*j*) in particular are not justified.
(ii) Error-free messages are slowed.
(iii) In any c.c.s. system there is the need to safeguard against major breakdown of a signalling link. Correction by retransmission has a significant merit relative to forward error correction in that it has an inherent ability to direct signalling traffic to an alternative signalling facility without loss of signal information.
(iv) With reasonable arrangements, correction by retransmission is superior to forward error correction in both throughput and undetected error rate.

It was concluded that forward error correction would have limitations in c.c.s. application and that correction by retransmission be adopted for the new c.c.s. system. This supported a previous conclusion in regard to system 6, but not the system 6 error-control method, which would not be suitable for the various reasons discussed in Section 11.5.4.

Guidelines for a preferred error-control method: From the analysis Section 11.5.4, it is considered that the main guidelines are:

(*a*) Messages should not get out of correct sequence in any working situations of the error control.
(*b*) Duplicated messages should not arise, but if they do they should be detected and dealt with by simple means at the signalling terminal, the processor function never being involved.
(*c*) Unnecessary and unrequested retransmissions should not arise.
(*d*) The processor function should not be involved in the error control function.

Guideline (*d*) requires all messages passed from the message transfer function to the processor function to be valid in all respects, except for errors undetected by the error control. Ideally, to achieve this, the error control method itself should be such that neither out of sequence nor duplicated messages occur, but this may not be conveniently realisable.

It would be illogical to require a signalling terminal to detect and correct for messages in incorrect sequence (*a*) as this would necessitate a measure of reasonableness checking and correction logic at the signalling terminal, to complicate the system. It is thought that the error-control method itself should be such that out of sequence messages do not occur. With sequence numbering (or the equivalent) of messages, it would be a simple matter for a signalling terminal to detect for duplicated messages (*b*) and to discard messages already received. Thus, if necessary, duplicated messages could be admitted on the message transfer function.

A philosophy of 'ignoring corruptions' in the error control could eliminate both unnecessary and unrequested retransmissions (*c*). Here, no action is taken on error detected, the system merely awaits the receipt of the next error free SU (or message) before taking decision as to the appropriate action to be taken. With this philosophy:

(*a*) corrupted SYUs (and corrupted messages which would have been unacceptable if error free) would not cause unnecessary retransmissions.
(*b*) to eliminate unrequested retransmissions, the acknowledgment indications are required to be in a continuous stream.

An error-control method based on the above philosophy precludes adoption of the system 6 error-control method because of the block transmission and the single acknowledgment (ACU).

'Ignore corruptions' error-control method: The error-control method for the new CCITT No. 7 c.c.s. system is not yet finalised by the CCITT, but present indications are that it will be based on the ignore corruptions philosophy. The following is a possible realisation of the philosophy (Fig. 11.10), is presented to demonstrate the principle, and should not be regarded as being the final agreement in the detail. Basic features are:

(*a*) Cyclic retransmission on error detected, meaning that the corrupted message SU (or message) and all the message SUs (or messages) following it in the retransmission store are retransmitted. Corrupted SYUs are not corrected and do not cause retransmission.

(*b*) Each SU (or message) contains both a forward sequence number (FSN) and a backward sequence number (BSN).

(*c*) SYUs transmitted between message SUs (or messages) all carry the same forward sequence number of the next message SU (or message) to be transmitted.

(*d*) Each SU (or message) contains a forward indicator bit (FIB) and a backward indicator bit (BIB), used to control retransmissions.

The philosophy may be applied with message SUs (LSUs and constituent SUs of MUMs) of fixed bit length or with messages of variable bit length. The FSN, BSN, FIB, BIB and check bits are 'housekeeping' bits, concerned with the control of message transfer, and are not passed to the processor. Drift problems do not arise.

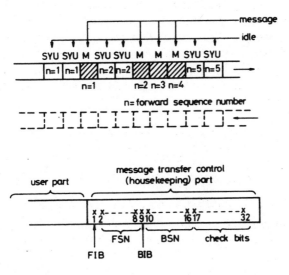

Fig. 11.10 *Typical realisation ignore corruptions error control*

FSN: Message SUs (or messages) are numbered sequentially to enable the correct sequence to be preserved during the message-transfer process, and to identify message SUs (or messages) when retransmission is required. The range of sequence numbers must be such that a retransmission store never contains two message SUs (or messages) bearing the same FSN and reflects the maximum number of message SUs (or messages) liable to exist in the error-control loop.

BSN: The BSN contained in a unit (or message) transmitted from a terminal indicates the FSN of the next message SU (or message) that terminal is prepared to accept. If a terminal has just accepted a message SU (or message) with a FSN of n, the next unit (or message) to be transmitted from that terminal will have a BSN of $n + 1$. When accompanied by a retransmission request (change in polarity of the BIB), the BSN indicates the point in the transmitted sequence at which the retransmission should start.

BIB: Is used by a receive terminal to signal back to the transmit terminal that a retransmission is required, the request being made by reversing the polarity of the next BIB to be transmitted. Once a receive terminal has requested a retransmission, no further reversals of polarity are made until a new retransmission is required. No action is taken by a terminal when a received unit (or message) is corrupted. The system waits (ignores corruptions) until an error-free unit (or message) is received. The polarity of the BIB is reversed to request retransmission when the FSN of this error free received unit (or message) is not one greater (cyclically) than the last correctly received, and accepted, message SU (or message). After requesting a retransmission, a terminal takes no further action until a change in polarity of a received FIB indicates the retransmission, the terminal then reverting to normal operation.

FIB: The polarity of the FIB in any particular unit (or message) transmitted by a terminal is always the same as that of the BIB contained in the last unit (or message) correctly received by that terminal. Since a change of polarity of the BIB is used to request a retransmission, a change of polarity of the FIB will indicate that the retransmission is taking place. The polarity of the FIB is reversed when a retransmission is started, and after a reversal the polarity is unchanged until the next retransmission is started. The FIB and BIB thus perform a 'handshaking' procedure.

Operation
Information is transferred in the two directions of a signalling link, but as the two directions are independent from the error-control point of view, it is only necessary to consider one of them (A-B).

(a) *Error-free:* A message SU (or message) is accepted at B provided it has a FSN which is one greater (cyclically) than the previous one accepted. The BSNs returned to A signal the progress made in the process of accepting correct message SUs (or messages). When received at A they enable the message SUs (or message) to be cleared from the retransmission store.

(*b*) *Error:* The corrupted unit (or message) is ignored. The system waits the receipt of the next error free unit (or message), the FSN of which will serve to indicate whether or not the corruption was a message SU (or message) or a SYU. If the FSN is the next in the sequence expected by B, the corruption must have been a SYU, in which case no further action is necessary and normal operation applies. The receipt of any other FSN serves to indicate that a message SU (or message) was corrupted and a retransmission necessary. B requests this by reversing the polarity of the BIBs sent back B to A and does not accept further message SUs (or messages) until it has detected that a retransmission has taken place.

The retransmission takes place when the BIB polarity is reversed and the reversal recognised at A, the first message SU (or message) of the retransmission being identified by the latest BSN received at A. The fact that the retransmission has taken place is indicated by reversed polarity of the FIBs transmitted A to B, commencing with the first message SU (or message) of the retransmission. B detects this FIB reversal to recognise the retransmission. If there are no further errors, the first message SU (or message) of the retransmission is accepted by B, which then reverts to normal operation.

If the first one (or more) message SU (or message) of the retransmission is corrupted, when B eventually receives an error free SU (or message) it will detect from the FIB reversal that a retransmission has taken place, but the FSN will be different from that B will accept. A second request for retransmission is made by again reversing the polarity of the BIBs sent back. B then rejects all further message SUs (or messages) until the next retransmission is detected by the reversals of the FIBs received.

Noncompelled error-control mode

The ignore-corruptions error control as described above is an example of this mode, messages being transmitted in a continuous stream. A number of messages may be present on the signalling link at any one time and the error control method is required to cater for this, which accounts for the sequence numbering in the example described. The mode is not dependent upon the magnitude of the error control loop delay. With noncompelled, the c.c.s. link will be very lightly loaded at high signalling bit rates in the telephony service, which gives possibility of repeat transmission of the information (typically by consecutive multiple transmission of a message SU (or message), or by repeated transmissions of the contents of the retransmission store) to minimize, but not replace, correction by retransmission.

Compelled error-control mode

Here, one message only is transmitted at a time, a following message not being transmitted until the previous message had been correctly received and the transmit end is aware of this, either implicitly or explicitly. The error control loop delay of the signalling link is occupied for the complete transmission of a single message, and relative to noncompelled, the c.c.s. link occupancy is significantly increased for a given signalling requirement. Compelled signalling is interleaved in the two directions. The mode reduces the number of speech circuits which may be served per c.c.s. link relative to noncompelled under otherwise equal conditions, the loading being dependent upon the error control loop delay and thus on the propagation time of the signalling link.

The compelled mode has potential for simplicity (one message only dealt with at a time, no sequence numbering, retransmission store limited to one message), and has possibility of being optional to noncompelled in the commonality concept of the c.c.s. system. While the mode may be suitable for the telephony service in certain restricted national network conditions, it is not thought suitable for the switched circuit data service in any conditions due to the excessive queueing delays which would arise, the data service having a high proportion of short duration calls, with consequential high c.c.s. link signalling requirement. It is thought that the application flexibility merit of the noncompelled mode in any service may well outweigh any merit which may be realisable with the compelled mode in restricted application.

11.11.4 Message structure

The message size, formatting and codings for the new No. 7 system are not yet finalised by the CCITT.* As the system is to be optimised for 64 kbit/s, and thus a c.c.s. link very lightly loaded in the telephony service, the bit length of the constituent bit fields of a message, and the total message bit length, can be reasonably long, and significantly longer than those in system 6, without penalty. The flexibility of the new c.c.s. system will permit, optionally, messages of fixed bit-length SUs, or messages of variable bit-length, as desired. For compatibility with the digital environment it is logical that the message length should

*Details of the system may be found in the Appendix to Chapter 11 on p. 377

be a multiple of the 8-bit octet, and to simplify the processing procedures, a byte structure adopted.

References

1 LUCAS, P., LEGARE, R., and DONDOUX, J.: 'Principes nouveaux pour la signalisation telephonique', *Commun. & Electron.*, 1967, **18,** pp. 7–23
2 CCITT: Green Book, **6,** Pt. 3, Recommendations Q252–295 'Specification of signalling system No. 6', ITU, Geneva, 1973, pp. 427–521
3 CREW, G. L.: 'CCITT system No. 6 – a common channel signalling scheme', *Telecommun. J. Australia,* 1968, **18,** 3, pp. 251–256
4 AKIMARU, H., TEKEDA, H., and ABU, M.: 'Common channel signalling system for DEX2 electronic switching system', *Rev. Electr. Commun. Lab.* (Japan), 1969, **17,** p. 11
5 DAHLBOM, C. A.: 'Common channel signalling – a new flexible interoffice signalling technique'. IEEE International Switching Symposium Record, Boston, USA, 1972
6 PETERSON, W. W., and BROWN, D. T.: 'Cyclic codes for error detection', *Proc. Inst. Radio Eng.,* 1961, **49,** Pt. 1, pp. 228–235
7 CCITT: Green Book, **6,** Pt. 3, Recommendation Q277 'System No. 6 error control', ITU, Geneva, 1973, pp. 497–499
8 CCITT: Green Book, **6,** Pt. 3, Recommendation Q267 'Unreasonable and superfluous messages', ITU, Geneva, 1973, pp. 477–480
9 CCITT: Green Book, **6,** Pt. 3, Recommendation Q274 'System No. 6 modulation method and modem requrements', ITU, Geneva, 1973, pp. 491–495
10 CCITT: Green Book, **6,** Pt. 3, Recommendations Q291 and Q293 'Security arrangements', ITU, Geneva, 1973, pp. 508–510 and 512–518
11 CCITT: Green Book, **6,** Pt. 3, Recmmendation Q271 'System No. 6 continuity check of the speech path', ITU, Geneva, 1973, pp. 484–487
12 WELCH, S.: 'Common channel signalling – a flexible approach'. International Switching Symposium Record, Kyoto, Japan, 1976
13 HAMMING, R. W.: 'Error detecting and error correcting codes', *Bell Syst. Tech. J.,* 1950, **29,** p. 147

Chapter 12

CCITT international
signalling systems

12.1 General

As signalling for international traffic crosses national boundaries, it
is a clear requirement that international signalling systems should
conform to standardised specifications, the production of which is
undertaken by the International Telegraph and Telephone Consultative
Committee (CCITT), a body of the International Telecommunications
Union (ITU), and the various administrations and operating organisations of the world participating in the work. The CCITT does not
specify the design detail of international signalling systems, this being
a matter for individual administration interpretation, and for this
reason design detail will not be discussed.

In general, the term international may be understood as implying
intercontinental, international within a continent, or regional, and
some systems have reasonably defined fields of application. In earlier
times of CCITT activity there was an implied distinction between
national and international signalling systems, due, perhaps, to national
systems having their own specific designs to meet particular network
conditions, and international systems sometimes being compromise
solutions to a variety of requirements. The basic concepts of particular
international signalling systems could, of course, be adopted for national
signalling application. While the CCITT has responsibility for international and not national signalling systems, the distinction between
national long-distance and international signalling systems is now
rapidly disappearing, this being particularly the case for the regional
systems R1 and R2, for c.c.s., and more particularly for c.c.s. in
the digital environment. This is thought to be a wholly logical process
as there is no basic reason why national long-distance signalling on
appropriate transmission plant should be significantly different

from that of international, national variants being adopted as required.

All international signalling systems are 4-wire and all international switching centres 4-wire switched. Preferably, the international exchange should function as an interface gateway exchange between the national and international networks to allow either network to evolve without reaction on the other.

International signalling systems 1, 2, 3, 4, 5, 5bis, 6, R1 and R2 are at present specified by the CCITT. A further system (probably CCITT 7) the optimised digital common channel signalling system, is currently being studied by the CCITT, but is not yet coded in the CCITT series as the study is incomplete. The range of CCITT specified systems reflects the evolution of international signalling requirements to meet the continually changing conditions of the international network. The early systems 1, 2 and 3 are of historical interest only and present interest in international signalling is confined to systems 4, 5, 5bis, 6, R1 and R2, and the new optimised c.c.s. system 7.

CCITT system 1   is a 500/20 Hz signalling system used in the international manual service.

CCITT system 2   is a 600/750 Hz 2 v.f. system for the international semiautomatic service, admitting 2-wire circuits, but was never used in the international service.

CCITT system 3   is a 2280 Hz 1 v.f. system with limited application (Section 12.2).

System 6 (Section 11.5), the new optimised c.c.s. system 7 (Section 11.11), R1 (Sections 8.7 and 10.6) and R2 (Sections 9.5 and 10.7) have been described in the respective sections stated.

12.2 CCITT signalling system 3

It is of interest to summarise the main features of this early system, specified by the CCITT in 1954, which were:

(*a*) 1. v.f., 2280 Hz, inband.

(*b*) Address signalling incorporated in the line-signalling system, separate interregister signalling not applicable.

(*c*) A 4-element binary code (50 ms pulses of 2280 Hz or 50 ms no tone) identifies the digit value, a pulse being present (mark) or absent (space). The mechanism is started, at the beginning of each digit signal, by a 50 ms pulse of 2280 Hz (the start signal) and stopped, after the 4 elements have been received, by the absence of tone (the stop signal). The digits do not correspond to their binary equivalents. This cycle is repeated for each address-digit signal transmitted.

The system is specified for unidirectional working only (bothway working not applying) and in semiautomatic and automatic terminal operation only. It has limited application in Europe and a few other places and will not be used for new international connections. It will not be discussed further.

12.3 CCITT signalling system 4

General

This system, specified by the CCITT in 1954, is 2 v.f. (2040 + 2400 Hz), compound, pulse, inband, end-to-end analogue signalling.[1,2] A digital version is not specified. Each digit of the address information is transmitted as a binary code of four elements, each element consisting of a pulse of either of the two signalling frequencies. The address signalling is incorporated with the line signalling, and a separate inter-register signalling system does not apply. The compound operation applies to the line signals only, the compound signal being usually, but not always, used as a preparatory signal element (the prefix) to the control signal element (the suffix) of single frequency. Thus the 2 v.f. receiver is required to operate in the compound- and simple- (single) frequency modes. A compound prefix is less likely to be imitated by speech than a single frequency of equal duration and serves to prepare the system for the reception of the suffix which follows. The guard features of the 2 v.f. receiver must take account of both the compound- and simple-frequency conditions (Section 8.2*d*). The prefix also brings about the receive-end line split to prevent the remaining part of the signal passing out of the section in which it is intended to be operative.

The end-to-end signalling would appear to be at variance with the preference for link-by-link signalling for line-signalling systems, but at the time the system was specified, the link-by-link preference was not so well established. The following points, however, may be made in regard to the end-to-end mode for system 4:

(i) It was preferred that the address signalling, which may be regarded as being interregister signalling although a separate interregister signalling system does not apply, be end-to-end for reasons such as potential for simple arrangements at transit registers, originating register control of the connection, etc. As address signalling is part of the line signalling in system 4, this dictated that the line (supervisory) signalling be end-to-end.

(ii) The maximum number of links (three) determined for the end-to-end signalling of system 4 was considered to be sufficiently modest as to avoid incurring undue penalty in adopting end-to-end signalling on international routes.

(iii) The end-to-end has some merit in the speed of transfer of certain important signals (answer, clear forward) on multilink connections.

While not intended, system 4 could be applied to satellite circuits, but the address-information signalling would be slow due to compelled digit-by-digit technique. It is not suitable for t.a.s.i. equipped circuits (Section 7.6). The system is specified for unidirectional working only (bothway working not applicable) for transit and terminal semi-automatic and automatic service. While the system was specified early (1954), and the signalling art has advanced since that time, it has considerable application, particularly in Europe, at present.

Signal code

Line signals (see Table 12.1)

Table 12.1 *CCITT system 4 line signals*

Signal	Code
Forward	
Terminal seizure	PX
Transit seizure	PY
Address information	Binary code (see Table 12.2)
Clear forward	PXX
Forward transfer	PYY
Backward	
Terminal proceed-to-send	X
Transit proceed-to-send	Y
Number received	P
Busy flash	PX
Answer	PY
Clear back	PX
Release guard	PYY
Blocking	PX
Unblocking (use of release guard)	PYY

$$x = 2040\,\text{Hz} \qquad y = 2400\,\text{Hz}$$

The symbols used in Table 12.1 have the following significance:

Prefix   P prefix signal of the two frequencies x and y compounded.

Suffixes
X	short signal element of the single frequency x
Y	short signal element of the single frequency y
XX	long signal element of the single frequency x
YY	long signal element of the single frequency y

The transmitted durations are:

$$
\begin{aligned}
\text{P} \quad & 150 \pm 30\,\text{ms} \\
\text{X and Y} \quad & 100 \pm 20\,\text{ms} \\
\text{XX and YY} \quad & 350 \pm 70\,\text{ms}
\end{aligned}
$$

The recognition times of the line signal elements as received from the d.c. output of the receiver are:

$$
\begin{aligned}
\text{P} \quad & 80 \pm 20\,\text{ms} \\
\text{X and Y} \quad & 40 \pm 10\,\text{ms} \\
\text{XX and YY} \quad & 200 \pm 40\,\text{ms}
\end{aligned}
$$

Notes on Table 12.1

(1) Two different types of seizure signal, which can perform switching functions, are provided:

(*a*) terminal seizure, which can be used at the incoming international exchange to seize equipment used exclusively for switching the call to the national network of the incoming country

(*b*) transit seizure, which can be used in the exchange at the incoming end of the international circuit to seize equipment used exclusively for switching the call to another international exchange.

(2) Two different proceed-to-send signals are provided:

(*a*) terminal proceed-to-send, used to invite the sending of information of use in a terminal exchange (the language digit, or the discriminating digit, plus the national significant number),

(*b*) transit proceed-to-send, used to invite the sending of information (beginning with the first digit of the country code) necessary for routing the call through the international transit exchange towards the incoming international exchange or to another international transit exchange.

(3) The number-received signal is sent from the incoming international exchange to the outgoing international exchange when the incoming

register has recognised that all the address information required for routing the call to the called party has been received. In automatic working the signal releases the registers and sets up speech conditions. The signal is used in:

(*a*) semiautomatic working, when the incoming register has received the end-of-pulsing signal ST,

(*b*) automatic working, when the incoming register recognises that all the digits of the national significant number have been received by the receipt of ST when given, or by checking the number of digits received in countries where the national significant number is always the same number of digits, or, in countries where this is not so either

(i) by receipt of the maximum number of digits used in the numbering plan of the country,

(ii) by analysing the early digit(s) of the national significant number to decide how many digits there are in the subscriber's number in the particular national numbering zone,

(iii) by using national end-of selection, or national called subscriber free, signal when given,

(iv) exceptionally, by observing that 5 s have elapsed since the last digit was received.

(4) In system 4, the number-received signal is sent when limited facility applies (e.g. called subscriber free signal not available) and also when full facility applies (e.g. called subscriber free signal available, but this is not sent on the international connection). Certain other signalling systems use the number-received signal when limited facility applies, but use the called line status signal instead when this is available.

(5) The busy flash signal is sent to the outgoing international exchange when the route, switching equipment, or (when detectable) the called subscriber is busy. The signal results in the release of the international connection by means of the clear forward/release guard sequence from the outgoing international exchange even though the caller is off-hook.

(6) On called-subscriber 'flash', repeated on-hook/off-hook, by the called party, the system sends a sequence of answer and clear-back signals. If the signalling system cannot follow the speed of the flash, the correct indication of the final position of the switchhook is always given.

(7) In automatic working, the international connection is released and the charge stopped, if, when 1–2 min after the clear-back signal, the caller has not cleared to send the clear forward. The release is in the direction from the originating international exchange, and is achieved by a clear-forward signal generated by the system.

(8) The clear-forward signal is sent end-to-end from the originating

international exchange and is recognised by all the international exchanges, including transit, on the connection. Each international link is released when the clear-forward signal is recognised on that link and the release-guard signal effective on that link. The release guard is thus on a per-link basis and is not end-to-end. A clear-forward signal is recognised, and is effective, in any condition of the circuit.

Address signals (see Table 12.2)

Unlike the more usual decadic address signalling in line-signalling systems, the coded address signalling in the line-signalling system 4 consists of an invariable number of signal elements (four) per digit, and thus permits a measure of error detection.

Table 12.2   *CCITT system 4 address information binary code*

Signal number	Signal	Combination elements			
		1	2	3	4
1	Digit 1	y	y	y	x
2	2	y	y	x	y
3	3	y	y	x	x
4	4	y	x	y	y
5	5	y	x	y	x
6	6	y	x	x	y
7	7	y	x	x	x
8	8	x	y	y	y
9	9	x	y	y	x
10	0	x	y	x	y
11	Call operator Code 11	x	y	x	x
12	Call operator Code 12	x	x	y	y
13	Spare	x	x	y	x
14	Incoming half-echo suppressor required	x	x	x	y
15	ST end-of-pulsing	x	x	x	x
16	Spare	y	y	y	y

Backward acknowledgment signal 1 — signal element x
Backward acknowledgment signal 2 — signal element y

This numerical signal binary code consists of four elements, each of a 35 ± 7 ms pulse of x or y frequency, and each separated from the next by a 35 ± 7 ms gap. The relation between the transmitted digits and the different combinations of the binary code is arrived at by assigning the value 8, 4, 2 or 1 to the presence of an element x in the 1st, 2nd, 3rd or 4th position in the code, e.g. $yxyx = y4y1 = 5$ = digit 5, and the transmitted digits correspond in the decadic and the binary form. The recognition time of the pulses and gaps as received from the d.c. output of the receiver is 10 ± 5 ms.

An acknowledgment signal is returned at the end of the reception of the 4th element of each numerical signal, the acknowledgment being a signal element 35 ± 7 ms of x or y frequency. Consequent on the transit seizure signal PY, a transit register returns the transit proceed-to-send signal Y to request the sending of the first digit of the address information it requires (the first digit of the country code in system 4). The transit register decides how many digits it needs for routing a call and the transfer of these digits from the outgoing to the transit register is controlled by the backward acknowledgment signals. Similarly, consequent on the terminal seizure signal PX, a terminal proceed-to-send signal X is returned by a terminal register to request the sending of the first digit of the address information it requires (the discriminating

Table 12.3 *CCITT system 4 control of address information transfer*

Backward signal	Signal	Interpretation at the outgoing exchange
X	Terminal proceed-to-send	Send discriminating (or language) digit
Y	Transit proceed-to-send	Send first digit of country code
x	Acknowledgment x	Acknowledgment of digit received with the alternative meaning according to the type of the last proceed-to-send signal received: (*a*) after X; 'send next digit' (*b*) after Y; 'stop the sending of digits'
y	Acknowledgment y	Acknowledgment of digit received; 'send next digit' (after Y only)

digit, or the language digit, in the case of system 4). The transfer of digits from the outgoing to the terminal register is then controlled by the backward acknowledgment signals (Table 12.3). The language digit is sent in semiautomatic working and the discriminating digit is sent instead in automatic working. These respective digits, located immediately after the country code, are inserted by the switching machine at the outgoing international exchange, and corresponding numerical digits are allocated for these digits as may be agreed upon by administrations.

Fig. 12.1 shows a typical signalling sequence with system 4.

Line splits

The transmit line split is applied 30–50 ms before a v.f. signal is sent, the split persists with signal and is ceased 30–50 ms after the end of sending a signal. The receive line split persists with received signal and ceases within 25 ms after the end of the signal; with a splitting time of 55 ms, this duration of signal will spill over to national networks.

While each system 4 link is equipped for both the transmit and receive line splits, these are inoperative at transit exchanges when end-to-end signalling is used. In the case of the end-to-end clear-forward signal, all international switching units i.e. outgoing, transit and incoming must release at the end of the clear-forward signal. The forward channel must not be split until the clear forward has completely ceased. As the various consequential release-guard signals are per-link, the transit and receive line splits on the backward channel on each link are operative to the release-guard signal on each link.

Normal release of registers

Register release switches the speech path through.

The outgoing register releases in any one of the following conditions:

(*a*) when the register has sent all the address information and has received an ST signal from the outgoing operator, or from the outgoing national network when this network gives ST

(*b*) on receipt of a number received signal, or a busy flash signal, from the incoming exchange.

The transit register releases:

(i) as soon as it has selected an outgoing circuit and sent a seizure signal forward

(ii) on congestion, after it has returned a busy flash signal.

The incoming (international terminal) register releases:

(*a*) when all the required address information has been received and

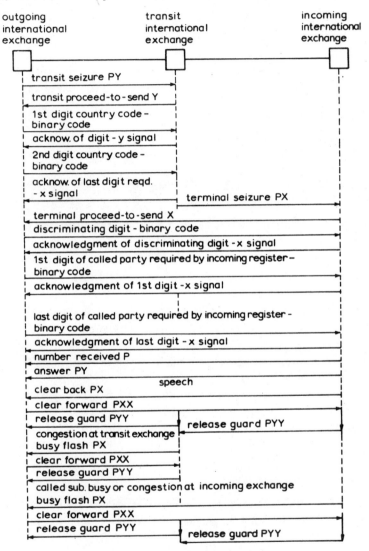

Fig. 12.1   *Typical signal sequence CCITT 4 system*

a number received signal returned, the incoming register determining when all the required address information is received

(*b*) on congestion, after it has returned a busy flash signal.

Abnormal release of registers

The outgoing register releases when any one of the following conditions arise:

(*a*) Semiautomatic operation, if within 10–20s from register seizure, or the receipt of the last digit, no further digit or the ST signal is received. Busy tone is returned.

(*b*) Automatic operation, if within 15–30s from register seizure, or the receipt of the last digit, the register is forced-released in any one of the following ways:

(i) seized, but no digits received

(ii) not all the digits necessary to determine the routing are received

(iii) digits necessary to determine the routing are received, but no further digits received

(iv) no number-received signal or busy flash signal received after relevant address information sent.

Busy tone is returned in the last two cases and the international connection released by a clear-forward signal.

In (*b*) above, the delay 15–30s covers the maximum period for receiving a number-received signal under the following most unfavourable conditions:

(*a*) address information received for which no routing has been provided

(*b*) proceed-to-send or busy flash signal not received within 10–30s following the sending of a seizure signal, or if not received within 15–30s following the sending to a transit exchange of the digits necessary to determine the routing

(*c*) an acknowledgment signal not received within 5–10s following the sending of a digit

(*d*) more than the maximum number (usually two) of transit proceed-to-send signals received.

The transit register releases, without returning any signal, under either of the following conditions:

(i) if the digits necessary to determine the routing are not received within 5–10s following the sending of a proceed-to-send signal to the outgoing exchange

(ii) if address information is received for which no routing has been provided.

The incoming register releases under any one of the following conditions:

(*a*) on incomplete number, no further digit received within 30–60s after receipt of the last digit (*b*) no digit received within 5–10 s following the return of a proceed-to-send signal

(*c*) address information received for which no routing exists, or incomplete number received followed by a ST signal.

In the first two cases, no signal is returned from the incoming register as the outgoing register remains in circuit and detects the abnormal condition. In the third case, a number received signal is returned before the incoming register releases, and is followed by appropriate tone indication (busy, number unobtainable).

Failure in sequence of signals

(*a*) An outgoing circuit is blocked and an alarm given under any one of the following conditions:

(i) if, after sending a seizure signal, a proceed-to-send signal or a busy flash signal is not received within 10–30 s

(ii) if a proceed-to-send or a busy flash signal is not received within 15–30 s of the sending to a transit exchange of the address information necessary to determine the routing

(iii) if, after sending a clear-forward signal, a release-guard signal is not received within 5–10 s.

(*b*) If a release-guard signal is recognised at a transit exchange without a clear-forward signal having been recognised, the transit register:

(i) returns the blocking signal to busy the outgoing end of the incoming circuit

(ii) immediately releases the circuit outgoing from the transit register. This prevents the receipt of a release-guard signal from giving a wrong indication that the circuit to the transit exchange is cleared.

Relevant data

Transmit:

Signal frequencies 2040 ± 6 Hz (x frequency)
2400 ± 6 Hz (y frequency)

Level. Absolute power level of each signal frequency at a zero relative-level point -9 dBm ± 1 dB
When compound, maximum difference in level between the two frequencies 0·5 dB

Receiver:

Operate   Signal frequencies 2040 ± 15 Hz, 2400 ± 15 Hz

Level. Absolute power level N of each received signal frequency to be within the limits

$$-18 + n \leqslant N \leqslant n \text{ dBm}$$

where n is the relative power level at the receiver input. This gives a margin of ± 9 dB on the nominal absolute level of each received signal at the receiver input.

The received level of the 2400 Hz signal frequency not more than 3 dB above, nor more than 6 dB below, the received level of the 2040 Hz signal

Nonoperate Not operate at the absolute operate level stated above when frequency differs by more than 150 Hz from the nominal 2040 Hz or 2400 Hz

When signal 2040 ± 15 Hz or 2400 ± 15 Hz, not operate at absolute power level at point of connection of receiver of $(-26 - 9 + n)$ dBm, n being the relative power level at receiver input. This limit is 26 dB below the nominal absolute power level of the signal at the receiver input

Comments on CCITT system 4

While it will be understood that system 4 was produced for a particular application i.e. international, national application not being in mind, it is of interest to examine the basic features of the system for national conditions. The following points arise in this consideration:

(i) The binary numerical signalling requires 2 v.f. and as the address information is incorporated in the system, this dictates that the system be 2 v.f. 1 v.f. line signalling is preferred in national networks for reasons of simplicity and cost.

(ii) While the availability of 2 v.f. conveniently allows compound-prefix signalling, modern 1 v.f. guarding is acceptably efficient from the signal-imitation aspect.

(iii) The v.f. binary signalling dictates that the line signalling be v.f. This does not allow the desired flexibility for national networks that line signalling be d.c., v.f., or outband, according to the transmission plant.

(iv) There is a broad similarity between system 4 and system R2 address signalling in that each address digit signal is acknowledged, and this acknowledgment carries an instruction. It is considered, however, that the R2 system achieves this more efficiently and with greater exploitation.

(v) Due to the 2 v.f. binary concept, the address signalling is relatively slow.

Thus while system 4 does not require a separate interregister signalling system, which in one regard may be considered as being a merit, it is thought that the system would not give the desired flexibility for line-signalling application, nor meet the requirements of national networks in a convenient way. Flexible line signalling with separate interregister m.f. signalling is considered to be a superior technique for national networks.

12.4 CCITT signalling system 5

General

This analogue system, specified by the CCITT in 1964, has separate line and interregister signalling. The line signalling is 2 v.f. (2400 + 2600 Hz), inband, compound and simple frequency continuous signalling and link-by-link. The interregister signalling is 2/6 m.f., link-by-link and forward signalling only. A digital version is not specified. Unidirectional, bothway, transit, terminal, semiautomatic and automatic operation may apply, the system being suitable for 3 kHz- and 4 kHz-spaced submarine, land cable, microwave radio and satellite circuits, with or without t.a.s.i.[3-6]

Initially, the system was a joint development between the UK Post Office and the Bell Laboratories (A.T. & T. USA) for dialling over the t.a.s.i. equipped transoceanic Atlantic cables, this being the first application of intercontinental dialling and of t.a.s.i. equipment.[7-9] The system was subsequently specified by the CCITT and has since found increasing application in other areas. Most of the intercontinental circuits of the world network (Atlantic, Pacific and Indian Oceans, etc. transoceanic cables and intercontinental satellite circuits) are equipped with system 5 at present.

It was decided at the beginning of the development that the system should be as simple and as robust as possible in view of the importance attached to the first intercontinental dialling service, and, in view of the high cost of transoceanic cables, that t.a.s.i. requirements should dictate the features of the system. In the result, the signals and facilities of system 5 were kept to a minimum consistent with the intercontinental dialling service.

Section 7.6 discusses the influence of t.a.s.i. equipment on signalling. For reasons of t.a.s.i. trunk-channel association, all line signals but one in system 5 are continuous-compelled, trunk-channel association always being assured in the actual time required for this function. The acknowledgment signalling in continuous-compelled increases the t.a.s.i. signalling time and activity, but some signals normally give rise to return signals (i.e. seizure/proceed-to-send, clear forward/release guard) and no penalty results from the adoption of continuous-compelled for these signals. Other signals (e.g. answer) do not normally require a return signal and adoption of continuous-compelled, instead of pulse, could be a penalty when t.a.s.i. is heavily loaded, but an advantage when light, and on balance, continuous-compelled was preferred for these signals.

The system 5 interregister m.f. signalling is pulse forward signalling only so as to minimise the t.a.s.i. signalling time and activity. The lack of backward signalling accounts for the reduced facilities given. The arrangement requires that a t.a.s.i. channel be prior-associated for address signalling to avoid clipping of the first address signal. As system 5 was to be the first intercontinental signalling system in use, some anxiety was felt as to the expensive intercontinental circuits being used ineffectively on incomplete dialled calls, and for this reason it was preferred to incorporate as much validity-checking of the address information as possible at the outgoing international exchange before seizing an intercontinental circuit. This implied complete *en bloc* of the address information to enable a measure of validity-checking to be performed, and thus *en bloc* transmission. In this mode, the t.a.s.i. speech detector hangover maintains trunk-channel association during the gaps between successive address pulse signals and thus clips do not occur. The alternative of overlap, in place of *en bloc*, mode would require some other arrangement to maintain t.a.s.i. trunk-channel association, such as lock tone, to avoid clipping of pulse-address signals (see CCITT system 5bis, Section 12.5).

Line signals (see Table 12.4)

Notes on Table 12.4
(1) By taking advantage of the fixed order of occurrence of specific signals, signals of the same frequency are used to characterise different functions, e.g. in the backward direction f_2 is used to indicate proceed-to-send, busy flash and clear back without conflict. The signalling equipment must operate in a sequential manner, retaining memory of the preceding signalling states and the direction of signalling to differentiate between signals of the same frequency. All signals except the forward transfer are acknowledged in continuous-compelled manner. The order of transmission of backward signals is subject to the following:
(i)   busy flash: only after a proceed-to-send signal and never after an answer signal
(ii) answer: never after a busy flash signal
(iii) clear back: only after an answer signal

Table 12.4 *CCITT system 5 line signals*

Signal	Direction	Frequency	Sending duration	Recognition time
				ms
Seizure	→	f_1	Continuous	40 ± 10
Proceed-to-send	←	f_2	Continuous	40 ± 10
Busy flash	←	f_2	Continuous	125 ± 25
Acknowledgment	→	f_1	Continuous	125 ± 25
Answer	←	f_1	Continuous	125 ± 25
Acknowledgment	→	f_1	Continuous	125 ± 25
Clear back	←	f_2	Continuous	125 ± 25
Acknowledgment	→	f_1	Continuous	125 ± 25
Forward transfer	→	f_2	850 ± 200 ms (pulse)	125 ± 25
Clear forward	→	$f_1 + f_2$ (compound)	Continuous	125 ± 25
Release guard	←	$f_1 + f_2$ (compound)	Continuous	125 ± 25

$f_1 = 2400$ Hz $\qquad f_2 = 2600$ Hz
→ forward signal
← backward signal, continuous-compelled mode

The receipt of the answer signal (f_1) permits discrimination between the busy flash and clear-back signals (both f_2).

(2) Except for the recognition time (40 ms) of the seizure/proceed-to-send signal sequence, which can be short as this sequence is not subject to signal imitation by speech and a rapid seizure is desired to minimise the postdialling delay and the probability of double seizure on bothway working, all recognition times are the same (125 ms). This simplifies signalling terminal design.

(3) The use of compound for the clear forward/release guard sequence improves the immunity to false release by signal imitation, this being particularly necessary as the recognition time (125 ms) of these important signals is relatively short for uniformity. The compound clear forward also discriminates against the f_2 forward transfer. The clear forward, which must always be acknowledged by a release-guard signal under all conditions, is completely overriding.

(4) The use of different frequencies for the seizure (f_1) and the proceed-to-send (f_2) facilitates double seizure detection on bothway working. The seizure signal continues until acknowledged by the proceed-to-send, which is returned when an incoming register is associated and continues until acknowledged by the cessation of the seizure signal. As there is no backward interregister signalling, the proceed-to-send signal must be

a line signal, which is convenient in system 5 to acknowledge the seizure signal.

(5) There is frequency discrimination between all the line signals except busy flash and clear back, which are both f_2 acknowledged by f_1. As the busy flash will be received without an answer signal having been received and the clear back will have been preceded by an answer signal, the answer signal is used to bring about a change of condition in the outgoing equipment to permit the discrimination.

(6) The forward transfer signal f_2 (a rarely used signal and used in semiautomatic operation only) is an unacknowledged pulse. The nonacknowledgment avoids possible confusion with the f_1 answer and the f_2 busy flash and clear back.

(7) The proceed-to-send ceases the seizure signal. The outgoing register pulses out the address information *en bloc* and a silent period of 80 ± 20 ms is arranged between the cessation of the seizure signal and the beginning of the register pulse out. The trunk-channel association is maintained by the t.a.s.i. speech-detector hangover during this interval. This, together with the speech-detector hangover maintaining t.a.s.i. trunk-channel association during the gaps between successive interregister address signals, avoids any requirement for locking the trunk-channel association (by lock tone or other means) during inter-register signalling.

(8) In transit operation, the line equipment at the transit exchange is informed by the interregister KP signal that the condition is transit. This facilitates the link-by-link transmission of line signals through the transit exchange without bringing about consequences appropriate to a terminal exchange.

(9) Should the called party flash his switchhook faster than the equipment can transmit a succession of clear-back (on-hook) and answer (off-hook) signals, the correct indication of the final position of the switchhook must always be given by the appropriate signal.

(10) If, after 1–2 min of receipt of the clear-back signal there is no clear-forward signal, the international connection is released by a system-manufactured clear-forward signal, and the measurement of call charge ceased.

(11) The clear-forward signal continues until acknowledged by the release guard, which, depending on the administration's choice, may be sent:

(*a*) on recognition of the clear-forward signal, and continues until acknowledged by the cessation of the clear-forward signal, or until the relevant incoming equipment at the international exchange has released, whichever occurs later, (*b*) in response to the clear-forward signal, to indicate that this signal has brought about the release of the relevant incoming equipment at the international exchange, the release-guard signal continuing until cessation of the clear forward is recognised.

On a bothway circuit, the outgoing access at the incoming end is maintained busy for 200–300 ms after the end of transmission of the release-guard signal to cover the responses to the cessation of the release guard at the outgoing end.

(12) Busy flash is transmitted for any of the following reasons:

(i) congestion at a transit or at an incoming international (terminal) exchange

(ii) error detected in the receipt of register signals

(iii) busy flash, if received, from interworking international signalling system(s) or from the incoming national network

(iv) time out of an incoming international register

Receipt of the busy-flash signal at the outgoing international exchange causes appropriate indication (e.g. tone) to be sent to the caller and the release of the international connection by means of a system-manufactured clear-forward signal.

(13) In inband v.f. signalling, a quick verbal answer may be clipped, partially or completely, by the line splitting on transmission of the electrical-answer signal. Nonrepetition of the verbal answer could result in either party, waiting for verbal response from the other, abandoning the call. Multilink connections increase the danger. Thus transmission of the electrical-answer signal should be as fast as possible in an effort to ensure that the line splits are terminated prior to verbal answer. As a contribution to this, the answer signal of system 5 is transmitted in the overlap-compelled, instead of normal-compelled, mode at transit exchanges. In this technique, the process of transmitting the answer signal from the transit exchange to a preceding exchange is initiated as soon as the transit signal receiver response to the incoming answer signal has caused the receive line split (35 ms maximum), the onward transmission not awaiting the full signal recognition time of the answer signal as would be the case in normal-compelled. The normal signal recognition time of the answer signal is still required at each transit exchange and the acknowledgment on a particular link would not be transmitted until signal recognition is complete. The answer signal on each link is ceased by its acknowledgment on that link. If the incoming answer signal duration is less than the signal recognition time, the transmission of the ongoing answer signal already instituted from that transit exchange is ceased. After signal recognition of the incoming answer, there is no control at the

transit exchange of the ongoing answer signal by the incoming answer signal.

The busy flash and clear-back signals are returned in the normal compelled manner mode at transit points, onward transmission not commencing until the incoming signal recognition is complete.

(14) The acknowledgments of the busy flash, answer and clear-back signals are sent after signal recognition $(125 \pm 25\,\text{ms})$ of the relevant primary signal. The primary signal is not ceased until the recognition time of the acknowledgment is complete $(125 \pm 25\,\text{ms})$. Signal cessation recognition time, primary or acknowledgment, is at least 40 ms.

Bothway operation

The extreme of t.a.s.i. trunk-channel association time (some 500 ms) combined with long propagation times and the various equipment response times, results in a relatively long unguarded interval in bothway operation. Double seizure is detected when the same frequency $(f_1$ seizure signal) is received as is being transmitted. On double-seizure detection, the transmitted seizure signal is ceased $850 \pm 200\,\text{ms}$ after commencement, an arrangement which ensures that both ends of a t.a.s.i. equipped bothway circuit will detect the double seizure.

The signalling equipment is released on cessation of both the outgoing and incoming seizure signals. A clear-forward signal is not sent. Either of the following arrangements may apply on double seizure, depending upon administration's choice:

(*a*) an automatic repeat attempt to set up the call,
(*b*) a reorder indication (tone) given to the callers.

The former is preferred. This does not require the repeat attempt to be limited to the circuit used for the first attempt, but if the first circuit is seized on the second search over the circuits, a minimum time of 100 ms elapses between the termination of the first-attempt outgoing seizure signal (or recognition of the cessation of the incoming seizure signal, whichever occurs later) and the commencement of the second-attempt seizure signal.

To minimise the probability of double seizure, the circuit selection at the two ends should be such that, as far as possible, double seizure can occur only when a single circuit remains free (e.g. by selection of circuits in opposite order at the two ends).

Line splits

Transmit line split: The exchange side of the circuit is disconnected 30–50 ms before a v.f. signal is transmitted. The split persists with signal and is terminated 30–50 ms following the end of sending of the v.f. signal.

Receive line split: The circuit is split at the international exchange when either a single or compound frequency is received so that the spillover does not exceed 35 ms. The split persists with signal and is terminated within 25 ms following the end of the signal. The splitting device may be a physical line disconnection, high-impedance device, insertion of signal frequency bandstop filter, etc. Leak current in the split condition should be at least 40 dB below the received signal level.

Relevant data: line signalling

Transmit:

f_1 2400 ± 6 Hz f_2 2600 ± 6 Hz

Transmitted level −9 dBm0 ± 1 dB per frequency

For compound the difference in transmitted level between f_1 and f_2 not to exceed 1 dB

The difference in the time between f_1 and f_2 of a compound signal on sending and ceasing not to exceed 5 ms

Receiver:

Operate f_1 2400 ± 15 Hz f_2 2600 ± 15 Hz

The absolute power level N of each received signal to be within the limits $(-16 + n) \leqslant N \leqslant (-2 + n)$ dBm, where n is the relative power level at the receiver input. These limits give a margin of ± 7 dB on the nominal absolute level of a received signal at the receiver input

The absolute level of the two frequencies of a compound signal not differing by more than 5 dB

Nonoperate   Receiver not to operate outside f_1 2400 ± $^{100}_{150}$ Hz f_2 2600 ± $^{150}_{100}$ Hz

Receiver not to operate on signal 2400 ± 15 Hz or 2600 ± 15 Hz whose absolute power level at point of receiver input is $(-17 - 9 + n)$ dBm. This limit is 17 dB below nominal absolute level of signal at receiver input.

Signals:

(*a*) Should the transmission of any seizure, busy flash, answer, clear back or clear-forward signal persist beyond a maximum of 10–20 s, the signal is terminated and the condition alarmed.

(*b*) If the transmission of any proceed-to-send, release guard, or other acknowledgment signal persist beyond a maximum of 4–9 s the signal is terminated and the condition alarmed.

(*c*) After signal recognition, interruptions of up to 15 ms in a signal or acknowledgment are to be ignored. Interruptions of more than 40 ms are recognised as the end of the appropriate signal (primary or acknowledgment) in the compelled sequence.

(*d*) Once the sending of a signal, pulse or continuous, has begun, it should be completed except when the clear-forward signal overrides.

(*e*) An interval of at least 100 ms should separate two successive signals in the same direction.

Interregister signals (see Table 12.5)

Automatic access to the international circuits is used for outgoing traffic, the address signals from the operator or subscriber being stored in an outgoing system 5 register before the international circuit is seized. As soon as ST (end-of-pulsing) is available to the outgoing register, a free international circuit on the appropriate route is selected and a seizure line signal sent. The seizure signal is terminated on receipt of a proceed-to-send, and KP (start of pulsing), address signals, and ST, transmitted by the register.

Both forward and backward interregister signalling is normally preferred in networks. This in turn, implies end-to-end signalling for reasons of reduced register holding time, reduced register provision, and leading register control. Backward interregister signalling, however, would increase t.a.s.i. signalling time and activity and would require some arrangement to assure t.a.s.i. trunk-channel association during the interregister signalling. It was assessed that t.a.s.i. efficiency was of first importance for system 5 and backward interregister signalling not adopted, the resultant reduction in facility exploitation due to this being accepted in the interests of t.a.s.i. This precluded the adoption of compelled signalling and also end-to-end signalling as no backward interregister signal is available to release registers. (Note. It is unacceptable to use the line answer signal to release registers and set up speech

Table 12.5 *CCITT system 5 interregister signals*

Signal	Pulse frequencies (compound)
	Hz
KP1 (terminal traffic)	1100 + 1700
KP2 (transit traffic)	1300 + 1700
Digit 1	700 + 900
2	700 + 1100
3	900 + 1100
4	700 + 1300
5	900 + 1300
6	1100 + 1300
7	700 + 1500
8	900 + 1500
9	1100 + 1500
0	1300 + 1500
Code 11 operator	700 + 1700
Code 12 operator	900 + 1700
ST (end of pulsing)	1500 + 1700

conditions as an answer signal is not always given, and further, the transmission path is required prior to answer to pass the ring tone.)

For the above reasons, the system 5 interregister signalling is link-by-link, pulse, forward signalling only, the registers releasing in sequence after transmission of ST. The signalling is 2/6 m.f. in the range 700–1700 Hz and 200 Hz spaced. The address information is always sent *en bloc* from the originating system 5 register and overlap from the transit and incoming system 5 registers.

The KP signal may be used to prepare the distant system 5 register on the link for the receipt of the subsequent address signals. It may also be used to discriminate between terminal and transit traffic as in the following:

Terminal KP (KP1): Used to create conditions at the next exchange so that equipment used exclusively for switching the call to the national network of the incoming country is brought into circuit.

Transit KP (KP2): Used to bring into circuit at the next exchange, equipment required to switch the call to another international exchange.

In system 5, the ST signal is transmitted from the register at the end of address signalling in both the automatic and semiautomatic operation. Both the outgoing and transit system 5 registers must determine the routing, and send the appropriate KP signal, by analysis of the early digits of the address information. The interregister signalling information in system 5 in the international automatic service comprises:

KP + country code (I digits) + characteristic digit (Z digit) + national significant number (N digits) + ST.

The characteristic digit is the discriminating digit (D) in automatic operation, or the language digit (L) in semiautomatic operation.

The country code, Z digit, and national number are composed of digit 1-0 signal Table 12.5, the allocation of digit signals for the Z digit being a matter of agreement between administrations.

Analysis of the country code of the destination country is generally sufficient to determine the forward routing, the code consisting in one, two or three digits. Exceptionally, early digit(s) of the national number may need to be included in the analysis to permit forward routing to any one of a number of locations in another country. In this event, the maximum number of digits to be analysed in a system 5 outgoing or transit register would be six, i.e. $I1$ $I2$ $I3$ Z $N1$ $N2$, which assumes the maximum three-digit country code. The country code will be the first digit(s) following KP2 received by a transit register. In system 5, the country code is not sent to the incoming (terminal) international register, and here the first digit received following KP1 will be the Z digit.

The 'Code 11' and 'Code 12' operators (so termed because the signals performing these functions are the 11th, and 12th, respectively of the 15 combinations provided by the CCITT number signalling code) are special international operators accessed by interregister signals.

Fig. 12.2 shows a typical signalling sequence with system 5.

System 5 register arrangements concerning ST

(*a*) *Semiautomatic operation:* The ST condition is determined by the receipt of the 'sending finished' signal from the operator.

(*b*) *Automatic operation:*
(1) When the ST signal is provided by the originating national network, this signal is transmitted to the outgoing system 5 register and no further arrangements are necessary.
(2) The outgoing system 5 register is required to determine the ST condition when this is not received from the originating national network. This determination may be on time delay or on digit count. When on time delay, ST is determined when cessation of the address

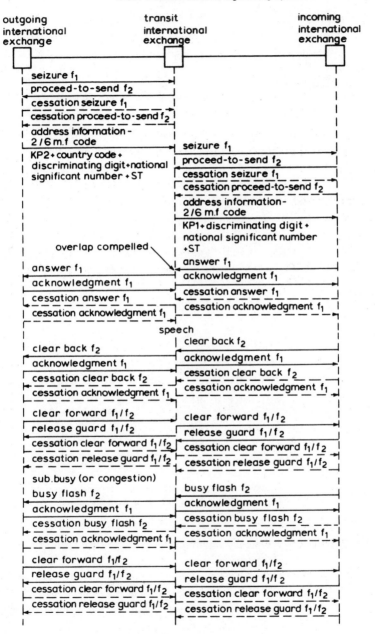

Fig. 12.2   *Typical signal sequence CCITT 5 system*

information input to the register exceeds a period of 4 s (5 ± 1) in either of the following conditions as preferred by the administration:
(i) after the minimum number of digits in the world numbering plan, or
(ii) after the minimum number of digits of the destination country numbering plan.

In (i) and (ii), prolonged cessation of the address information input before the minimum number of digits results in time-out release of the register without production of the ST condition.

An immediate ST condition may be produced by digit count to avoid the 4 s delay when the destination country numbering plan has a fixed number of digits, or when the maximum number of digits of the numbering plan of the destination country has been received.

Under all circumstances, the outgoing international circuit is not seized until the ST condition is available in the register. Thus, when operative, the 4 s delay to determine ST, while increasing the post-dialling delay, does not react on t.a.s.i. signalling time and activity.

Relevant data: interregister signalling

Transmit:

Frequencies 700, 900, 1100, 1300, 1500, 1700 Hz, tolerance ± 6 Hz

Transmitted level − 7 dBm0 ± 1 dB per frequency

Difference in transmitted level between the two frequencies of a signal not to exceed 1 dB

Signal duration: KP1 and KP2 100 ± 10 ms
All other signals 55 ± 5 ms
Interval between all signals 55 ± 5 ms

Interval between cessation of seizure line signal and the transmission of KP interregister signal 80 ± 20 ms

Receiver:
Operate

Frequency variation ± 15 Hz of the nominal
Receive level. The absolute power level N of each signal to be within the limits $(-14 + n) \leqslant N \leqslant n$ dBm, where n is the relative power level at the receiver input. These limits give a margin of ± 7 dB on the nominal absolute level of each signal at the receiver input.

The absolute levels of the two frequencies comprising a signal not to differ from each other by more than 4 dB

Minimum signal duration 30 ms

Minimum interval between successive signals 30 ms

Nonoperate   Receiver not to operate to a signal whose absolute power level at its input is $(-17-7+n)$ dBm. This limit is 17 dB below the nominal absolute power level of the signal at the receiver input.

Receiver not to operate to a signal of 10 ms duration or less.

Interruptions to signal, and intervals between successive signals, of 10 ms or less ignored.

Release of system 5 registers

Normal release
(*a*) An outgoing system 5 register releases when it has transmitted ST.
(*b*) An incoming system 5 register releases in either of the two cases:
(i) on transmitting ST, on receipt of a number received condition from the destination national network, etc., depending on the arrangement adopted by an administration
(ii) when the busy flash signal is returned.
(*c*) A transit system 5 register releases in either of the two cases:
(i) when it has transmitted ST
(ii) when the busy flash signal is returned.

Abnormal release
(*a*) An outgoing system 5 register releases, and clears the forward connection in any one of the following cases:
(i) after a 15–30 s time-out if, after seizure, none, or less than the minimum number of address signals, received
(ii) proceed-to-send signal is not received within the 10–20 s time-out of the seizure signal
(iii) proceed-to-send received, but, due to fault, the outgoing register has not pulsed out. The outgoing register will be released by the clear forward/release guard sequence prompted by the busy flash signal sent from the incoming end on nonreceipt of register signals within 4–9 s. This assumes that the busy flash signal is received at the outgoing end before the termination of any forced release delay that administrations may wish to incorporate at the outgoing register.
(*b*) An incoming system 5 register releases in any one of the following cases:
(i) no interregister signals received within 4–9 s after the start of sending the proceed-to-send signal

(ii) ST not received within 20–40 s after the start of sending the proceed-to-send signal

(iii) on return of the busy flash signal from the incoming end when an error is detected in the receipt of interregister signals.

(*c*) A transit system 5 register releases in any one of the cases stated for the release of the outgoing and incoming (terminal) registers.

12.5 CCITT signalling system 5bis

General

The importance given in system 5 (Section 12.4) to minimising t.a.s.i. signalling time and activity resulted in:

(*a*) complete *en bloc* of the address information

(*b*) the requirement to determine the ST end-of-pulsing condition under all conditions

(*c*) no backward interregister signalling.

(*a*) and (*b*) increase the postdialling delay. (*c*) limits facility exploitation.

As it was thought unlikely that t.a.s.i. would be applied to satellite circuits and perhaps not to all transoceanic cables, the CCITT considered it advisable to produce an intercontinental analogue signalling system to take advantage of non-t.a.s.i. conditions, should administrations so desire. The system would also be suitable for t.a.s.i. application, but with the penalty of increased t.a.s.i. signalling time and activity, but which could be applied should this penalty be acceptable to particular administrations. To this end, system 5bis was specified by the CCITT in 1968 as a variant of system 5.[10] The system operates in the overlap mode at all, including the outgoing, system 5bis registers to reduce the postdialling delay. It does not require ST as a mandatory condition in automatic operation, and incorporates both forward and backward interregister signalling to permit greater facility exploitation.

The line signalling is the same as that of system 5 (Section 12.4), but the interregister signalling is different. The field of application is the same as that of system 5 with which system 5bis readily interworks. Like system 5, a digital version of system 5bis is not specified.

As far as is known, system 5bis has not found application to date. As mentioned, most intercontinental dialling circuits of the world network are equipped with system 5, with satisfactory i.s.d. service, and there is no reason to replace existing system 5 equipment by system 5bis. In regard to new provision, it is thought that the evolution of the signalling art has overtaken requirement for system 5bis as common channel signalling will undoubtedly be the future system for application to intercontinental, and other, circuits.

T.A.S.I. guard and locking frequency
The system 5bis interregister signalling is 2/6 m.f. forward and backward signalling, pulse, link-by-link, all registers, including transit, remaining line-associated to pass the link-by-link signalling, which increases register holding time and register provision.

Partial *en bloc* of the address information applies at the outgoing system 5bis register, an initial address block normally consisting of seven address signals, and thus an intercontinental circuit can be taken unnecessarily in some situations of incomplete dialling. The remaining address signals are transmitted when available and therefore in the overlap mode. As in system 5, t.a.s.i. trunk-channel association is established by the seizure signal and is maintained by the t.a.s.i. speech-detector hangover during the interval (80 ± 20 ms) between cessation of the seizure signal and the transmission of the first interregister signal. This trunk-channel association, however, cannot be assured for the transmission of the later address signals transmitted overlap as these signals could arrive with relatively long intervals. For this reason, a t.a.s.i. locking frequency (1850 Hz) is applied in the forward signalling direction after the transmission of the first signal of the initial address block, the locking frequency being interrupted to send address signals and is transmitted until register release.

Backward interregister signalling can occur at any time during connection set-up. Thus means must be incorporated to assure t.a.s.i. trunk-channel association in the backward direction to avoid t.a.s.i. prefixing of the backward signals. For this reason, cessation of the proceed-to-send signal is followed within 80 ± 20 ms by a backward t.a.s.i. locking frequency of 1850 Hz, which is interrupted to pass backward signals and is transmitted until register release.

With this arrangement, energy, signal or locking frequency, is present in each direction during almost the whole time registers are associated, which accounts for the increase in t.a.s.i. signalling time and activity. Such continuous signalling at the more usual signal level would overload transmission systems and to avoid this, the level of the forward locking frequency is initially -7 dBm0 ± 1 dB, continuing at this level until a new m.f. signal is sent or until the expiration of 90 ± 30 ms, when the signal level is reduced to -15 dBm0 ± 1 dB. The backward locking frequency is at a uniform low level of -15 dBm0 ± 1 dB.

As energy is continuous during register association, signal cessation

CCITT international signalling systems

has no meaning and thus false interruption to signal, or to t.a.s.i. locking frequency, has no signal meaning, which improves the reliability of interregister signalling. Furthermore, receipt of the locking frequency can be exploited to improve reliability, e.g. should interfering frequencies be received in the presence of the locking frequency, the presence of the locking frequency would allow logic to enable such interference to be ignored. Thus the t.a.s.i. locking frequency can be exploited to give a 'guard' function and for this reason the frequency is termed 'guard and t.a.s.i. locking frequency', which indicates the dual function. The exploitation of the frequency for guard is a design matter and is not specified by the CCITT, being left for individual administrations to adopt their own arrangements.

It will be noted that this guard function of the locking frequency has some similarity to the guard tone of the UK MF2 interregister signalling system (Section 10.8.2). There is, however, a difference in detail in that UK MF2 prefers a 2/6 m.f. combination as the guard to avoid a separate guard frequency receiver. All detection is thus 2-and-2 only, and a prefix, as distinct from continuous, guard to avoid change from high to low signal level.

Line signalling
System 5bis uses the same line-signalling system as system 5 (Section 12.4), but with the following qualifications owing to the expanded interregister signalling:

(*a*) As congestion signals are included in the system 5bis backward interregister signalling, the busy flash signal is treated slightly differently, this line signal being sent only if the release of the incoming register has already occurred. In the event that a busy flash signal is received from a subsequent part of the connection prior to the release of the incoming register, an interregister congestion signal is sent.
(*b*) The answer signal will not start the charge nor the measurement of call duration if, prior to this line signal, a no-charge interregister signal is also received.
(*c*) In addition to the release conditions stated for system 5, in system 5bis the clear-forward signal is sent after the receipt of a backward interregister signal calling for release of the connection by the outgoing system 5bis register.

System 5bis bothway operation is the same as that for system 5.

Interregister signalling
This is 2/6 m.f. forward and backward signalling, pulse, link-by-link, with a t.a.s.i. guard and locking frequency in each direction being

used to assure trunk-channel association during the signalling. As all international switching units and circuits are 4-wire, advantage is taken of this to simplify system 5bis and the same signalling frequencies 700, 900, 1100, 1300, 1500, 1700 Hz (200 Hz spaced and the same as those of system No. 5) are used in each signalling direction. This is different from CCITT system R2 which has different frequencies in the two directions owing to the possibility of 2-wire switching, or 2-wire circuits, which could arise in national application of R2. National application of system 5bis is not visualised.

Forward interregister signals (see Tables 12.6, 12.7 and 12.8): The forward sequence begins with an *en bloc* initial address block which normally takes any one of the following forms:

$$X \ I_1 \ Z \ N_1 \ N_2 \ N_3 \ N_4$$
$$X \ I_1 \ I_2 \ Z \ N_1 \ N_2 \ N_3$$
$$X \ I_1 \ I_2 \ I_3 \ Z \ N_1 \ N_2$$

and thus normally of seven digit signals. Any further address digits are sent overlap as soon as they are available.

X = instruction signals to indicate switching control information in regard to echo suppressors and routing constraints imposed by satellite circuits (Table 12.6)

I = digits (one, two or three) of the country code (Table 12.8)

Z = characteristic digit (D) to indicate a subscriber dialled call or the language digit (L) in semiautomatic service (Table 12.7)

N = early digits of the national significant number (Table 12.8).

The outgoing international circuit is not selected and a seizure signal not sent until the initial address block is formed at the outgoing system 5bis register. This initial address block enables analysis to be made at system 5bis registers to determine the forward routing. The block is transmitted to system 5bis transit and incoming registers which function in the overlap mode when the forward routing has been determined and the distant register is ready to receive address signals. In enabling complete routing to be performed on the international part of the connection, the initial *en bloc* block partially adopts the approach of system 5 to ease the t.a.s.i. problem on interregister signalling. The alternative of overlap for this initial part of the address would have reduced t.a.s.i. efficiency.

The type of X signal (signals 1–4 Table 12.6) may be changed when transmitting the block forward from a transit register depending on

Table 12.6   *CCITT system 5bis forward interregister signals:
content of the X signals*

Signal number	Frequency (compound)	Meaning
	Hz	
1	700 + 900	Incoming half-echo suppressor not required and no satellite link included in preceding routing
2	700 + 1100	Incoming half-echo suppressor required and no satellite link included
3	900 + 1100	Incoming half-echo suppressor required and a satellite link included in preceding routing
4	700 + 1300	Outgoing half-echo suppressor required and no satellite link included
5	900 + 1300	
6	1100 + 1300	
7	700 + 1500	
8	900 + 1500	Spare (note 1)
9	1100 + 1500	
10	1300 + 1500	
11	700 + 1700	
12	900 + 1700	
13	1100 + 1700	Reserved to facilitate inter-working with system No. 5
14	1300 + 1700	
15	1500 + 1700	Unavailable to avoid conflict (note 2)

Notes: (1) The spare signals are for future use:
(*a*) to extend the precision of echo-suppressor control
(*b*) to monitor the number of t.a.s.i. circuits on a connection, to give potential for t.a.s.i. routing restriction should this be a future requirement.
(2) Forward X signal 15 is not available, as backward signal 15 (Table 12.9) is a substitute for an erroneous pulse. If X signals were to be sent backwards within a route-monitoring block the the use of X = 15 would produce a conflict. If it is decided in the future that an X signal be sent backwards as part of a route-monitoring block, it will be the same X signal which was transferred forward on the last link.

Table 12.7 *CCITT system 5bis forward interregister signals: content of the Z signals*

Signal number	Meaning
1	French ⎫
2	English ⎪
3	German ⎬ language digit (L)
4	Russian ⎪ (semiautomatic calls)
5	Spanish ⎪
6	Spare ⎭
7	Calls requiring access to test equipment
8	Spare language digit
9	Spare
10	Digit 0; sub. without priority ⎫ discriminating
11	Subscriber with priority ⎬ digits (D) for
12	Data transmission ⎭ automatic calls
13	Spare
14	Spare
15	See note

Note: Signal 15 is reserved to have the same meaning (ST-end of pulsing) as in the N digit Table 12.8.

the action taken at the transit exchange in regard to the echo-suppressor provision and type of outgoing circuit (satellite or nonsatellite) when it received the X signal from the previous exchange.

Discrete transit and terminal indicators are not provided in system 5bis. An exchange will function as an international terminal when it receives its own country code, and as an international transit on receipt of a country code other than its own.

If the total address comprise less than the seven signals of the normal initial block, as may arise on calls to incoming operators, a suffix ST signal is sent to indicate the shorter length and also that forward transmission is complete. ST is always sent in international semiautomatic operation and it is also sent should it be received from the outgoing national network in automatic operation.

Backward interregister signalling (see Table 12.9): Normally, one backward interregister signal only, from the list shown in Table 12.9, is returned on each link, the signal releasing the registers; and exceptionally two on error-detected (see 'error detection'). No acknowledgment

Table 12.8   *CCITT system 5bis forward inter-register signals: content of the N signals*

Signal number	Meaning
1	Digit 1 ⎫
2	2 ⎪
3	3 ⎪
4	4 ⎪
5	5 ⎬ See note
6	6 ⎪
7	7 ⎪
8	8 ⎪
9	9 ⎪
10	0 ⎭
11	Code 11 operator
12	Code 12 operator or component of the test code pattern when associated with $Z = 7$ (see Table 12.7)
13	Repeated transmission prefix
14	Spare
15	ST end of pulsing

Note: The signals 1–10 are used for the country code and the national number digits.

signalling, proceed-to-send signalling or control of address-information-transfer signalling applies. Normally it is preferred that proceed-to-send signals be backward interregister signals, but it is line signalling in system 5bis for t.a.s.i. reasons and also to retain the same line signalling as system 5.

An incoming system 5bis register recognises that all the required address information has been received on the same criteria as for CCITT system 4 (note 3 on Table 12.1 Section 12.3). The appropriate called subscriber status signal (sub. free, etc) is returned when this type of information is available to the incoming system 5bis register. The address-complete signal is returned when limited facility conditions apply, status indication not being available.

Table 12.9   *CCITT system 5bis backward interregister signals*

Signal number	Meaning (note 1)
1	International switching equipt. ⎫
2	International circuit group ⎬ congestion
3	National network ⎭
4	Spare ⎫
5	Address complete – charge │
6	Sub. free – charge called
7	Sub. free – no charge party's line
8	Sub. busy condition
9	Number unobtainable (status)
10	Sub. transferred (changed number) ⎭
11	Error detected
12	Spare
13	Route monitoring prefix (note 2)
14	Spare (note 2)
15	Substitute for erroneous pulse

Notes: (1) Any of the backward signals, except signal 15, may be given meanings independent of those listed if they appear in the route-monitoring block.

(2) In the event that route-monitoring blocks of different lengths are used in the future, signal 14 is available as an additional route-monitoring prefix.

Error detection

The error-detected signal (signal 11 Table 12.9) is returned on a given link on error (failure to satisfy the 2-and-2 only signal recognition) in the forward interregister signalling, error correction being by retransmission. Should a first error-detected signal result in a successful retransmission, and assuming that a second error does not occur on the same link and that connection set-up proceeds in normal manner, then a backward signal relating to normal register dismissal will eventually follow in normal course. Thus two backward interregister signals will be returned in this case, one being the error-detected signal.

Should a second error occur on the same link, a second error-detected signal is returned on that link and no further backward interregister signalling takes place. The action taken then depends on the type of register function, outgoing or transit, receiving the two error-detected signals:

(*a*) An outgoing system 5bis register clears the connection forward by the clear forward/release guard sequence and causes a circuit group congestion signal to be returned to the preceding part of the connection.
(*b*) A transit system 5bis register sends the congestion signal back and releases, the congestion signal resulting in forward connection clear down from the outgoing system 5bis register and the return of the congestion signal to the preceding part of the connection.

The error situation is thus limited to two detections per link per call in the automatic procedures.

Retransmission for error correction commences with the 'repeated transmission prefix' (signal 13 Table 12.8), followed by the initial address block and any subsequent address digits previously sent. If the error had occurred in any of the first five signals of the initial address block, the return of the error-detected signal is delayed until the fifth signal of the block is received. This procedure avoids possible conflict between signal $Z = 13$ (Table 12.7) and signal $N = 13$ (Table 12.8). If the link-by-link address transmission has commenced on the link immediately following the link engaged in error correction, the correction process should be such that it does not retransmit valid information already sent on the subsequent link.

Potential is specified in system 5bis to allow the backward inter-register signal sequences to be expanded for a route-monitoring feature, should this be a future requirement. In this event, a backward sequence headed by the route-monitoring prefix (signal 13 Table 12.9) will be sent from each system 5bis-equipped international exchange except the outgoing. This 'route-monitoring block' backward sequence would be of fixed length and transmitted immediately upon receipt of the forward X signal, the block being repeated link-by-link on the international part of the connection. Should an error-detected signal require to be transmitted back, this would be transmitted after the backward route-monitoring block, but not until the first five signals of the forward initial address block have been received.

Error-correction retransmission does not occur on error-detected in the backward signalling. A system 5bis transit register (and a system 5bis incoming register when arrangements permit) returns the 'substitute for erroneous pulse' signal (signal 15 Table 12.9) on the incoming link as a substitute for the corrupted signal received on the outgoing link, and error-detected. The eventual receipt of this substitute signal by the outgoing system 5bis register releases the connection forward by the clear forward/release guard sequence, except when the substitute occurs within a route-monitoring block (excluding the route-monitoring prefix itself), the connection not being released in this circumstance. The same action is taken by the outgoing system 5bis register on error-detected in the backward signalling it receives, the substitute signal not being received.

Fig. 12.3 shows a typical signal sequence with system 5bis.

Relevant data

Line signalling: Same as for line signalling CCITT system 5 (Section 12.4).

Interregister signalling

Transmit:

Frequencies (in both directions) 700, 900, 1100, 1300, 1500, 1700 Hz, tolerance ± 6 Hz

Transmitted level -7 dBm0 ± 1 dB per frequency

Difference in transmitted level between the two frequencies of a signal not to exceed 1 dB

Signal duration 55 ± 5 ms

Interval between pulse signals 55 ± 5 ms

Interval between cessation of the seizure line signal and the transmission of the first register signal 80 ± 20 ms

Guard and t.a.s.i. locking frequency:

Frequency 1850 ± 6 Hz

Transmitted level in forward direction initially -7 dBm0 ± 1 dB continuing until a new m.f. pulse is sent or until expiration of 90 ± 30 ms; after this interval the level reduced to -15 dBm0 ± 1 dB

Transmitted level in backward direction uniformly -15 dBm0 ± 1 dB

Interval between cessation of the proceed-to-send line signal and the transmission of the backward guard and t.a.s.i. locking frequency 80 ± 20 ms

Interval between cessation of forward X signal and the first application of the forward guard and t.a.s.i. locking frequency 1 ± 1 ms

After initial application in both directions, the guard and t.a.s.i. locking frequency sent during the intervals

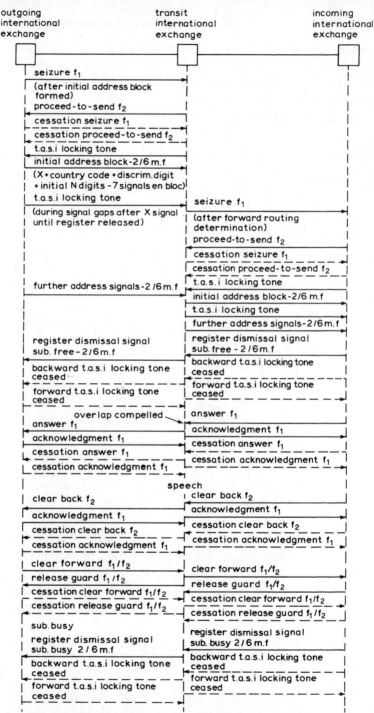

after sub. busy status signal received and registers released, the clear forward signal is automatically sent from the outgoing exchange and the international connection cleared down by the clear forward/release guard sequence.

Fig. 12.3   *Typical signal sequence CCITT 5bis system*

between m.f. signals, the frequency persisting until register dismissal. Interval between m.f. signals and the guard and t.a.s.i. locking frequency 1 ± 1 ms

Receiver:

Operate   Frequency variation ± 15 Hz of the nominal

Receive level. The absolute power level N of each signal to be within the limits $(-14 + n) \leqslant N \leqslant n$ dBm, where n is the relative power level at the receiver input. These limits give a margin of ± 7 dB on the nominal absolute level of each signal at the receiver input.

The absolute levels of the two frequencies comprising a signal not to differ from each other by more than 4 dB

Minimum signal duration 30 ms

Minimum interval between successive signals 30 ms

Nonoperate   Receiver not to operate to a signal whose absolute power level at its input is $(-17 - 7 + n)$ dBm. This limit is 17 dB below the nominal absolute power level of the signal at the receiver input.

Receiver not to operate to a signal of 10 ms duration or less

Interruptions to signal, and intervals between successive signals, of 10 ms or less ignored

Not to operate to a 2/6 m.f. signal in the presence of the guard and t.a.s.i. locking frequency

Guard and t.a.s.i. locking frequency detector:

Operate   Frequency variation ± 15 Hz of the nominal

Receive level in the range 0 to -22 dBm0

Minimum signal duration 15–25 ms

Release   When a signal satisfying the above operate limits is reduced to a level of -37 dBm0 or below for 10–20 ms

Nonoperate   Not to operate to a frequency outside the range 1850 ± 100 Hz

Not to operate to a signal of level below -32 dBm0

Error determination

In addition to error detected by failure to satisfy the 2-and-2 only signal recognition (typically 20 ms), error is detected, if, for an interval of 20–30 ms in the forward signalling direction, no valid signal is present at the output of either the m.f. receiver or the guard-frequency detector. This condition prompts the return of the error-detected signal unless the error occurs after reception of ST when given. No provision is made for detecting and acting upon interruptions in the guard frequency in the backward signalling direction.

Release of system 5bis registers

Normal release
(*a*) An outgoing system 5bis register releases when it has received an appropriate backward interregister dismissal signal (Table 12.9), or when the connection is cleared earlier by the outgoing side.
(*b*) An incoming system 5bis register releases when it has transmitted a backward interregister dismissal signal, or when the connection is cleared earlier by the outgoing side.
(*c*) A transit system 5bis register releases when it has sent a backward interregister dismissal signal, or when the connection is cleared earlier by the outgoing side.

Abnormal release
(*a*) An outgoing system 5bis register releases and clears the forward connection:
(i) if 10–20 s after the start of the seizure line signal no proceed-to-send line signal is received
(ii) if 20–30 s after the reception of the proceed-to-send signal the outgoing register has not pulsed out due to fault
(iii) if 20–30 s after sending the last valid forward interregister signal no register dismissal signal is received
(iv) if a second error-detected signal is received
(v) if a 'substitute for erroneous pulse' signal is received, provided it is not part of the route monitoring block.
(*b*) An incoming system 5bis register releases and clears the connection forward, if no digits, or insufficient digits to establish a connection, have been received and 15–30 s have elapsed since the end of the proceed-to-send signal or the receipt of the last forward interregister signal. Signal 2 Table 12.9 (circuit-group congestion) is transmitted backward prior to release. This results in clear down of the international connection from the outgoing system 5bis equipped exchange.
(*c*) A transit system 5bis register releases in any one of the cases stated

for the abnormal release of outgoing and incoming (terminal) system No. 5bis registers. When releasing because of reception of a second error-detected signal, the transit register sends the circuit-group congestion signal backwards prior to release. This results in clear down of the international connection from the outgoing system 5bis equipped exchange.

12.6 International signalling: sending sequence of numerical information

The numerical information transferred forward between registers on the international network varies depending on the particular international signalling system, transit or terminal traffic, and on the type of call (automatic, semiautomatic, test equipment, etc.).[11] A caller dials the following sequence in international automatic working:

i.s.d. prefix + country code + national (significant) number

The i.s.d. prefix is used only in the outgoing national network to enable a caller to access the centralised international equipment in that network, and is not transferred over the international network. In addition to the transfer of the destination country code and the national significant number of the called party, the international signalling system transfers other information generated by the international equipment, and used to control the connection set-up on the international network. This additional information may give the following indications to the equipment on the international part of the connection:

> characteristic digit
> nature of call (satellite, etc.)
> echo suppressor control
> transit or terminal indication
> ST (end-of-pulsing) when applicable

The general term characteristic digit concerns the discriminating digit or the language digit (English, French, etc.). The language digit is sent in semiautomatic working to facilitate the bringing into circuit of an appropriate language-speaking incoming operator. The discriminating digit is sent, instead of the language digit, in automatic working. Receipt of the appropriate digit, language or discriminating, by an international register indicates the type of working (semiautomatic or automatic) and enables appropriate equipment arrangements to be made. The characteristic digit has a position in the transferred sequence to enable it to be located by the receiving register, usually immediately following the country code when this is sent, and in a corresponding sequence when the country code is not sent.

The nature of call indicator gives information concerning previous routing of the call to a receiving register to enable the register to impose any appropriate restriction on the forward routing. In the main, it concerns restriction of the number of satellite links permitted in a connection (Section 7.5), but may also include t.a.s.i. routing restriction (Section 7.6).

CCITT systems 4, 5bis, 6 and R2 pass signals to ensure appropriate echo-suppressor association. An outgoing international exchange so equipped takes decision with respect to its echo-suppressor requirement at the time the outgoing circuit is selected, and passes an appropriate signal forward to guide subsequent international exchanges as to echo-suppressor provision action to be taken (Section 7.3). Systems 5 and R1 do not include echo-suppressor signals. The normal application of system 5 would be to long circuits normally requiring echo suppressors, the circuits being so equipped. Regional control procedures which would not require signals are applicable for the regional system R1.

Some international signalling systems (4 and 5) have discrete transit and terminal indicators to give direct instruction. In the others (5bis, 6, R1 and R2) transit and terminal indications are interpreted from the forward interregister signalling information received.

The new CCITT optimised c.c.s. system 7 will follow the principle of system 6 with regard to the above points.

The ST end-of-pulsing signal is always transferred on international semiautomatic working. An international equipment generated ST may or may not be transferred on automatic working depending upon the signalling system, but ST is always transferred over the international network on automatic working when this signal is available from the outgoing national network.

Table 12.10 summarises the interregister numerical information transferred forward over the international network by the various international signalling systems in the typical case of automatic working.

Notes on Table 12.10:
(1) Direct transit or terminal indication is given by the line seizure signal.
(2) Used only by multilateral or bilateral agreement for echo suppressor control, and, when used, Code 14 is sent as the first signal in the numerical signal sequence in response to each proceed-to-send line signal received.

Table 12.10   *Forward interregister numerical information transferred: automatic working*

System	To international transit exchange	To international (terminal) incoming exchange
4[1]	Code 14[2] Country code[4]	Code 14[2] Discriminating digit National (sig) number ST[3]
5	KP2 Country code[4] Discriminating digit National (sig) number ST	KP1 Discriminating digit National (sig) number ST
5bis	X digit Country code[5] Discriminating digit National (sig) number ST[3]	X digit Country code[5] Discriminating digit National (sig) number ST[3]
6	Routing information[6] Calling party's category[7] Country code[4,8] National (sig) number ST[3]	Routing information[6] Calling party's category[7] National (sig) number ST[3]
R1	KP Country code[5,10] Discriminating digit[10] National (sig) number ST	KP Country code[5,10] Discriminating digit[10] National (sig) number[9] ST
R2	Country code indicator[11] Country code[4]	Discriminating digit National (sig) number

(3) On automatic calls, the ST signal (Code 15) sent only when this signal is available from the outgoing national network.

(4) The country code is not sent to the incoming (terminal) international exchange.

(5) Discrete transit and terminal signal indicators are not given. The received country code is analysed by a receiving international register to obtain transit or terminal indication by interpretation.

(6) The routing information includes the satellite and echo suppressor detail.

(7) The discriminating digit information is included in the calling party's category.

(8) The country code is not sent to the incoming (terminal) international exchange. Transit or terminal indication is given by 'country code included' or 'country code not included' indicators in the routing information bit field.

(9) The trunk (area) code is not sent to the called numbering area (NPA – number plan area) of a country in an integrated numbering plan (i.e. N. America).

(10) The characteristic digit (discriminating or language) or equivalent information and the country code are not sent within an integrated numbering area (i.e. N. America).

References

1 CCITT: Green Book, 6, Pt. 2, Recommendations Q120-Q139 'Specification of signalling system No. 4', ITU, Geneva, 1973, pp. 259–291

2 MILES, J. V., and TURNBULL, M. G.: 'Switching arrangements for international subscriber dialling of calls to Europe', *Post Off. Electr. Eng. J.*, 1964, 57, 2, pp. 75–78

3 CCITT: Green Book, 6, Pt. 2, Recommendations Q140-Q164 'Specification of signalling system No. 5', ITU, Geneva, 1973, pp. 305–337

4 WELCH, S.: 'Signalling systems for dialling over transoceanic telephone cables', Inst. Post Off. Electr. Eng. Printed Paper No. 225, 1964

5 WELCH, S.: 'Line and interregister signalling systems for dialling over transoceanic telephone cables', *Post Off. Electr. Eng. J.*, 1963, 56, 2, pp. 111–118

6 WELCH, S.: 'Signalling arrangements for dialling over transatlantic telephone cables', *ibid.*, 1960, 53, pp. 201–205

7 BULLINGTON, K., and FRASER, J. M.: 'Engineering aspects of t.a.s.i.', *Bell Syst. Tech. J.*, 1959, 38, p. 353

8 LEOPOLD, G. R.: 'T.A.S.I. – B system for restoration and expansion of overseas circuits', *Bell Lab. Rec.*, 1970, 48, p. 299

9 CLINCH, C. E. E.: 'Time assignment speech interpolation (TASI)', *Post Off. Electr. Eng. J.*, 1960, 53, pp. 197–200

10 CCITT: Green Book, 6, Pt. 2, Recommendations Q200-Q221 'Specification of signalling system No. 5bis', ITU, Geneva, 1973, pp. 373–391

11 CCITT: Green Book, 6, Pt. 2, Recommendation Q107 'Sending sequence of numerical (or address) information', ITU, Geneva, 1973, pp. 240–242

Glossary of Terms

Terms adequately explained in the text are omitted from the glossary. Some terms are not yet defined and definitions for some others are in process of change. Such terms are included in the glossary with explanation of their meanings. The glossary is limited to the main signalling and associated terms.

1 General

Address information: The totality of digits at a point in a network which identify the called party (subscriber, operator, etc.) or which define the onward routing.

Address signal: A forward signal containing one or more digits of the address information.

Analogue signalling: A method of signalling in which signals are sent by specified waveforms in an analogue transmission path.

Channel associated signalling: A method of signalling in which the signals relating to each traffic circuit are accommodated within the transmission path comprising the circuit (channel bandwidth, channel time slot, etc.), or, are carried by a segregated path which forms an integral part of the transmission system providing the traffic circuit (e.g. a time slot in a p.c.m. system carrying the signals relating to the traffic circuits provided only by other time slots in the same system).

The method implies that each traffic circuit is provided with signalling means dedicated exclusively to that circuit.

Compelled signalling: A method of signalling in which every signal must be acknowledged in the opposite direction before another signal may be sent. The signalling may be continuous or pulse.

Compound signal: In a.c. signalling. A signal in which more than one frequency is transmitted at a time.

Decadic signal: A signal consisting of a sequence of pulses, the number of which conveys one digit of the address information. In telephony, the radix of the digit numbering notation is 10.

Digital signalling: A method of signalling in which signals are sent by binary bit(s) in a digital transmission path.

Earth-battery signalling: A method of d.c. signalling in which signals are sent by alternate application of earth and battery to a single wire earth-return circuit.
The method implies that the send-end impedance is low, and substantially the same, during both the earth and battery application conditions.

En bloc signalling: A method of signalling in which address signals are assembled into one or more blocks for onward transmission, each block containing sufficient information to enable onward routing to be determined by switching centres which receive it.

End-to-end signalling: A method of signalling in which signals are passed from one end to the other of a multilink connection without signal processing at intermediate switching points.

Inband signalling: A method of signalling in which signals are sent over the same path as the user's communication and within the same frequency band as that provided for the user.

Link-by-link signalling: A method of signalling in which signals are passed from one end to the other of a multilink connection, the signals being received and passed on at each intermediate switching point.

Loop-disconnect signalling: A method of d.c. signalling in which signals are sent by the alternate making and breaking of a loop circuit. The method implies that the send end impedance is low during the loop, and high during the disconnect, conditions.

Loop-battery signalling: A method of d.c. signalling in which signals are sent by the alternate application of loop and battery to a loop circuit. The method implies that the send-end impedance is low, and substantially the same, during both the loop and battery application conditions.

Multifrequency (m.f.) signalling: A method of a.c. signalling in which each signal is sent as a compound signal consisting of n frequencies from a set of m frequencies.

Noncompelled signalling: A method of signalling in which every signal need not be acknowledged in the opposite direction before another signal may be sent.

Outband signalling: A method of signalling in which signals are sent over the same path as the user's communication, but in a different frequency band from that provided for the user.

Off-hook signal: A signal directly or indirectly derived from the off-hook (active) condition of a subscriber's telephone set, or the equivalent.

On-hook signal: A signal directly or indirectly derived from the on-hook (quiescent) condition of a subscriber's telephone set, or the equivalent.

Overlap signalling: A method of signalling in which the reception and onward transmission of signals are overlapping processes at a switching centre.

Signal code: The rules for sending the logic states of every signal conveyed by a signalling system and for identifying every received signal.

Signalling channel: On a 4-wire circuit. A one-way transmission path.

Signalling link: On a 4-wire circuit. A combination of two signalling channels, in opposite directions, which operate together in a single signalling system.

Simple signal: In a.c. signalling. A signal in which only one frequency is transmitted at a time.

Voice frequency (v.f.) signalling: A method of a.c. signalling in which signals are sent by an alternating current (or currents) of frequency (or frequencies) within the telephone speech band.

2 Signals

Address complete: A backward signal sent from a terminal exchange to indicate that all the address information required for routing the call to the called party has been received.

The signal may be qualified as charge, no charge, or coin box, to indicate the appropriate action to be taken on receipt of the subsequent answer signal.

Receipt of this signal also indicates that no called party line condition (status) signals will be sent.

This signal is a register dismissal signal.

Address incomplete: A backward signal sent from a terminal exchange to indicate that the connection should be released, or other appropriate action taken, because insufficient address information has been received to enable the call to be set up.

Answer: A backward signal sent to indicate that the called party has answered.

In automatic working, the signal may be used to start the charge, this function sometimes being indicated by an answer (charge) signal. If charging is to be inhibited, an answer (no charge) signal may be sent, or, if such a signal is not included in the repertoire, the answer signal is suppressed.

In semiautomatic working, the answer signal has a supervisory function.

Blocking: A backward signal sent on an idle circuit to the outgoing end to cause engaged conditions (blocking) to be applied to the circuit to guard it against seizure.

Busy flash: A backward signal sent to indicate called subscriber busy or plant congestion conditions.

In some signalling systems, the busy flash signal is replaced by discrete congestion and subscriber busy (electrical) signals.

In automatic working, receipt of the busy flash signal at an outgoing end results in busy tone to the caller, and, in some systems, release of the connection.

In semiautomatic working, the busy flash signal causes an appropriate indication to be given to the operator.

Calling party's category indicator: A forward signal sent to indicate the category of the calling party (operator, ordinary subscriber, priority subscriber, data call, test call, etc.).

In semiautomatic working of some international signalling systems, the signal may indicate the service language to be spoken by an incoming operator.

Clear back: A backward signal sent to indicate that the called party has cleared.

In automatic working, the signal is sometimes used to stop the charging process if the caller has not cleared, and, depending upon the release philosophy of the system, may or may not initiate release of the connection.

In semiautomatic working the clear back has a supervisory function.

Clear forward: A forward signal sent to terminate a call or call attempt and to release the connection.

Code 11: A forward signal sent in international semiautomatic working to obtain an incoming (Code 11) operator at the incoming international exchange for the completion of calls which cannot be routed automatically at the incoming international exchange.

The Code 11 signal is the 11th code combination provided by the international interregister signal code.

Code 12: A forward signal sent in international semiautomatic working to obtain a delay (Code 12) operator at the incoming international exchange.

The Code 12 signal is the 12th code combination provided by the international interregister signal code.

Congestion: A backward signal sent to indicate that there is insufficient plant to handle the call.

In some signalling systems, discrete signals indicate switching equipment, circuit group, or, in international service, distant national network congestion.

Depending upon the arrangments adopted, receipt of a congestion signal at an outgoing end may result in busy tone to the caller, connection release, or reroute.

Country code indicator: A forward signal sent in international signalling systems to indicate whether or not the country code of the called country is included in the address information.

Echo-suppressor indicator: A forward signal sent to indicate whether or not an outgoing half-echo suppressor is included in the connection.

Forward transfer: A forward signal sent by an outgoing operator to request the assistance of an operator at an incoming exchange, if the call is automatically set up at that exchange.

When a call is completed via an incoming operator, the signal should preferably cause this operator to be recalled.

KP (start-of-pulsing): A forward signal sent to indicate that transmission of the address signals is about to start.

It may be used to prepare an incoming register for the receipt of address signals.

When used in international working, this signal may be qualified to convey terminal (KP1) call or transit (KP2) call indication to a receiving international exchange.

Line out of service: A backward signal sent to indicate that the called party's line is out of service or faulty.

Nature of circuit indicator: A forward signal sent to indicate the nature of the circuit or any preceding circuit(s) already engaged on the connection (e.g. satellite or no satellite circuit).

A receiving exchange uses this information to determine the nature of the outgoing circuit to be selected.

Number received: A backward signal sent from a register to indicate that sufficient address information has been received to enable the connection to be set up forward, but does not necessarily imply address complete.

Usually the number received is a register dismissal signal.

Proceed-to-send: A backward signal, sent following the receipt of a seizure signal, to indicate that the incoming equipment is ready to receive address information.

In some signalling systems, the proceed-to-send signal may be qualified as transit or terminal to indicate which address digits should be sent.

Reanswer: A backward signal sent to indicate that the called party, after having on-hook cleared, again lifts his receiver, or in some other way, reproduces the off-hook answer condition, e.g. switch-hook flashing.

Release guard: A backward signal sent to an outgoing end in response to a clear forward signal to indicate that the latter has been fully effective in releasing the incoming equipment and that the circuit is free at its incoming end.

It serves as an acknowledgment to the clear forward signal and to protect a circuit against subsequent seizure at its outgoing end until the incoming equipment is fully released.

Seizure: A signal sent forward from the outgoing end of a circuit at the start of a call to prepare the incoming end of a circuit for reception of subsequent signals.

The signal may have subsidiary functions such as switching, causing a register to be associated, or busying the distant outgoing end of a bothway circuit.

ST (end-of-pulsing): A forward address signal indicating that there are no more address signals to follow.

Subscriber busy (electrical): A backward signal sent from the exchange nearest to the called party to indicate that the line(s) connecting the called party to the exchange are busy.

This signal is a register dismissal signal.

Busy tone is returned when no provision for the subscriber free electrical signal is made in the signalling system.

Subscriber free: charge: A backward signal sent as an alternative to the address complete (charge) signal to indicate address complete, that the called party's line is free, and that the call should be charged on answer.

This signal is a register dismissal signal.

Subscriber free: no charge: A backward signal sent as an alternative to the address complete (no charge) signal to indicate address complete, that the called party's line is free, and that the call should not be charged on answer.

This signal is a register dismissal signal.

Subscriber free: coin box: A backward signal sent as an alternative to the address complete (coin box) signal to indicate address complete, that the called party's line is free, that the called line is a coin box station, and that the call should be charged on answer.

This signal is a register dismissal signal.

Subscriber transferred (changed number): A backward signal sent to indicate that the subscriber number received has ceased to be used and that the subscriber to whom it was allocated must be reached by another number.

This signal is a register dismissal signal.

Vacant number (spare code): A backward signal sent to indicate that the received address information identifies a number which is not in use.

Number unobtainable, or other appropriate tone, is returned when no provision is made for the vacant number signal in the signalling system.

Unblocking: A signal sent to the exchange at the other end of a circuit to cancel at that exchange the engaged condition of that circuit caused by a previous blocking signal.

When the blocking signal is continuous, cessation of the blocking signal performs the unblocking function and a discrete unblocking signal is not used. When the blocking signal is pulse, a discrete unblocking signal is used.

3 P.C.M. signalling

In-slot signalling: In p.c.m. A method of channel associated signalling in which the signals for the associated traffic circuit are transmitted in a bit time slot permanently or periodically allocated in the channel time slot of that circuit.

Out-slot signalling: In p.c.m. A method of channel associated signalling in which the signals for the associated traffic circuit are transmitted in a separate time slot not within the channel time slot of that circuit.

Time-assigned (time addressed) signalling: In p.c.m. A method of time divided signalling in which the traffic circuit to which a signal refers is indicated by the time slot conveying the signal.

4 Common channel signalling

(a) Signalling network

Common channel signalling: A method of signalling in which the signals relating to a number of traffic circuits are carried on a transmission path dedicated exclusively to signalling, and in which an address label sent with a signal identifies the traffic circuit to which the signal refers.

Associated signalling: A mode of operation in which the signals carried by the system relate to a group of traffic circuits between two switching centres, the circuits terminating at the same switching centres as the signalling system.

Nonassociated signalling: A mode of operation in which the signals for a group of traffic circuits between two switching centres are sent over two or more signalling links in tandem, the signals being processed for onward routing and forwarded to the next link by equipment at one or more signal transfer points.

Fully dissociated signalling: A form of nonassociated signalling in which the routes the signals may take through the network are restricted only by the rules and configuration of the signalling network.

Quasiassociated signalling: A form of nonassociated signalling in which the route the signals may take through the network is prescribed.

Signal transfer point: A signal relay centre handling and transferring signal messages from one signalling link to another in a nonassociated mode of operation. There is no processing of the signal information content of messages at signal transfer points.

Signalling module: A group of two or more signalling links between two switching centres, which group provides signalling facilities for a group of traffic circuits, secures these facilities within itself, and has a self contained set of address labels.

(*b*) *Signal units and messages*

Signal unit (SU): The smallest defined group of bits handled as a composite entity and used for the transfer of signal information.

Signal message: Signal information pertaining to a call, management action, etc., sent at one time on the signalling channel.
A message may consist of SU(s) or be of variable bit length.

Lone signal unit (LSU): A signal message which consists of one signal unit only.

Multiunit message (MUM): A signal message which consists of more than one signal unit.

Initial address message (IAM): The first message sent in a call set up process and which contains sufficient address information to enable the routing of the call to start.

Initial signal unit (ISU): The first signal unit of a multiunit message.

Subsequent address message (SAM): An address message, which may be either a one unit or a multiunit message, sent following the initial address message.

Subsequent signal unit (SSU): A signal unit, other than the first, in a multiunit message. In some systems, the last unit in a multiunit message is not referred to as being a subsequent signal unit.

Final signal unit (FSU): The last signal unit in a multiunit message. In some systems, it is referred to as being a subsequent signal unit.

Acknowledgment signal unit (ACU): A discrete signal unit which carries information appertaining to the correct or erroneous reception of one or more signal units or signal messages.

Discrete acknowledgment signal units are not used in systems which have acknowledgment bit(s) contained in the normal signal units.

Synchronisation signal units (SYU): A signal unit containing a bit pattern and information designed to facilitate rapid synchronisation of signalling on a channel. Synchronisation signal units are also sent, as idle units, when no signal messages are available for transmission.

Address (circuit) label: The coded field of bits within a signal message used to identify the particular traffic circuit with which the message is associated.

The label may be subdivided into a band number and a circuit number.

Band number: The subdivision of the address label used to identify the circuit group (or band) containing the particular traffic circuit with which the message is associated, or the equivalent group for a nontraffic circuit related message.

Translation of the band number may be used for routing a message at a signal transfer point.

Circuit number: The subdivision of the address label used to identify the particular traffic circuit within a circuit group (or band) with which the message is associated.

Byte: A subdivision of a signal unit, containing a constant number of bits, used as an entity during processing and possibly used to define the bit field boundaries of the format.

Queueing delay: The delay incurred by a signal message as a result of the sequential transmission of signal units (or messages) on the signalling channel.

5 Numbering plans

Country code: A digit, or combination of digits, characterising a country in international operation.

International number: The number dialled, following the international prefix, to call a subscriber in another country. It consists of the country code of the required country followed by the national (significant) number of the called subscriber.

International prefix: The combination of digits dialled by a caller making a call to a subscriber in another country, to obtain access to the automatic outgoing international equipment.

National number: The number dialled, including the trunk prefix when applicable, to call a subscriber in the same country, but outside the caller's local network or numbering area.

National (significant) number: The number dialled, following the trunk prefix, to call a subscriber in the same country, but outside the caller's local network or numbering area. It consists of the trunk code followed by the called subscriber's number and is thus the national number with the omission of the trunk prefix.

Numbering plan area (NPA) code: A form of trunk code identifying a geographical area to which the called subscriber belongs and within which subscribers can call one another by their subscriber numbers.

This is adopted by the Bell system, the NPA being a 3-digit code, e.g. New York City area NPA code 212.

Subscriber number: The number dialled to call a subscriber in the same local network or numbering area.

Trunk code: A digit, or combination of digits, not including the trunk prefix, characterising the called numbering area within a country (or group of countries included in one integrated numbering plan).

Trunk prefix: A digit, or combination of digits, dialled to obtain access to the automatic outgoing trunk equipment when making a call to another subscriber in the same country, but outside the caller's local network or numbering area.

6 Routing

Alternative routing: When a group of circuits over which overflow traffic is routed involves at least one exchange not involved in the previous route choice.

Overflow: The process of routing a call automatically via another (second choice) group of circuits from the same exchange when a call cannot find a free circuit in a first choice group of circuits. There may be also overflow, at the same exchange, from a second choice group of circuits to a third choice, and so on.

Reroute: The process of rerouting a call automatically from an outgoing exchange when congestion occurs at a transit exchange on a previous routing, the congestion condition being signalled from the transit to the outgoing exchange.

Appendix to Chapter 11
Format principles CCITT No. 7 c.c.s. system message signals

Section 11.11.4 makes general comment on the message structure approaches for the new CCITT No.7 c.c.s. system, and it is of interest to make further comment on the continuing CCITT work in this area to indicate the trend. At the time of writing, the CCITT study is not yet completed.

Significant importance is now given to the desirability for variable bit length message signals for system 7, as distinct from the MUM and standard size SU philosophy of system 6. As mentioned in Section 11.11, a complete message signal consists of two parts:

(i)   That part passed to the processor function.
(ii) The message transfer control 'housekeeping' part concerned with the error control and not passed to the processor.

The present interest is in (i).

It is proposed that a message signal consists of an integral number of octets. The first octet is a service information bit field used to associate the signal information of a message with a particular user subsystem of the processor function (Section 11.11.1 and Fig. 11.9). This service information field is coded to indicate the service (e.g. telephony), whether the message is international or national, and is only used with message signals (it would not be used with signalling system control signals for instance).

The signal information of a message signal (the part of the message dealt with by a particular user subsystem of the processor and conveniently called the 'user information') is of variable length consisting of an integral number of octets. It basically contains the circuit label, a heading code, and one or more signals and/or indications. The user information is divided into a number of subfields, which may be either of fixed or of variable length. When of fixed length, they contain the same number of bits in all messages of that type. When of variable length, the size is indicated in an immediately preceding fixed length subfield.

The circuit label (following the service information octet) is conveyed by the first five octets of the user information, the 40 bits being divided into:

(a)   Destination point code (14 bits) indicating the signalling point for which the message is intended.

(b)   Originating point code (14 bits) indicating the signalling point which is the immediate source of the message.

(c)   The circuit identification code (12 bits) indicating one speech circuit among those directly interconnecting the originating and destination points. This code field may be further subdivided in the circuit number/band number philosophy to give the speech circuit time slot in a p.c.m. system and the identity of the system, or an f.d.m. channel in an f.d.m. transmission system.

This circuit label arrangement may be varied for national networks.

All telephone message signals contain a heading octet of two parts H0 (4 bits) and H1 (4 bits). The H0 coding identifies message groups (e.g. forward address messages group, call supervision messages group). The H1 coding identifies a particular message in the relevant group. Two typical cases are:

(a)   The H0 code group 'call supervision messages', with H1 codes for the specific message signals – answer (charge), answer (no charge), clear back, clear forward, reanswer, forward transfer.

(b)   The H0 code group 'circuit supervision messages', with H1 codes for the specific message signals – release guard, blocking, blocking acknowledgment, unblocking, unblocking acknowledgment, continuity check request, reset circuit.

Code H1 is either a signal code in an H0 group (e.g. clear forward signal in the call supervision messages group), or in the case of more complex messages, identifies the format of these messages (e.g. initial address message in the forward address messages group). Thus, indication as to variable or fixed length is given by the H0 codings for some message groups, and by the combination of the H0 and H1 codings for others, typically:

(a)   H0 codes 'call supervision messages' and 'circuit supervision messages' imply fixed length messages. The user information part of these message signals consists of the circuit label and the heading octet (H0 and H1) bit fields. The H1 code is a signal code.

(b)   H0 code 'forward address messages', the H1 code identifies the format and implies variable or fixed length. When variable, the actual length indication is given by a bit field in the message.

This H0 and H1 arrangement, giving potential for 16 H0 message groups and 16 H1 specific message signals in each group, is more than adequate to cater for message signals presently defined. There is ample spare capacity for further international and national message signals not presently defined.

To illustrate the application of the principles, it is useful to describe the user information part (following the service information octet) of a few typical telephone message signals. The bits stated are in the transmitted sequence of the user information part.

(i)   *Initial address message (variable length)*

40 bits	label
8 bits	heading octet (H0 4 bits, H1 4 bits)
	H0 code indicates 'forward address messages group'. H1 code indicates 'initial address message',
	implying variable length, and identifies the format.
6 bits	calling party category (operator language digit, ord. subscriber, priority subscriber, test call, etc.)
2 bits	spare
12 bits	message indicators
	nature of address message (local, national, international number)
	nature of circuit (satellite, echo suppressor, etc.)
	continuity check requirement, etc.
4 bits	number of address signals in message (length indication)
n x 8 bits	address signals (successive 4-bit fields)

(ii)   *Subsequent address message (variable length)*

40 bits	label

8 bits   heading octet (H0 4 bits, H1 4 bits)

H0 code indicates 'forward address messages group'. H1 code indicates 'subsequent address message of two or more address signals', thus variable length, and identifies the format.

4 bits   filler 0000

4 bits   number of address signals in message (length indication)

n x   8 bits   address signals (successive 4-bit fields)

(iii) Subsequent address message (fixed length, one-address signal)

40 bits   label

8 bits   heading octet (H0 4 bits, H1 4 bits)

H0 code indicates 'forward address messages group'. H1 code indicates 'subsequent address message, one-address signal', thus fixed length, and identifies the format.

4 bits   address signal

4 bits   filler 0000

(iv) Clear forward message signal (fixed length)

40 bits   label

8 bits   heading octet (H0 4 bits, H1 4 bits)

H0 code indicates 'call supervision messages group', implies fixed length and that the H1 codes are signal codes. H1 code indicates 'clear forward signal'.

For complete message signal, the user information detailed above for each example is headed with the service information octet and terminated with the error-control message transfer-control bit field.

Index

Acknowledged signalling, 97
Acknowledgment signal unit (ACU),
      c.c.s., 279, 284, 285,
      289–293, 310
Address complete, 218, 241, 354
Address information, 2, 17, 18, 27,
      30, 32, 201
Address, transfer control, 216–219
Aperiodic guard, v.f., 122, 123
Associated signalling p.c.m., *see*
      Time-assigned signalling
Audio 4-wire circuit, 8, 73
Audio 2-wire circuit, 8, 16, 73
Auto-to-Auto relay set, 30
Automatic repeat clearing, 161

Backward busy, *see* Blocking
Backward interregister m.f. signalling,
      preference for, 208, 209,
      212–216
Basic signalling requirement, 2, 16–19
Blocking, 65, 70, 72, 73, 79, 163,
      185, 198
Block structure, c.c.s., 269, 280, 284,
      285
Bothway operation, 66, 67, 71, 79,
      103–107, 152, 153, 186,
      195, 199, 278, 340, 341
Buffer amplifier, v.f., 117, 159, 165
Built-in p.c.m. signalling, 26
Bunched signalling p.c.m., 91, 92
Busying, outgoing circuit, 33, 71, 72,
      162

CCITT, 21, 322
CCITT signalling systems
    No. 1, 323
    No. 2, 323
    No. 3, 323, 324
    No. 4, 324–335

    No. 5, 335–348
    No. 5bis, 348–361
    No. 6, 269–296
    No. 7, 311-320, 323, 377-380
    R1, 137–157, 220–223
    R2, 190–199, 223–240
C.C.S.
    advantages, disadvantages,
      260–262
    basics, 260–262
    reasons for, 259, 260
C.C.S. general features, 263, 264
C.C.S. modes,
    associated, 264
    nonassociated, 264
    dissociated, 264
    quasiassociated, 266
C.C.S. network configuration (Bell),
      302, 303
C.C.S. rationalised system, 311–313
C.C.S. system CCITT 6,
    basic concepts, 269
    error control, 282–291
    formats and codings, 270–281
C.C.S. system CCITT optimised
      digital, No. 7,
    basic concepts, 311–313
    error control. 314–319
    message structure, 320
    reasons for, 311, 312
CEPT, 41
Changeback, c.c.s., 308, 309
Changeover criteria, c.c.s., 306, 307
Characteristic digit, 344, 351, 361
Characteristic impedance, 161
Check 2-and-2 only, 21, 219, 221
Check tone, v.f., 162, 163
Circuit label, c.c.s., 262, 263, 266,
      271, 298, 299

Circuit label band, c.c.s., 293, 294
Class of service, 21, 208, 245–248
Codes 11 and 12, 227, 328, 344, 354, 369
Coded signalling, interregister, 201–204
Coin and fee check, 46
Comma free codes, p.c.m., 86
Common control switching, direct, 48
    indirect 5, 7, 25, 26, 201
Common channel signalling (c.c.s.), 13, 14, 269–291, 311–320
Compandors, 102, 144
Compelled error control, c.c.s., 98, 320
Compelled signalling sequence, 224, 225
Compound signal, 21, 118, 123, 124, 324, 337
Continuity check, speech path, 263, 309
Continuous/pulse signalling analysis, d.c., 43, 44
    v.f., 154–157, 169, 170
Continuous signalling 2-state, 19, 32, 33, 43, 44, 73, 79, 82, 90, 105, 114, 115, 174, 191, 196
Control signals, c.c.s., 279
Country code, 226, 227, 234, 334, 361
    indicator, 226, 227, 233, 234
CX d.c. system (Bell), 73–75
Cyclic retransmission, c.c.s., 316

D.C. signalling, 9
DC2 system (UK), 61–67
DC3 system (UK), 67–73
Decadic address
    general, 17, 25, 27, 36, 41, 44, 60, 61, 73, 78, 88, 90, 113, 143, 151, 152, 175
    direct control switching, 6, 23, 28 47, 48, 53, 153, 179, 182, 259
    interregister, 7, 9, 10, 27, 48, 53, 201
Delay dialling, 42, 140, 141
Dial telephone, 17–19
Dial characteristics, 18, 53
Digital
    switching, 13, 32
    line signalling
        R1, 157
        R2, 196–199

Direct control switching, 5, 6, 26–32, 50, 51
Discrimination, signalling, 46, 47
Discriminating digit, 226, 236, 326, 344, 351, 361
Double seizure, 43, 46, 67, 71, 79, 105–107, 152, 153, 195, 199, 278, 340
    procedures, 105–107
Duplex signalling, 2, 12, 32, 63, 81, 95, 117, 158, 172
Duplicated messages, c.c.s., 286, 289, 315, 316
DX d.c. system (Bell), 75–80

E and M lead control, 73, 75, 94, 95, 107–115, 138, 144–147, 173, 174, 188
E and M Bell,
    type I, 107, 108, 115
    type II, 107, 111, 115
    type III, 107, 111, 112
    repeaters and convertors, 111–113
    signalling limits, 112
Earth-battery d.c. signalling, 51, 73
Earth potential difference (e.p.d.), 33, 65, 78, 80
    compensation, 73, 75, 78
Echo suppressors, 95, 228–233, 275, 351, 362
    location in signalling path, 95
En bloc, 101, 207, 208, 212, 216, 217, 221, 336, 348, 349
End of selection, 327
End-to-end signalling,
    line, 117, 124, 324, 325
    interregister, 103, 208, 211–217
    compandors, 102
Error control,
    CCITT 5bis, 355–357
    R2, 234
    UK MF2, 249, 250
Error control CCITT 6,
    analysis of method, 289, 290
    check polynomial, 283
    drift compensation, 287, 290
    error correction, 284, 285
    error detection, 283, 284
    retransmission procedures, 285, 286
Error control CCITT optimised digital c.c.s., No. 7
    guidelines for preferred method, 315, 316
    ignore corruptions method, 316–319

Facilities, influence on signalling, 10
Facility enhancement, 7, 201, 202,
    223
F.D.M. transmission outband arrange-
    ments, 172–174
Forward error correction, 314, 315
Forward retest, *see* Retest
4-wire circuits, 8, 12, 19, 94, 95, 117
4-wire/2-wire termination, 94, 95,
    132, 159, 182

Glare, *see* Double seizure
Glossary of terms, 365–376
Group acknowledgement, 205, 241
Guard coefficient, v.f., 121, 122
Guard tone, interregister m.f.,
    240–244, 248, 249

Heading code, c.c.s., 271–273
History of routing, 98

Idle unit, c.c.s., *see* Synchronising SU
Inband signalling, 20, 117, 202, 203
Independent signalling paths, 94
Initial address message (IAM), c.c.s.,
    274
Initial signal unit (ISU), c.c.s., 273,
    274
In-slot signalling, p.c.m., 85–86
Integrated digital networks, 13, 14
Interdigital pause, 18, 30, 48, 59, 72
International signalling, *see* CCITT
    signalling systems
Interregister m.f. signalling, 9, 10, 18
    187, 190, 202, 203, 219
    link-by-link/end-to-end analysis,
        212–216
Interregister m.f. signalling modes,
    analysis, 203–212
    choice, 211
    coded continuous, 203, 204
    fully compelled, 204–206, 224,
        225
    noncompelled, 204
    pulse, 204, 205, 220, 236
    reliability assessment, 210, 211
    semicompelled, 204, 206, 224
Interregister m.f. signalling systems,
    CCITT R1, 220–223
    CCITT R2, 223–240
    UK MF2, 218, 240–253
    Socotel, 218, 253–257
Interruption control R2, 194, 195
Interworking R2, 237, 238

Junction signalling,

direct control switching, 5, 6, 27,
    47, 48
common control switching, 5, 6,
    27, 48, 49
Junction d.c. signalling,
    battery-earth pulsing and
        supervision, 37
    continuous/pulse analysis, 43, 44
    high/low resistance, 34, 35
    loop, 27, 32, 33
    loop-disconnect, 27, 30–32
    loop reverse battery, 27–30
    pulse, 34
    single wire, 33, 34

Keyphone, *see* Pushbutton telephone
KP signal, 221, 241, 338, 342, 343,
    363

Language digit, 226, 236, 326, 344,
    351, 361
Limited facilities, *see* Number
    received
Line and interregister signalling,
    separation of, 6, 49, 201
Line split(s),
    v.f., 117, 123–126, 148–150,
        161, 165, 330, 341
    types, 125, 126
Line (supervisory) signalling, 4
Link-by-link signalling
    line, 33, 117, 124, 125, 158,
        173, 264, 265
    interregister, 208, 212–216,
        335, 343
Loading, c.c.s., 296–299
Load sharing, c.c.s., 304–306
Lone signal unit (LSU), c.c.s.,
    271–273
Long-distance d.c., 60, 61
Loop-battery d.c. signalling, 51, 63
Loop-disconnect d.c. signalling, 19,
    50, 60
Low-frequency a.c. signalling, 37–40,
    44

Management signals, c.c.s., 279
Manual hold, 46
Memory logic, 135–137, 157, 169
Metering over junctions, 4, 33, 38,
    46, 87, 182
Meter pulses, trunk network, 177
Modem, 260, 294, 295
Multiframe, p.c.m., 86–90
Multiunit message (MUM), c.c.s., 264,
    271–274

National network configuration, 11
Nature of call indicator, 362
Network centralised services, 267–269
Noncompelled error control, c.c.s., 98, 297, 319
Number-complete determination, 235
Number plan area (NPA), 216, 217, 364
Number received, 218, 241, 248, 327, 331
Numbering scheme, 206, 208
Numerical information international, sending sequence, 361– 364

Off-hook, 1, 2, 16
On-hook, 1, 2, 16
Operational amplifiers, 128–131
Out-of-sequence messages, c.c.s., 286, 289, 315, 316
Outband line signalling systems,
    Australian T, 289, 190
    Bell, 188
    CCITT R2,
        analogue, 190–196,
        digital, 196–199
    two frequency, 190
    UK AC8, 180–187
    UK AC12, 187, 188
Outband line signalling, 172–175
    application constraints, 177–178
    modes, 174–177, 188
    simplicity, 175
Outpulsing from registers, control of,
    see Proceed-to-send
Out-slot signalling, p.c.m., 85, 91
Overlap, 207, 216, 336, 349

Parity check bit field, c.c.s., 263, 270 283, 284
Partial *en bloc,* 217
P.C.M. signalling,
    general, 12, 13, 26
    24-channel Bell D1, 86
    24-channel Bell D2, 88–91
    24-channel UK, 86–88
    30-channel CEPT, 91–93
    R2 line, 196–199
Phantom, 8, 61
Point-to-point p.c.m., 26, 92
Postdialling delay, 3–7, 49, 97, 101, 201, 203, 207, 209, 223, 238, 240
Power level, signal, 22, 102, 103
Prefix, 123, 244, 324
Primary/secondary meanings, m.f.

signalling, 225, 227
Proceed-to-send, 40–43
Propagation time, 97, 98, 100, 105, 107, 239
Pulse code modulation (p.c.m.), 83
Pulse correction,
    general, 56–59
    target diagram, 57, 58
Pulse outband signalling, 174, 188–190
Pulse regeneration, 59
Pulse repetition distortion, 52–54
    influence of battery voltage, 56
    influence of line capacitance, 55
    influence of line leakance, 55
    influence of line resistance, 54
    influence of relay adjustment, 55, 56
Pulse signalling, v.f., preference for, 169, 170
Pulsing circuit, 51–53
Pushbutton telephone, 20–23

Queueing delay, c.c.s., 93, 263, 296, 297

Reasonableness checks, processor, c.c.s., 286, 287, 289
Receiver sensitivity, v.f., 121, 122
Register dismissal signal, 217, 218
Register outpulsing, control of, 40–42
Registers, release of, 222, 236, 237, 330–333, 347, 348, 360, 361
Release guard, 64, 65, 70–72, 155, 161, 162, 183, 192, 198
Retest, 70, 161, 162, 184
Retrieval, c.c.s., 307, 308
Routing restriction, 98, 102, 275, 351, 352, 362
R2 interregister pulse signals, 236
R2 potential for rationalisation, 239, 240

Satellite circuits, 98, 238
Satellite loop delay, 97, 98
Security, c.c.s.,
    associated signalling module, 300, 301
    exploitation of signalling modes, 302, 303
    general, 299, 300
    inbuilt associated signalling, 300
    quasiassociated, 301
    signal transfer point, 302

Selection, *see* Address information
Sequence numbering,
    of blocks, c.c.s., 269, 281, 284,
        285, 288, 289
    of messages, c.c.s., 316–319
Sequenced release, *see* Release guard
SF 1 v.f. signalling, Bell E and F
        types, 138, 149
Signal imitation, 20, 21, 117–123,
        125
Signal integrity, check of line, 43,
        109, 140
Signal interference, v.f., 121
Signal priority, c.c.s., 309, 310
Signal repertoire, 33, 45–47
Signal transfer point (s.t.p.), c.c.s.,
        266, 267, 299, 302, 303
Signal unit (SU), c.c.s., 263, 269
Signalling bit stream, derivation of,
        c.c.s., 294–296
Signalling channels, p.c.m., 86–92
Signalling functions,
    supervisory, 3, 25, 27
    selection, 3, 27
    operational, 4
Signalling module, c.c.s., 300–302
Signalling sequence, network, 232,
        235, 282, 331, 345, 358
Simple signal, v.f., 123, 124, 157,
        324
Simplex signalling, 4-wire circuits,
        81, 95, 172
Socotel, *see* Interregister signalling
        systems
Speech immunity, *see* Signal
        imitation, v.f.
Speech path signalling systems,
        limitations, 259
Speech period, signalling during, 38
Spill over, v.f., 125
ST signal, 221, 241, 327, 338,
        342–344, 346, 348,
        353, 361, 362
Status signal, called line, 218, 241,
        247, 248, 327, 354
Stored program control (s.p.c.), 5, 7,
        259, 260
Subsequent address message (SAM),
        c.c.s., 278
Subsequent signal unit (SSU), c.c.s.,
        274
Subscriber line circuit, 24
Subscriber line signalling, 1, 16–27
Suffix, 123, 324

Supervisory (line) signals, 3, 4, 19,
        25, 27, 201
Supervisory unit, 25–27, 33
Switching, influence on signalling,
        5–7
SX d.c. system (Bell), 80, 81
Symmetrical signalling terminals, 75,
        82, 141, 156
Synchronisation, c.c.s.,
    block, 293
    general, 290, 291
    initial, 291, 292
    multiblock, 293, 294
    resynchronisation, 292, 293
Synchronisation signal unit (SYU),
        c.c.s., 269, 280, 281
System X, 311
Tandem, 30
T.A.S.I. guard/lock tone, 349, 350
Through-channel unit, E and M, 175
Time assignment speech inter-
        polation (t.a.s.i.), 97–102,
        105, 335–338, 342, 348
Time-assigned signalling, 12, 85, 88,
        91, 93, 262
Time-division multiplex, 85
Time-shared signalling, 262, 263
Tone-on idle, 136, 138, 174, 175,
        180, 188, 191
Tone-on-idle v.f., influence on
    type line split, 148–150
    receiver signal-guard, 150–152
Tone-off idle, 174, 175, 180
Tone-on/tone-off idle outband,
        analysis, 175–177
Touchtone, *see* Pushbutton telephone
Transmission, influence on signalling,
        7–10
Transmission bridge, 25, 30, 32, 33,
        36, 182
Trunk circuit equipment, 108, 111
Trunk (toll) network line
        signalling,
    preference for outband, 177
    why v.f. widely applied, 178
Trunk offering, 46
Tuned guard, v.f., 122, 123
2 v.f. signalling, 123, 124, 157, 158

Undetected errors, c.c.s., 270, 284
Unidirectional (one-way) operation,
        104, 105
Unnecessary retransmissions, c.c.s.,
        289, 315

Unrequested retransmissions, c.c.s.,
    289, 290, 315

V.F. receiver,
    guard circuit, 120–122
    signal-guard, 126–132, 150–152
Voice frequency (v.f.) signalling,
    choice of frequency, 119, 120
    continuous/pulse, analysis,
        154–157, 169, 170
    general arrangements, 9, 132, 134
    preference for 1 v.f., 118, 119
V.F. signalling systems,
    Bell SF analogue, 137–157

Bell SF digital, 157
F.R. Germany, 167
France, 167, 168
UK AC9, 158–164
UK AC9M, 164, 165
UK AC11, 166–167
V.F. signal type
    continuous compelled, 135–137
    2-state continuous, 135–138, 156,
        169, 170
    pulse, 137, 157, 158, 169, 170

Wink, 42, 141
Wired logic control switching, 7, 13